Brownfields

Redeveloping Environmentally Distressed Properties

Brownfields

Redeveloping Environmentally Distressed Properties

Harold J. Rafson
Robert N. Rafson

McGraw-Hill

New York   San Francisco   Washington, D.C.   Auckland   Bogotá
Caracas   Lisbon   London   Madrid   Mexico City   Milan
Montreal   New Delhi   San Juan   Singapore
Sydney   Tokyo   Toronto

McGraw-Hill

A Division of The **McGraw·Hill** Companies

1 2 3 4 5 6 7 8 9 0   DOC/DOC   9 0 9 8 7 6 5 4 3 2 1 0 9

ISBN 0-07-052768-7

*The sponsoring editor for this book was Bob Esposito and the production supervisor
was Pamela Pelton. It was set in Palatino by North Market Street Graphics.*

Printed and bound by R. R. Donnelley & Sons Company

McGraw-Hill books are available at special quantity discounts to use as premiums and sales promotions, or for use in corporate training programs. For
more information please write to the Director of Special Sales, McGraw-Hill,
11 West 19th Street, New York, NY 10011. Or contact your local bookstore.

Contents

OFFICE OF THE MAYOR

CITY OF CHICAGO

RICHARD M. DALEY
MAYOR

Dear Readers:

As Mayor of the City of Chicago, I have long recognized that the key to revitalizing our urban communities is the redevelopment of the abandoned, underutilized and contaminated industrial properties known as "brownfields." Chicago has committed an extraordinary amount of resources to not only overcoming the barriers to brownfields redevelopment but to implementing a plan of action which has resulted in new development, job creation and tax generation.

The City's plan focuses its resources into revitalizing our industrial corridors. Over the past seven years, the Chicago Brownfields Redevelopment Initiative has moved aggressively in acquiring underutilized land, assembling large tracts, cleaning and then marketing it to industrial users. This strategy makes sense because it concentrates development efforts in areas with existing infrastructure, ready access to transportation and a skilled labor pool.

America's urban centers cannot allow their industrial districts, which are essentially the lifeblood of cities such as Chicago, to remain idle. It is critical that we position our industrial districts to meet employers' needs now and into the 21st century. Brownfields redevelopment is a crucial part of accomplishing that goal.

I am pleased to continue to offer my support to efforts designed to stimulate discussion and inform others about the brownfields process, and by doing so encourage brownfields redevelopment.

Sincerely,

Mayor

Acknowledgments

We would like to thank all the contributors for their thought and effort in preparing their sections of this book. The whole is a great deal more than the sum of the parts.

We would especially like to thank our patient editor, Robert Esposito, and the editor in charge of publication.

We would also like to thank Dorothy Rafson for her help, kind words, and support in producing the final manuscript.

Finally, we would like to comment that the city of Chicago, where we live, is an excellent illustration of the problems and opportunities for brownfields remediation, and has provided us access to many experts in the field.

Thanks to all,

—HAROLD J. RAFSON
—ROBERT N. RAFSON

Introduction

This book is about brownfields redevelopment, and it aims to present the viewpoints of all parties involved. It addresses the question of what is necessary to get the job done. A broad overview is given of efforts around the nation, but the book concentrates on Chicago (and Illinois and Cook County) as an example of aggressive efforts to accomplish brownfields remediation. Chicago is an interesting case because of its own efforts, as well as those of private developers. A large section relates to the issues faced by the private developer. This is an evolving field, and throughout the text and in the conclusion there are comments and reflections on the improvements needed to further encourage revitalization of brownfields.

Chicago has been described by Carl Sandburg as the "city of broad shoulders," where Daniel Hudson Burnham, who helped plan the city, decreed that one should "make no small plans"; it has also been described as "the city that works." Chicago faces the enormous problem of abandoned industrial properties, of which it has more than its share because of its history as an industrial center. Chicago recognizes the need to attract and retain manufacturers and jobs, and sees that progress toward solving the problem has been very slow. With Mayor Daley's leadership, Chicago is striding forward to solve these problems itself. While other cities may have begun brownfield redevelopment programs, none of these have been on the scale of the effort in Chicago.

One of the goals of this book is to help the private developer (and all other parties involved) through the myriad of difficulties of brownfields development. It is certainly true that the past record of private development of brownfields has been meager.

To solve some of those issues, the State of Illinois has modified the regulations realistically with the Tiered Approach to Corrective Action Objectives (TACO). The state, county, and city have added incentives to help make brownfields redevelopment more viable. Private development of brownfields is now (spring 1999) rapidly expanding.

Simultaneously, the city is going down its own path of brownfields redevelopment. It will buy and assemble properties and do all the same things developers do, and more. The city enjoys numerous advantages: it has the right of eminent domain, it can negate tax delinquencies, it can provide future tax benefits, it has extensive and inexpensive funding, and it is not driven only by the profit motive.

At this time, both the city and private developers are charging forward to work on brownfields redevelopment. In a few cases efforts may overlap, but for the most part the city is taking on the larger jobs, particularly of large site assemblies the private developers could not have done alone. Certainly, there are enough brownfields sites in Chicago to keep both the city and private developers busy for years to come.

In this time of change, of laws and incentives, the city and private developers have to learn to work together. The city has the intentions and the monetary and legal resources to pursue many redevelopments. It does not presently have the staff to handle the ambitious amount of work to be done, and it will probably solve this with outsourcing. Private developers have the expertise to get jobs completed. The next five years will be a period of greatly expanding activity, through the mutual efforts of the city and private developers, that the city will be proud of.

This is an exciting time for Chicago. Many other cities and developers will do well to keep advised of Chicago's experiences in the next years. The authors will try to help in this regard, and this coming period is certainly going to be active and interesting.

Let us first define what we mean by *brownfields*.

Region 5 of the United States Environmental Protection Agency (USEPA) defines brownfields as "abandoned, idled, or underused industrial and commercial sites where expansion or redevelopment is complicated by real or perceived environmental contamination that can add cost, time and uncertainty to a redevelopment project."[1] It is also possible to expand this definition to cases in which the site development is hindered by additional factors; neighborhood decline, poor infrastructure, and poor locations.[2] Brownfields, while having similarities, can also display great diversity, ranging from corner gas stations to multiacre abandoned chemical plants or facilities constructed over contaminated land.

Brownfields encompass those projects that are economically viable when remediated as well as those that are not, and every level of desirability in between. From a practical point of view, these can be classified as follows:

1. Sites that, despite needed remediation, remain economically viable due to sufficient market demand and value
2. Sites that have some development potential, provided financial assistance or other incentives are available
3. Sites that have extremely limited market potential, even after remediation
4. Currently operating sites that are in danger of becoming brownfields, because historical contamination will ultimately discourage new reinvestment and lending[3]

This definition is affected by changes in the regulations specifying what environmental remediation is required. These regulations are subject to variation in time, location, and use. As the cleanup standards loosen, the set of definitions just listed shifts toward greater economic viability.

The reader of this book no doubt has in mind a particular piece of property, whether it is one he or she aims to develop, one owned by the reader or his or her company, or an area to be developed by a community. That piece of property or area has unique characteristics, and the reader should keep these in mind, as a more generalized discussion is presented in this book.

This book is directed toward the *process* of brownfields redevelopment, from beginning to end. Many interests are involved as a site is investigated and developed. This book aims to reflect the viewpoints of each of those interests. It is meant to be a practical book and to point out the pitfalls as well as the rewards. Since the redevelopment path is somewhat different for the city, as compared to the developer, these discussions have been separated and grouped in Chaps. 4 through 8.

Before each section, a short introduction is inserted that describes the content of the section and why those issues are discussed. Individual sections may be discussed by experts, who often look at a subject from their own unique perspectives. The editor has taken the liberty, in a few cases, of adding comments intended to clarify the perspectives of those sections.

Since the reader's interest may be limited to a unique project, or a specific aspect of many sites, he or she may skip certain sections as not being applicable. We urge the reader to go back and read all the sections at some time, since there is a substantial interdependence between the various steps and the various disciplines involved in reviewing and accomplishing a project.

As will be discussed further and in more detail in the book, brownfields are a major national problem, and estimates state that there are more than

500,000 brownfields sites in the U.S. There may be more if we use the expanded definition to which this book subscribes. These brownfields may be viewed as a negative factor to society, but they are also an opportunity. Over the past generation the revitalization of brownfields has been virtually nil, and in fact the number of brownfields has expanded as industries have idled or abandoned sites. In the past five years, conditions and perceptions have been changing. It is a necessary change, not only in order to use these assets but to eliminate contamination, to reverse urban blight, to reinvigorate urban growth, and to make better use of existing infrastructure. The revitalization of brownfields will be one of the significant positive movements of the next generation, and it is hoped that this book will provide useful information to assist to that end.

We have identified the authors of the various chapters and sections because they each bring to their subject their own experience and attitudes, and the reader should recognize the various viewpoints. The editors have tried to allow the authors latitude in discussing their specialized subjects and to mesh these perspectives in the developers' chapters as well as introductory and concluding sections.

References

1. USEPA Region 5 Office of Public Affairs, *Basic Brownfield Fact Sheet.* (1996)
2. U.S. Office of Technology Assessment (USOTA), *State of the State on Brownfields: Programs for Cleanup and Reuse of Contaminated Sites.* (1995)
3. Davis, T. S. and Margolis, K. D., *Defining the Brownfields Problem, A Comprehensive Guide to Redeveloping Contaminated Property.* Chicago: American Bar Association. (1997)

Brownfields

Redeveloping Environmentally Distressed Properties

1
Background

In the first chapter we present background information for the book. There is a discussion of the development of brownfields, and information about the Superfund law. The liability provisions of the law caused a halt in brownfields redevelopment. Changes in the regulatory laws will be discussed in subsequent chapters.

One of the significant issues is the recognition of environmental remediation liabilities on corporate financial statements. As Gary Ballesteros comments in Sec. 1.3, there are "sleeping dogs" that are being awakened.

There follows an overview of the various stakeholders who are impacted by the brownfield situation. This is discussed in further detail by Donna Ducharme in Sec. 1.5.

1.1 Historical Perspectives

Harold J. Rafson

Progress is not a straight line. It has its dips and turns, and even its retrogressions. Looking at the historical background to a discussion of brownfields, there are many simultaneous tracks: the development of cities, the growth of and changes in industry, major changes in transportation, historical changes of the economy, the depression and the two world wars; the growth and changes of the environmental movement, and the changing perspectives and attitudes toward risk.

Let us begin the retrospective review at the beginning of the twentieth century, both because that scope would encompass virtually all the buildings that would fall into the brownfield category and because we write this at the end of the century.

The beginning of this century was a period of great optimism. The 1890s had seen the completion of westward expansion, the war with Spain and the global spread of American interests, and the continuation of a wave of immigration. Chicago had the World's Columbian Exposition, the growth of skyscrapers, the annexation of land that increased Chicago's area five-fold, and the completion of the Loop on the El. In 1909, the Chicago Plan was originated by Daniel Hudson Burnham.

The history of the environmental movement has several branches. With increasing and controlled expansion into the West, the first objective was to protect irreplaceable national treasures. John Muir was a leader in this field; his efforts brought about the establishment of the first national park at Yellowstone in 1872. With Gifford Pinchot at the Department of Interior, supported by President Theodore Roosevelt, this effort toward conservation was the first major direction of the environmental movement.

This direction was radically changed (or rather expanded) with Rachel Carson's influential book *Silent Spring* (1952), which brought to light the ecological impacts of chemical pesticides. This caused a rethinking that led to the idea that we should be concerned not only with conservation against encroachment by development but with protection against contamination. Contamination can be transported by air, water, or soil, and pollutants are introduced into the ecological system by the manufacture and the use of industrial products.

This consumer concern extended not only to the health of the environment, but the health of the human population as well. In the 1950s, there was concern about the safety of food additives. One federal law that was passed was the Delaney amendment, which stated that no compound that can cause cancer could be added to food. This well-intentioned outright abolition was unrealistic, and it launched the rethinking of how much added risk was socially acceptable. This approach applies to food additives, and to air and water pollution as well.

There was a lot of fuzzy activist, political, and scientific thinking, which muddied the atmosphere. A clarifying work called *Acceptable Risk* was written by William W. Lowrance in 1956. Lowrance's thesis (which we agree with wholeheartedly) is crystalized in the words *acceptable* and *risk*. Lowrance's point is that risk is a scientific concept—it can be measured or estimated (even if there are arguments as to how correctly this is done). What is acceptable is a social question, not to be determined by scientists (except as they, too, are citizens) but by the voice of the people. The scientists have made greater progress in defining and estimating risk than society has in organizing to indicate what is acceptable.

Recent movements to face these issues that began in the 1980s have referred to this concept as *comparative risk*. This is a noteworthy effort, and must go forward to arrive at answers to these questions, however imprecise or questionable those answers may be. The comparative risk movement also arose following the same kinds of obstacles that faced brownfields redevelopment. Well-meaning laws that aimed toward a "clean" environment (and specified the ways such remediation would be paid for) had the same kind of stifling effect on cleaning up environmental contamination as the Delaney amendment had had a generation before on the introduction of new food additives that were beneficial. Very simply, there is nothing cut and dried about these issues; there is no black or white. There must be an agreement that progress must be made. The comparative risk movement continues to wrestle with the problem of defining acceptability, which is very important in terms of allocation of natural resources and getting the most protection for the buck.

We now face new agendas that try to define risk-based objectives. These still have great shortcomings (as far as this author is concerned) because they do not try to answer the question of what is acceptable to society. Nevertheless, they do a more realistic job of defining what is clean, not only as virgin clean but in relationship to realistic use and health risks. The acceptable limits vary from state to state, and in Illinois, for example, they can be established in several ways (tiers). These tiers are a conservative approach that aims for a 1 in 1 million risk; a variation that can be recalculated for each situation and may come up with a less stringent limit; or a third approach that allows for a more open-ended proof of risk.

The point the author wishes to make is that these problems are not new, are not easy, and have not been answered. We continue to grope for answers, and there is a need for continuing reevaluation and change as we work toward a realistic approach to a safer environment.

Let us focus more specifically on the changes that have occurred in Chicago over the past century. As a sign that improvement in the environment is a continuing effort, we should note that in 1900 Chicago reversed the flow of the Chicago and Calumet Rivers. Chicago is virtually flat with poor drainage. Sewage flowed into the Chicago River, which discharged into Lake Michigan, from which water was drawn for the principal water supply. In 1850 and 1889, thousands died of cholera. First, efforts were made to raise the level of the city and improve drainage. In 1889, the Sanitary Authority was established. Chicago, in its typical bold manner, in ensuing years dug the Sanitary Ship Canal and moved a greater amount of earth than was dug for the the Panama Canal. Chicago's first approach, after the river reversals, was to flush the city's waste into the Mississippi with a great deal of the Lake Michigan water. This was later limited, and in 1930 wastewater treatment plants were established to treat wastes before discharge into the river. Chicago's bold planning continues to this day, as enormous underground piping and reservoirs are built as deep tunnels to avoid stormwater flows that exceed plant treatment capacities.

In 1902, the world's largest industrial park (the Central Manufacturing District) was established in Chicago, and it was remarkably successful. Portions of it remain, and some of the facilities now qualify as brownfields.

This is an elegant example of brownfields issues. The Central Manufacturing District was ideal planning in its time, served by railroads and trolley cars. New facilities were designed and built to their specific industries' requirements and according to the latest thinking. In a hundred years (or less), conditions have changed radically. Trucks have replaced railroads. Larger trucks have placed new demands on the design of buildings and overpasses. Automobiles have replaced trolley cars. The needs of the agricultural equipment and steel and metalworking industries have changed, and industries have relocated. The manufacturing district, still a wonder of development with excellent buildings, location, and infrastructure, had to readjust. The district has deteriorated, as have the nearby industrial areas. This decline occured even before it was ever considered that some of these buildings and sites were contaminated because of operating practices during earlier periods of less environmental concern.

Should this area be redeveloped? Yes, to maintain the viability of the city. Should these individual buildings be retained and improved? That is a case-by-case decision, but wherever it is possible to find the client who can put a building to good use, with reasonable modifications, yes. Obviously it requires effort by the city, the developer, the user, and the bank to make this work. It is much easier for a company to move out, to tear down, to build new.

In the 1960s and 1970s, asbestos removal became a common part of any redevelopment of an older industrial facility. Concerns about previous contamination and liabilities were the first considerations in negotiations. Stricter air quality standards in the 1970s and growing citizen complaints made site acquisition in suburban areas and industrial parks more attractive, because environmental problems could be avoided.

In the 1980s, stringent environmental cleanup standards and attachment of liabilities with ownership virtually halted the purchase of contaminated sites. Uncertain cleanup costs, vague or extremely stringent cleanup standards, lack of financing, liability, and the costs in time and money kept developers and manufacturers away from contaminated sites redevelopment. These sites became blights on their communities.

Banks also would not finance projects that had little contamination for fear of becoming attached to the liabilities of the site. These liabilities could extend not just to the cleanup costs, but also to any environmental legal actions including toxic tort claims. Banks would call loans if they suspected that a business had contaminated the collateral supporting the loans. This made a bad situation worse for the owners of the contaminated properties.

From the developers' perspective, the uncertainty of cleanup costs, as well as the time and expense, made brownfields redevelopment impossible. The developers shared the concerns of the banks, worrying that they too would be linked to liabilities, costs of cleanup, and lawsuits.

In the 1990s there were several major changes in the business and environmental landscape that made brownfields redevelopment possible. The most important change has been the reduction and clarification of environmental contaminant residual standards. In addition, lender liability protection, improved cleanup methods, better estimation of remediation costs, and financial incentives have paved the way for brownfield redevelopment.

Older industrial sites that had been abandoned were often in prime locations for redevelopment. These sites were often built close to cities and transportation, and the land had been built on, which meant that the soil conditions and other site issues were favorable. Developers recognized the opportunities but were still cautious because remediation costs were high.

Chicago and other cities have made it clear that the cleanup standards did not need to consider the pathway for migration to groundwater. Chicago legislated the elimination of groundwater wells within the city limits, and its sole source of water is Lake Michigan. This has completely changed the cost and complexity of cleanup. The cleanups deal only with the contaminants within the buildings and within the soil. These changes have led to better estimates of costs and time, since groundwater cleanup had always been uncertain, expensive, and lengthy.

With many of the uncertainties limited, manufacturers and other businesses call more frequently on developers to consider brownfields sites. Banks are now more willing to put up both construction loans and permanent financing when sites are shown to be clean to the standards set by the state environmental agency, the Illinois Environmental Protection Agency (IEPA).

In 1997 the Illinois legislature passed the Tiered Approach to Corrective Action Objectives (TACO) as the new state cleanup standard. These standards are based on health risk and are clear, definite, and understandable. It is widely believed that these standards will be in effect for a long time. This allows the developers to enjoy a greater level of comfort that the remediation of any industrial site can be accurately determined and budgeted.

IEPA now also may provide no further remedial action planned (NFRP) letters for the specified use when the remedial action is completed. NFRP letters and TACO give lenders comfort that these sites will not have continuing cleanup liabilities and that the sites, once clean according to the regulations, will be resalable.

Harold J. Rafson received his B.Ch.E. and M.Ch.E. from the Polytechnic Institute of Brooklyn. He is now an independent consultant working in the area of brownfields redevelopment, environmental assessments, and planning and management of redevelopment. He is nationally rec-ognized in the field of control of odors and toxic and volatile com-pounds. He is editor and author of Odor and VOC Control Handbook, published by McGraw-Hill. He has published and presented over 150 technical papers and speeches over the past 40 years. Mr. Rafson holds numerous U.S. and international patents concerning gas emission con-trol technology and soil remediation. He has been an active member of the Odor and Emission Control Committees of both the Water Environ-ment Federation (WEF) and the Air & Water Management Association (AWMA). As president of Quad Environmental Technologies, Mr. Rafson developed mist scrubbing technology over a period of 20 years. This method has been brought from a concept to a traditionally accepted technology. Mr. Rafson has been directly involved in the design, sale, fabrication, installation, and start-up of over 300 gaseous emission control projects. His technical expertise in air emissions includes test-ing methods and all control technologies and their impacts on commu-nities. Prior to his work in air pollution control, Mr. Rafson was a food process engineer. He is a member of Tau Beta Pi and Sigma Xi.

1.2 The Superfund Statute's Common Law Foundation

Carey S. Rosemarin and Steven M. Siros*

Introduction

Contamination created brownfields, but the "brownfields problem" was created by the Comprehensive Environmental Response, Compensation and Liability Act (CERCLA or Superfund).[1] The universal interpretation of this statute, passed by Congress in 1980, is that persons who purchase contaminated property can be held liable to clean it up, irrespective of whether they had any part in contaminating it. Consequently, would-be purchasers have found potentially contaminated property less attractive, fearing the imposition of liability for exorbitant costs of remediation. The result is that otherwise valuable real estate cannot be sold and instead sits idle—exactly the opposite of the intended effect of CERCLA, which was enacted to force the cleanup of property at the expense of those who prof-ited by the contamination.

---

* The material in Chaps. 1.2, 5.2, and 7.2 was prepared by Carey S. Rosemarin and Steven M. Siros for informational purposes only and does not constitute legal advice. This informa-tion is not intended to create, and receipt of it does not constitute, a lawyer-client relation-ship. Readers should not act upon the information provided without seeking professional counsel.

The legal sections of this book provide a broad perspective of this problem, and discuss the means that have been devised to rectify it. The present chapter explains the historical roots of CERCLA, and shows that many of the concepts embodied in the statute had their origin in the common law (law as it has been developed by judicial precedent, as opposed to statutes). Section 5.2 discusses the current state of the law under Superfund, primarily as it pertains to the liability of owners. Finally, Sec. 7.2 addresses various legal devices that have been used to overcome the possibility of CERCLA liability being imposed, and thus facilitate the transfer of contaminated property.

From the developer's point of view, the passage of CERCLA in 1980 all but turned property redevelopment on its head. The concept that developers could be faced with potentially devastating costs of environmental remediation on their own property and the property of others—merely on the basis of ownership of contaminated property—was met in the real estate industry with shock and anger. The rules of joint and several liability (the imposition of liability on one liable party for the damages caused by other liable parties) and strict liability (liability without regard to fault) for environmental contamination seemed grossly unjust and contrary to the doctrines of fairness and equity upon which the American judicial system was founded.

The negative emotional reaction that CERCLA often evokes is understandable. The CERCLA statute effected major changes in the way in which persons could be held accountable for environmental injuries. Nonetheless, the concept that one person could be held liable for the acts of others long predated CERCLA in American jurisprudence. In reacting to the demand for a cleaner environment, as manifested by the environmental movement of the 1960s and 1970s, Congress gave new powers to the federal government and private plaintiffs, but drew upon preexisting common law concepts to achieve that end. Thus, CERCLA's authorization of the government to require a nonculpable property owner to remediate environmental damages on its own property and that of others may be seen as draconian. On the other hand, as discussed in this chapter, it can also be seen as a logical extension of the combination of a number of familiar common law concepts.

The source of most of these concepts is the law of torts, which is concerned with compensating persons for losses suffered at the expense of a wrongdoer.[2] At its most basic level, a tort is defined as a "civil wrong, other than breach of contract, for which the court will provide a remedy in the form of an action for damages."[3] A tort arises from the existence of a duty to avoid causing harm to others, through acts of omission as well as commission. Every person has a duty of care for the personal and property rights of others while engaged in everyday activities. Carelessness in exercising this responsibility may give rise to tort liability. The duty owed is noncontractual, which means that it does not arise from an agreement between the parties. Well-known torts include assault, battery, negligence, trespass, and nuisance. At common law, the torts most often utilized by property owners seeking redress for environmental harms are trespass and nuisance. The following discussion focuses on these concepts and others to show that Superfund has its roots in the common law and the demand for environmental quality in the mid-twentieth century.

Damage to the Property of Others

One point on which CERCLA and the common law appear to coincide is the situation in which a condition maintained on the property of one person affects the property of another person. Under CERCLA, a property owner may be held liable for the harms that environmental conditions located on the owner's property cause to the property of others.[4] This concept is a direct outgrowth of the common law torts of trespass and nuisance, and is consistent with the precept that a person is responsible for the adverse consequences of his or her actions.

Trespass

At common law, a person commits a trespass where the person:

(1) enters land in the possession of another, or causes a thing or a third person to do so;

(2) remains on the land; and

(3) fails to remove from the land a thing which he is under a duty to remove.[5]

The tort of trespass reflects the importance of the ownership of land in western culture, and prior to the passage of CERCLA in 1980, was an important tool for property owners seeking redress for environmental harms. For example, in *Rushing v. Hooper-McDonald, Inc.*, 300 So.2d 94 (Ala. 1974), a property owner was found liable for trespass where pieces of asphalt from the the property owner's property fell into a stream and were subsequently deposited onto another person's property.[6] Thus, in those situations where a physical object from one person's property enters onto the property of another, the law of trespass may impose liability on the property owner to either pay for the damages or remove the offending object from the other person's property.

Nuisance

The tort of nuisance also provides a basis for holding a property owner liable for environmental harms caused to adjacent property owners, but nuisance, unlike trespass, does not require damage to the plaintiff's property. The offensive act can occur entirely on the defendant's property and still constitute a nuisance. In order to prevail on a nuisance claim, a plaintiff must demonstrate the following:

(1) the defendant acted with the intent of interfering with the use and enjoyment of the land by those entitled to that use;

(2) there was some interference with the use and enjoyment of the land of the kind intended;

(3) the interference that resulted, and the physical harm, if any, from that interference, proved to be substantial; and

(4) the interference was of such a nature, duration, or amount as to constitute unreasonable interference with the use and enjoyment of the land.[7]

The intent required to prevail on a nuisance claim may be demonstrated by evidence that the defendant took no action to abate the interference once the defendant was made aware of such interference.[8] Additionally, the nuisance must tangibly affect the physical or mental health of ordinary people under normal circumstances or conditions.[9]

As noted above, nuisance differs from trespass in that nuisance does not require a demonstration of a physical invasion of another's property, while trespass does. The distinction between an action for trespass and an action for nuisance tends to blur in cases of environmental pollution. The law of nuisance developed well in advance of technology that could detect objects not visible to the human eye, and therefore, in the past, interferences caused by odor, smoke, or noise were likely to be redressed under the law of nuisance.[10] Currently it is recognized that even smoke or odors are made up of very small objects that, in order to cause an interference, must invade an adjacent landowner's property to cause a nuisance.[11]

A single set of facts may often be characterized as both a nuisance and a trespass. A common scenario in which a property owner could face liability under a trespass theory would be where the property owner installed an underground petroleum storage tank that subsequently began to leak. If the petroleum product were to leak onto an adjacent property, the owner of the adjacent property could institute an action for trespass against the other property owner.[12] The same scenario could also support a nuisance action.[13]

Thus, the common law imposed liability on property owners for the harm caused by conditions on their properties to adjacent properties and their owners. To that extent, the corresponding CERCLA liability was not a new concept. However, CERCLA also imposes liability on property owners not only for harms caused by those property owners, but for injuries caused by others. As discussed next, under certain conditions the common law also imposed liability on some parties for the actions of others.

Liability for the Actions of Others

Liability for the Actions of Prior Landowners

Under CERCLA, a property owner may be held responsible for remediation of property even if the environmental condition was in existence prior to the current owner's purchase of the property.[14] Under the common law, the current owner is in part liable for the actions of others. For example, under the law of trespass, an owner who purchased property from which a contaminant was continuing to migrate onto an adjacent property was liable for the trespass, but only to the extent that the trespass continued after the property owner was made aware of the trespass.[15]

In this respect, the law of nuisance is similar. A purchaser of property where a nuisance exists is liable for damages and abatement of the nuisance, even in those situations in which the purchaser had no connection with the activities that created the nuisance in the first place. For example, in *New York v. Shore Realty Corp.*,[16] in which Shore purchased property that already been contaminated by another party, the court reiterated the concept that landowner liability for a nuisance is not based on responsibility for creation of the nuisance, but rather on the fact that the landowner had exclusive control over the land and the things on it, and should have the responsibility for remediating conditions on the land that were a source of harm to others. It did not matter that persons other than Shore had placed the chemical on the site; rather, the court noted that Shore was liable for the maintenance of the nuisance regardless of Shore's negligence or fault.[17]

Liability for the Actions of Joint Tortfeasors

There may also be situations in which CERCLA requires the present owner to remediate the property, but in which the property was contaminated either by third parties or by the present owner and others. An example of such a situation might be the spillage of hazardous substances by several vendors in the course of delivery of chemicals to a manufacturer.

Again, in some instances the common law also provided for the imposition of liability on a single party, even though the injury was caused by
persons acting in concert (joint tortfeasors).

Where the conduct of two or more persons combines to create one indivisible harm, either defendant can be held responsible for the entire
harm.[18] This rule is known as *joint and several liability*. As numerous courts
have noted, fairness and public policy dictate that a wronged party should
not be deprived of the right to recover full damages merely because one of
the negligent parties is unable to pay.[19] However, where a defendant is
required to pay more than his or her fair share of the plaintiff's damages,
then that defendant generally has the right to sue other liable tortfeasors
for the excess in an action for "contribution."[20]

Similarly, Sec. 107 of CERCLA, 42 U.S.C. § 9607, imposes joint and several liability on four categories of "potentially responsible parties" (PRPs)
for remediation of environmental contamination. One of these categories
includes current property owners.[21] Section 113 of CERCLA, 42 U.S.C.
§ 9613, provides a right of contribution where a PRP has been required to
pay more than its fair share of liability. The difference between Sec. 107 of
CERCLA and the common law concept of joint and several liability is that
pursuant to the common law, joint and several liability can only attach to
joint tortfeasors. Under CERCLA, joint and several liability attaches without regard to fault; and liability is based on ownership of the property.

Thus, the provisions of CERCLA that impose liability on current owners
for the actions of former owners and third parties are similar to some
aspects of the common law. Under theories of trespass and nuisance, liability could be imposed on the current owner for conditions created by the
former owner (from the time the new owner became aware of those conditions). Also, joint and several liability could be imposed on a single
party where there existed joint tortfeasors and the harm was indivisible.
Yet, CERCLA also imposes joint and several liability on current owners for
remediation of contaminated properties regardless of the degree of fault of
the current property owner. Even this concept was not foreign to the common law. Although the common law was premised on the rule of fault-
based liability, there existed situations in the common law in which
liability was imposed on persons whose acts caused injuries, despite their
lack of fault. This issue is addressed in the next section.

Strict Liability: Liability
Without Regard to Fault

As just noted, under CERCLA liability for property contamination may be
imposed on the owner of property based solely on his or her status as a

property owner.[22] Similarly, under the common law concept of strict liability, a person could also be held liable for acts that were legal and were undertaken with all due care, that is, acts that did not involve fault on the part of the defendant. An example of such instances where persons are held liable on a strict liability basis is "ultrahazardous activities."[23] The common law also imposes strict liability where a person undertakes an "inherently dangerous activity." Some courts have concluded that the storage of hazardous substances constitutes an inherently dangerous activity.[24]

Strict liability does not depend on the intent of the actor. For example, in *United States v. Earth Sciences, Inc.,*[25] the defendants were conducting a gold leaching operation that discharged cyanide into the Colorado River. The United States brought an action seeking to compel Earth Sciences to remediate the contamination. One of the theories of the United States' action was based on strict liability because the United States argued that the gold leaching operation was an inherently dangerous activity. Earth Sciences argued that it was not liable because there had been no intentional discharge of cyanide into the Colorado River. The court ruled that intent was irrelevant, and Earth Sciences was held strictly liable for the discharge of cyanide into the river.[26]

Therefore, the aspect of CERCLA that imposes liability without regard to fault—strict liability—has a substantial foundation in common law. However, liability under CERCLA, unlike strict liability, does not require a current property owner to conduct an inherently dangerous activity. Status as an owner of contaminated property suffices, as discussed earlier. Additionally, most common law cases dealing with strict liability involve private plaintiffs seeking redress for personal injury or property damage. On the other hand, governmental authorities figure prominently among plaintiffs in CERCLA cases. Nonetheless, as illustrated by *Earth Sciences,* governmental authorities, seeking to protect the public interest, have sued private parties under common law principles to hold them liable for causing environmental damage. This point is addressed in the final section of this chapter.

Government Enforcement

Under CERCLA, an action to force a current property owner to remediate contaminated property can be brought directly by the United States, even if no adjacent property owner is complaining of damages due to the offending hazardous substances, or even if the hazardous substances have not left the confines of the owner's property.[27] The concept that the government can force a property owner to clean up his or her own property can find its foundation in the common law concept of a public nuisance.

A public nuisance differs from a private nuisance in the harm that each causes. A private nuisance causes harm to a particular landowner, while a public nuisance causes harm to the general public—this can include interference with the public health and safety.[28] Thus, the pollution of a stream may constitute a private nuisance if it merely causes harm to riparian owners, but this same action will constitute a public nuisance if it kills fish or contaminates the drinking water.[29] The general rule is that a private party does not have the right to bring an action to abate a public nuisance. Such a right is generally viewed as being possessed by the state.[30]

Thus, the common law concept of public nuisance provided a basis for the government's ability to compel a private landowner to remediate environmental conditions on his or her own property. Moreover, even if the current landowner did not participate in the activities that resulted in contamination of the property, if the condition on the property constituted a public nuisance, the government could compel a current property owner to abate such a public nuisance, similar to its ability to do so in an order issued by the Environmental Protection Agency pursuant to Sec. 106 of CERCLA, 42 U.S.C. § 9606.

Conclusion

Many characteristics of CERCLA have a solid basis in the common law. The common law doctrines of nuisance, trespass, strict liability, and joint and several liability provide the foundation for the CERCLA statute. In enacting CERCLA, Congress extended these concepts, but did not create new powers of whole cloth. In doing so, the legislature provided private and public entities with powerful new tools to force the remediation of contaminated property.

References and Notes

1. 42 U.S.C. § 9601 *et seq.*
2. Prosser and Keeton on Torts, 5th ed. (1984), at 7
3. Prosser and Keeton on Torts, 5th ed. (1984), at 2
4. *See* 42 U.S.C. § 9607(a); *Westfarm Associates Ltd. Partnership v. Washington Suburban Sanitary District,* 66 F.3d 669 (4th Cir. 1995) (noting that a property owner could be liable for damage to property of others caused by release of hazardous substances)
5. Restatement (Second) of Torts 2d, § 158
6. *See also Martin v. Reynolds Metals Co.,* 342 P.2d 790 (Ore. 1959), *cert. denied,* 362

U.S. 918 (defendant liable for trespass where fluoride gases and particulate matter contaminated plaintiff's property)

7. Restatement (Second) § 822; Prosser and Keeton on Torts, 5th ed. (1984), at 2, § 87, at 622–23; *Williams v. Amoco Production Co.,* 734 P.2d 1113 (Kan. 1987) (describing the necessary elements to recover for a private nuisance)

8. *See Williams,* 734 P.2d at 1124. *See also New York v. Shore Realty Corp.,* 759 F.2d 1032 (2d Cir. 1985) (landowner subject to liability for private nuisance on its property upon learning of the nuisance and having the opportunity to abate it)

9. Restatement (Second) of Torts, § 822. *See also Jost v. Dairyland Power Cooperative,* 172 N.W.2d 97 (Wis. 1970) (emission of sulfur fumes from a neighboring property which destroyed vegetation on plaintiff's property constituted a nuisance); *Boomer v. Atlantic Cement Co.,* 257 N.E.2d 870 (N.Y. 1970) (emissions of particulate matter and fumes from cement plant constituted a nuisance to neighboring landowners)

10. *Meat Producers Inc. v. McFarland,* 476 S.W.2d 406 (Tex. 1972) (pollution of the air constitutes a nuisance rather than a trespass)

11. *See Boomer,* 257 N.E.2d 870 (particulate matter treated as nuisance even though particulate matter would clearly qualify as a physical object)

12. *See Hudson v. Peavey Oil Co.,* 566 P.2d 175 (Ore. 1975) (property owner liable for trespass from underground storage tank onto adjacent property)

13. *See Hauser v. Calawa,* 366 A.2d 489 (N. Ham. 1976) (landowner liable for nuisance where septic system backed up and caused damage to adjacent landowners' property)

14. *Nurad Inc. v. W.E. Hooper & Sons, Inc.,* 966 F.2d 837, at 846 (noting that liability under CERCLA is triggered by ownership of the site and not by responsibility for the contamination)

15. *Williams v. Baton Rouge,* 1998 WL 289747, at *11 (La. App. Ct. April 30, 1998) (where a person allows an object to remain upon the land of a property owner and fails to remove it, that person commits a continuing trespass and may be ordered to remove the offending object from plaintiff's property); *Rudd v. Electrolux Corp.,* 982 F. Supp. 355 (M.D. N. Car. 1997) (once an owner of property becomes aware of environmental contamination on his property, the owner is responsible for any continued migration of the contamination onto adjacent property under a trespass theory regardless of whether the owner of the property actually was responsible for the contamination). *See also Bassilakis v. Saland Corp.,* 1998 WL 197627 (Conn. 1998) (noting that injunctive relief is available to compel abatement of continuing trespass regardless of when release occurred).

16. *New York v. Shore Realty Corp.,* 759 F.2d 1032 (2d Cir. 1985)

17. *Shore,* 759 F.2d at 1037

18. *See Wisconsin Natural Gas Co. v. Ford, Bacon & Davis Construction Corp.,* 291 N.W.2d 825 (Wis. 1980) (noting that to the extent that there are multiple tortfeasors whose harm cannot be easily divided, each tortfeasor is liable for the entire harm)

19. *Martinez v. Stefanich*, 577 P.2d 1099 (Color. 1978)

20. *See Cordier v. Stetson-Ross Inc.*, 604 P.2d 86 (Mont. 1979) (where a defendant is subjected to joint and several liability, the right to contribution exists as to the remaining tortfeasors)

21. *See Sun Company, Inc. v. Brown & Ferris, Inc.*, 124 F.3d 1187, 1191 (10th Cir. 1997)

22. *See Nurad, Inc. v. W.E. Hooper & Sons, Inc.*, 966 F.2d 837, at 846 (stating that the trigger to liability under CERCLA is ownership or operation of a facility at the time of disposal, not culpability or responsibility for the contamination)

23. *See Poole v. Lowell Dunn Co.*, 573 So.2d 51 (Fla. App. Ct. 1990) (blasting qualifies as ultrahazardous activity)

24. *See State of New Jersey v. Arlington Warehouse*, 495 A.2d 882 (N.J. 1985) (storage of hazardous substances subjects landowner to strict liability); *see also Sterling v. Velsicol Chem. Corp.*, 647 F. Supp. 303 (W.D. Tenn. 1986), *aff'd in part, rev'd in part*, 855 F.2d 1188 (6th Cir. 1988) (defendants subject to strict liability for creating, maintaining, and operating chemical waste site); *Branch v. Western Petroleum, Inc.*, 657 P.2d 267 (Utah 1982) (landowner liable under doctrine of imposed liability for contamination of adjacent landowner water wells due to storage of waste oil on property). *See also McAndrews v. Collerd*, 42 N.J.L. 189 (1880) (storing explosives on property subjects one to strict liability for all actual injuries caused by the explosives).

25. *United States v. Earth Sciences, Inc.*, 599 F.2d 368 (10th Cir. 1979)

26. *Earth Sciences*, 599 F.2d at 372

27. *See* Section 106 of CERCLA, 42 U.S.C. § 9606; *United States v. Rohm & Haas*, 2 F.3d 1265 (3d Cir.1993) (pursuant to CERCLA Section 106, government can order private party to initiate a site cleanup)

28. Prosser and Keeton on Torts, 5th ed. (1984), at 6

29. Prosser and Keeton on Torts, 5th ed. (1984), at 645

30. *See Indiana Limestone Co. v. Staggs*, 672 N.E.2d 1377 (Ind. App. Ct. 1996) (noting the general rule that a private party does not have standing to seek abatement of a public nuisance); *Maples v. Quinn*, 64 So.2d 711 (Miss. 1953) (state had the power to order the abatement of certain operations at a fish processing plant because odors constituted a public nuisance). *See also Shore Realty*, 759 F.2d at 1035 (state has standing to bring suit to abate nuisance in its role as "guardian of the environment").

Carey S. Rosemarin is a partner in the Chicago office of Jenner & Block. Mr. Rosemarin has practiced in the environmental discipline throughout his entire career. Prior to joining Jenner & Block, he served as assistant regional counsel for the U.S. Environmental Protection Agency (Region V) for approximately six years, and later practiced as a partner in a major Chicago law firm. Previously he was employed by Union Carbide Corp. at the Oak Ridge National Laboratory.

Mr. Rosemarin is a past chairman of the Environmental Law Committee of the Chicago Bar Association. He lectures frequently on envi-

ronmental issues and has authored articles on various subjects of interest to environmental professionals. He has substantial expertise in environmental regulation, litigation, and enforcement actions involving air, water, and hazardous substances, and corporate and real estate transactions requiring environmental counseling.

Mr. Rosemarin holds a B.S. from the University of Michigan School of Natural Resources and an M.S. from the Pennsylvania State University. He received his J.D. in 1978 from the University of Tennessee College of Law.

Steven M. Siros is an associate at the Chicago office of Jenner & Block. Mr. Siros is a 1994 magna cum laude graduate of the Northern Illinois University College of Law. He received a B.A. in Political Science and French from Illinois State University.

While in law school, Mr. Siros was articles editor for the Law Review *as well as president of the International Law Society. Prior to joining Jenner & Block, Mr. Siros served two years as a law clerk for the Honorable Justice Lawrence D. Inglis of the Second District Illinois Appellate Court.*

Mr. Siros focuses his practice on environmental law. He has extensive experience in related air issues as well as Superfund cost recovery actions, RCRA citizen suit actions, and regulatory and due diligence counseling.

Mr. Siros has written several articles concerning environmental law. His articles have appeared in the Northern Illinois University Law Review *and the* Southern Illinois University Law Review.

1.3 Recognition of Environmental Remediation Liabilities

Ernest Di Monte

In January 1993, the American Institute of Certified Public Accountants (AICPA) held an environmental issues discussion. The main objectives of the discussion were to examine practice problems in applying generally accepted accounting principles to environment-related financial statement assertions; to detect environmental issues that may need authoritative accounting and auditing guidance; and to make inroads toward the development of guidance on applying existing accounting and auditing standards to environment-related matters. Out of this discussion came Statement of Position 96-1 from AICPA.[1]

Statement of Position 96-1 is provided to improve and narrow authoritative literature of existing principles as applied by entities to the specific circumstance of environmental liabilities. This may include the recognizing, measuring, and disclosing of environmental remediation liabilities in the financial statements. For purposes of this book we will be discussing the aspects of recognition, measurement, and disclosure of environment remediation liabilities.

Recognition is the determination of *when* amounts should be disclosed in financial statements for the purpose of reporting environmental liabilities. *Measurement* has to do with the *amount* to be reported in the financial statements. According to the Financial Accounting Standards Board (FASB) Statement of Financial Accounting Standards No. 5, "Accounting for Contingencies," the accrual of a liability is required if (1) information available prior to issuance of the financial statements indicates that it is probable that an asset has been impaired or a liability has been incurred at the date of the financial statements and (2) the amount of the loss can be reasonably estimated.

Once an entity has determined that it will probably incur costs for the remediation of an environmental liability, the entity should estimate the liability it figures to bear. This estimate should be based on available information. The entity should include its allocable share of the liability for a specific site. Many sites that are contaminated have been contaminated by a few different entities, for example a waste disposal site that has been used as a dumping ground for many companies. This estimate should then be recognized as a liability on the balance sheet of the entity and as a charge to income. The liability should also be disclosed in notes to the financial statements; it should be described along with any additional appropriate information.

Estimates formed in the early stages of remediation can vary significantly. Many times, early estimates require major revision. Many factors are essential to forming cost estimates, such as the extent and types of hazardous substances at a given site and the given technologies that can be applied to abatement of the site. Also to be considered when developing estimates are the number of potentially responsible parties and their financial ability to pay their share of the environmental cleanup costs.

In disclosing notes to the financial statements, entities are encouraged to report the nature and a brief description of the environmental remediation liability. Included in the description may be the estimated time frame of disbursements for expenses, the estimated time frame for probable recoveries (i.e., insurance proceeds, other responsible party recoveries), the accounting principles used, and other pertinent information. Disclosures to the financial statements themselves include the amount of the liability to be incurred. This amount should only be incurred on the financial statements if the liability is probable. If the liability is reasonably possible, it should be disclosed in the notes to the financial statements but not accrued on the financial statements. Also to be included on the financial statements are any future receivables, such as receivables from other responsible parties, recoveries from insurers, or recoveries from prior owners, that are related to the environmental remediation liability. Remediation expenses should be disclosed as a charge against operating income, since the costs are considered a part of the normal operation of a company.

Reference

1. American Institute of Certfied Public Accountants (AICPA), Statement of Position 96-1, "Environmental Remediation Liabilities." New York: AICPA. (October 10, 1996)

Ernest R. Di Monte passed the CPA exam in May 1987 after graduating from Northern Illinois University with a B.S. in accounting. He began his career at KPMG Peat Marwick in Chicago as a member of the real estate division of the tax department. There he gained invaluable experience in the tax aspects of complicated real estate transactions. In 1990 he formed the CPA firm Ernest Di Monte & Associates with his father. The firm specializes in real estate and construction accounting. Its clients include some of the most prominent developers in the Midwest.

1.4   What Is at Stake for Everyone?

Harold J. Rafson

More is at stake than money. Consider the example of an abandoned site, in an urban area, that has contaminated a building, the land, and the groundwater.

Who Is Affected—and in What Ways?

1. The owner is affected directly by the loss of the property, by the environmental liabilities attached to the site, and by the need to acknowledge the existence of these environmental liabilities. This can impact the corporate financial statement, the market value of shares, and relationships with stockholders, bankers, or other lenders. The property owner's relationship with neighbors, community leaders, city officials, and local and state regulators can be harmed. There may be an ongoing stigma attached to the property longer than is warranted. The owner may become liable to a variety of lawsuits, including third-party lawsuits and toxic torts.

2. Neighboring property values can also be impacted by the existence of contamination on the site. Neighbors will experience increased health risks. This is especially true where groundwater is contaminated; groundwater is mobile and can affect broad areas, sometimes miles from the source of contamination. The value of the neighboring property may also be diminished.

3. The community is negatively affected by the urban blight caused by abandoned property, as well as by the loss of jobs from the local labor pool.

4. The city suffers a loss of tax revenues, of jobs, and possibly of population. It finds its investments in infrastructure inadequately used. There may be added costs of police and fire protection caused by these degenerated properties and neighborhoods.

5. The lender's loans to the company are at risk. Possibly a company may declare bankruptcy, or it may move out of state, resulting in a loss of business for the bank. The property may have been used as collateral, and its value is diminished by contamination.

6. Suppliers to the company may suffer losses as a result of bankruptcy. Many of the suppliers are local businesses that are affected if the company moves to the suburbs or out of state.

7. All taxpayers—at some time, everyone—is going to have to clean up the mess left by a polluter. Whatever the mechanism—a tax break or a federal cleanup program—all costs ultimately get back to the citizens. Even if the company pays for a cleanup through a price increase, this ultimately is paid for by the consumer. The person in the street is the payer of last resort, resulting in a transfer of funds from the taxpayer to the stockholder. It is in the taxpayer's interest not to be taken advantage of by corporate irresponsibility or mismanagement.

What If Brownfields Development Is Not Done?

There will be a continuation of all the negative effects—urban blight; community deterioration; flight of workers and jobs to the suburbs or other states or countries.

Every business owner with a brownfield property, and every developer, must be aware of changes in the conditions that deter or foster redevelopment. The decision made last year not to pursue a brownfield redevelopment may be wrong today. All of the conditions are improving—financing, insurance, liability, incentives, real estate values, remediation standards, and methods. All who are involved can affect the viability of the project by their actions. There are continuing negative factors as well—inertia, uncertainty, increasing incentives to move out, unimproved urban issues.

This book is about this dynamic, evolving process. In these pages we highlight some cases where the authors believe extra efforts are required, or changes in attitudes are needed, to create a better climate for revitalization.

For biographical information on Harold Rafson, see Sec. 1.1.

2
National Lessons and Trends

Charles Bartsch provides a broad-ranging perspective on many of the issues involved with brownfields projects in states throughout the nation. Many of these issues will be revisited in later chapters. The book focuses mainly on Chicago's efforts in this field. This chapter offers insights into other areas and programs for dealing with brownfields.

2.1   Coping with Contamination

Charles Bartsch

Brownfields have emerged as the preeminent economic development issue of the 1990s. Communities all across the country have had to address the legacy of their past in the context of contamination, complexity, and uncertainty. Brownfields are like fingerprints; no two are alike. As such, they pose significant challenges for local elected officials and economic development agencies. Redeveloping these sites can be a costly proposition. The complicated process and the legal hurdles of acquiring, cleaning, and reusing the sites can be expensive in terms of site preparation expenses and fees, and costly in terms of time delays. Site evaluation processes, testing, possible legal liabilities, and other factors serve to deter private participation in activities geared toward bringing old industrial sites back to productive use. In many situations, the private development and financial sectors are not able or willing to act on their own to ensure that the full economic potential of site reuse will be achieved.

As of early 1998, more than three dozen states had put so-called voluntary cleanup programs (VCPs) in place. VCPs have gone a long way toward bringing certainty to the cleanup process, defining the necessary extent of cleanup, what types and levels of contamination can be left on site, and how contamination will be addressed and controlled. These advances have taken place even in the absence of federal Superfund reauthorization. In practice, they have helped sort out the liability process and bring some finality to it, which has made brownfields reuse more economically viable for prospective new site users. Across the country, thousands of sites have gone through state VCPs and gained some measure of closure.

Now, critical funding gaps are, in fact, the primary deterrent to site and facility reuse. The financing situation is especially gloomy for start-up firms or small companies with little collateral outside the business. How this is addressed is an emerging trend in brownfields reuse strategies. Clearly, governments at all levels can find creative ways to help enterprises overcome the obstacles that environmental contamination brings to the economics of the site reuse process; such actions range from regulatory clarification for liability stemming from loan workouts to direct financial assistance programs. For decades, federal, state, and local governments have used or sponsored public finance mechanisms to stimulate economic activity in certain geographic areas or industries. Now, publicly driven economic development initiatives are reaching into new sectors and incorporating new concerns, such as environmental improvement. Brownfields reuse strategies and techniques are rapidly evolving.

Redeveloping Contaminated Sites—Financing Barriers in Brief

In most areas, adequate private financing to carry out both cleanup and redevelopment activities is simply not available. The costs of preparing financing packages have tripled since 1980 because of environmental requirements. In practice, whether sites are cleaned and reused or not boils down to dollars and cents; even if an old industrial facility has only small amounts of contamination, site assessment and cleanup add considerably to the cost of a redevelopment project, making the project's economics much harder to justify.

In large part, fear of Superfund liability has made lenders wary, even though Congress in 1996 clarified lender liability uncertainties sparked by several court cases; this fear makes lenders reluctant to provide the resources needed to carry out site reuse projects. Even though only a handful of lenders have actually been held liable for the impacts of contamination on projects in which they have participated, there is no question that environmental concerns have affected banking practices. Trade groups such as the American Bankers' Association and regulatory agencies such as the Federal Home Loan Bank Board have outlined the types of risk that could emerge from transactions involving environmentally contaminated property. In sum, these risks include:

1. Reduced value of collateral

2. Borrowers' inability to repay loans if they must also cover site cleanup costs

3. Potential for the bank to become liable for the cost of site cleanup if it forecloses on the property, or to forgo its collateral interest and not foreclose in the face of significant cleanup costs

4. Preemption of a mortgage loan security by a cleanup lien imposed under a state "super lien" law

5. Possibility that the borrower would not maintain the facility in an environmentally sound manner (which could trigger any of the other risks)

In the face of these concerns, lenders have changed the way they deal with projects that even remotely involve hazardous wastes in response to these risks—real or perceived. This, in turn, affects the reuse potential of specific sites as well as the broader economic development climate in many areas. In practice, financial institutions grappling with concerns over environmental liability and contaminated project sites are either sharply curtailing their level of lending, especially to manufacturing, or simply cutting off financing for certain types of businesses, such as those that routinely handle toxic substances—service companies such as dry cleaners and auto body shops, as well as manufacturers such as high-technology metal fabricators, semiconductor makers, and tool and die shops. In addition, previously used sites, brownfields or not, must often shoulder significantly increased transaction costs because additional documentation and thorough environmental assessments (which can easily cost $50,000 or more) are required; often, cleanup must be completed as a condition of loan approval.

Money to undertake site investigations and carry out cleanups is the hardest piece of the financing puzzle to fit together. Considerable attention is being focused on this problem, and solving it will be one of the next brownfields reuse process issues. Increasingly, the public sector is stepping up to the plate to provide seed money for these purposes; this is becoming a necessary ingredient to establish a climate in which brownfields reuse can flourish.

Promoting Reuse: Goals
of Public Sector Incentives

The public sector can do much to help level the economic playing field
between greenfields and brownfields sites. Creatively crafted and carefully
targeted incentives and assistance can help advance cleanup and reuse
activities. Such strategies must recognize, however, that brownfields projects differ considerably in terms of barriers to investment and opportunities to redevelopment. Therefore, no one "best" public sector approach will
fit all needs. Clearly, a variety of incentives can make the most effective use
of public sector assistance, as well as improve the climate that invites private investment in brownfields. These incentives, used separately or in
combination, are being adopted to meet any of several goals, including:

- Reducing the lender's risk, making capital more available by providing
  incentives or legal clarification for lending institutions to help companies or projects at sites deemed riskier because of their prior uses

- Reducing the borrower's cost of financing, for example by making capital more affordable by subsidizing the interest charged on brownfields
  loans, or by establishing policies that reduce loan underwriting and
  documentation costs

- Easing the developer's or site user's financial situation by providing
  incentives, such as tax credits, that can help improve the project's cash
  flow

State and local governments, in many respects, are the brownfields inno-
vators. (A complete listing of current state-level brownfields project activity
is found in Table 2-1.) Typically, reuse success stories are found in places that
have adopted their own site characterization and reuse tools and are creatively built on the foundation provided by federal programs and policies.

Yet as important as these initial successes are, the potential exists for even
greater activity. Many jurisdictions are starting to explore ways to help
prospective reusers overcome the difficulties that contamination can bring
to the redevelopment process, setting up finance programs to ease the cost
or terms of borrowing, augmenting private funds, or filling funding gaps
that the private sector will not bridge. Moreover, public sector support does
not have to be limited to helping specific companies; other related activities
can be financed that help improve the broader brownfields investment climate. For example, states and localities can assume some of the responsibilities for site preparation and cleanup, recovering some of their costs during
subsequent site sale or development. Jurisdictions can support such activities by earmarking tax revenues, loan repayments from other programs,

and other sources of funds to pay for necessary project activities such as site testing or soil removal.

State Initiatives in Brownfields Reuse

As can be seen in Table 2-1, 38 states have established formal voluntary cleanup programs (VCPs) and several others have put similar procedures in place to help bring considerable certainty and finality to the remediation process. Some states have gone further, though, directing financial assistance to support cleanup and reuse activities. More than two dozen states have launched some type of financing initiatives or incentives linked to state voluntary site cleanup programs. These focus on brownfields reuse situations, and many are targeted to small and midsize companies that go through state voluntary cleanup programs. Most of these programs are less than two years old—very new in the economic development context—although initial results seem promising.

Characteristics of State VCPs

Three dozen plus programs in place today:

- Twenty-five enacted in 1993 or later
- Nearly a dozen older programs changed significantly in 1997
- Many new initiatives already introduced into state legislatures in 1998

Eligibility:

- Typically, open to any contaminated site except landfills, sites on Superfund's National Priority List (NPL), or sites subject to corrective action under other federal environmental programs (notably the Resource Conservation and Recovery Act (RCRA) and leaking underground storage tank (LUST) sites)
- A few states have additional limitations or targeting requirements

Oversight:

- Older programs—state sign-off on remediation plans, state review of cleanup activities
- Recent approaches—state oversight varies by level of cleanup required, type of site (i.e., orphan or prospective purchaser), involvement of private sector

Table 2-1. Brownfields: Where the States Are

This table attempts to summarize the brownfield program state of the art as 1998 commences. A few points: eligible sites typically are limited to volunteers and include all types of contaminated sites except for Superfund, RCRA, or LUST sites, as well as landfills. Exceptions are noted. The most common types of assurance provided, as noted, are no further action (NFA) letters and covenants not to sue (CNTSs). A few states are developing generic cleanup standards pegged to types of site use, but virtually all take future site use into consideration when running their voluntary cleanup programs. Identifying financing programs targeted to brownfields and tax and other incentives to attract private investment to brownfields proved challenging; typically, in practice, the applicability of specific programs comes down to agency interpretation of brownfields-type site activities as eligible within the program's scope. This chart includes programs directly available through state voluntary cleanup programs, as well as those identified by state agency staff as having consistent applicability to brownfields reuse efforts. The most common types of assistance include grants and revolving loan fund (RLF) initiatives.

This information is based on telephone interviews with state environmental and economic development staff in all 50 states, conducted in late summer 1997, and additional information submitted by states as part of their review processes. In addition, state officials were provided with a copy of this matrix for a final fact check in late 1997.

State	Financing program targeted to brownfield situations	Voluntary cleanup programs and assurances provided	Incentives to attract private investment to brownfields
Alabama	• Industrial development grants, up to $375,000; can be adapted for brownfields purposes	• State voluntary cleanup program (1996)—offers NFA letter or notice of completion	
Alaska	• Cleanup grants for underground storage tanks • Contaminated Sites Remediation Program—response fund (not limited to brownfields) available when companies not able to clean up	• Pilot voluntary cleanup program (1996), final program anticipated in late 1998—offers NFA letter after streamlined cleanup process with less regulatory oversight; no sites with groundwater contamination eligible	

	Voluntary Cleanup Program	Financial Incentives	Related Incentives
Arizona	• VCP (1997) with three components —agency-wide voluntary remediation program and a Water Quality Assurance Revolving Fund (WQARF), with state oversight of the cleanup and subsequent letter of completion (prospective purchaser agreements available for sites with WQARF liability issues); and a Greenfields Pilot Program with soil cleanup oversight by a "certified remediation specialist" who drafts an NFA document that releases owner from any further action by state DEQ; all sites eligible except those subject to enforcement action		• Related incentives include credits on income tax for average wage × 100 if the firm is in the state enterprise zone program • Refund on sales and use taxes on machinery and building material for EZ participating firms
Arkansas	• Voluntary cleanup program (1995)—offers CNTS and comfort letter to lenders; limited to prospective purchasers of abandoned industrial, commercial, or agricultural properties	• Low-interest loan program for brownfields projects, created within existing state LLF program	
California	• Voluntary cleanup program (1993)—offers NFA letters for sites needing no remediation; CCC once cleanup is completed; prospective purchaser agreement policy adopted July 1996	• Mello-Roos Districts—designation allows community to abate property taxes and issue bonds to capitalize RLFs for site assessment and cleanup	• Mello-Roos designation allows property tax abatements

Table 2-1. (*Continued*)

State	Financing program targeted to brownfield situations	Voluntary cleanup programs and assurances provided	Incentives to attract private investment to brownfields
Colorado		• Voluntary cleanup program (1994)—offers NFA letter; geared (but not limited to) current owners of contaminated sites	
Connecticut	• Urban Sites Remedial Action Program—capitalized with $30.5 million in state bond funds for assessment/remediation of sites in "distressed municipalities" and "targeted investment communities"; Department of Environmental Protection (DEP) can clean up and acquire site if it chooses, recovering cost from future users (nearly all money already allocated) • Dry Cleaner Establishment Remediation Fund for financing (maximum $50,000/year) soil and groundwater remediation and prevention • Special Contaminated Property Remediation and Insurance Fund—loans to municipalities and private entities for Phase II and III investigations and demolition costs	• Urban Sites Remedial Action Program (1992)—offers CNTSs to new owners, or to current owners not associated with contamination who agree to clean and "productively" use the property; sites may be Superfund, RCRA, or LUST sites, although eligibility may be restricted where ground-water is used for drinking; site owners may use state "licensed environmental professional" to verify remediation	• Enterprise zone program incentives—tax abatement of 5 years and 80% of local property taxes on real estate improvements; 10 years/50% tax credit; 7-year minimum deferral of increased taxes resulting from property value rise after remediation has been completed

Delaware	• Grants for 50% of site assessment costs, up to $25,000 • Low-interest loan program—up to $250,000 for 90% of cleanup costs • Existing RLF for wastewater treatment may be extended to include brownfields activities	• Voluntary cleanup program (1994)—site's proximity to drinking or surface water supply or residential population places constraints on eligibility; offers NFA letters; prospective purchasers may sign a consent decree for contribution protection, and new owners of remediated sites may receive a CNTS	• Quasi-public Riverfront Development Corp has $25 million to acquire, investigate, and redevelop sites • Tax credits of $650/year per new job created related to cleanup and redevelopment ($900/year in poverty areas)
Florida	• $3 million for disbursement to USEPA pilot communities or pilot applicants	• Brownfield Redevelopment Program (1997)—offers NFA letters; PRPs are eligible and liability protection is offered under certain circumstances; state incorporates risk-based corrective action principles	• $2500 tax "bonus refund" per job created at a remediated site for certain industries • 35% tax credit, up to $250,000 per site, for costs of voluntary cleanup activity integral to site rehabilitation; credits can be transfered once to a new owner • Revolving loan trust fund provides low-interest loans to local governments, community redevelopment agencies, or non-profit corporations, for the purchase of outstanding, unresolved contractor liens, tax certificates, or other liens or claims • Guarantee program provides up to 5 years of loan guarantees or loan loss reserves for primary lender loans for redevelopment projects in brownfields areas

Table 2-1. (*Continued*)

State	Financing program targeted to brownfield situations	Voluntary cleanup programs and assurances provided	Incentives to attract private investment to brownfields
Georgia	•Hazardous Waste Trust Fund—provides local governments with money for site investigation and remediation at solid waste disposal facilities; no more than $2 million per site	•Hazardous Site Reuse and Redevelopment Act program (1996, amended 1998)—offers limited liability relief via a certification mechanism; site must be listed on state's hazardous waste inventory; 1998 amendments expanded and clarified lender liability and prospective purchaser relief	
Hawaii		•Voluntary Response Program (1997)—offers letter of completion giving future liability exemption to prospective purchasers and to current owners, which is transferable to prospective purchasers; state oversight costs paid by participants	
Idaho		•Voluntary cleanup program (1996)—offers COC and CNTSs; sites not subject to existing regulations may participate	•Idaho Land Remediation Act—sites receiving a CNTS may qualify for a 7-year, 50% tax break on the property's appreciation due to remediation

Illinois	•Brownfields Redevelopment Grant Program—$1.2 million available each year to municipalities (limit of $120,000 per city) to coordinate activities related to reuse (but not pay for cleanup) •Bank Participation Loan Program (in Chicago)—up to $250,000 or $350,000 for commercial and industrial loans (respectively) that are matched by banks at 75% of prime rate, for terms from 3 to 15 years	•Illinois Pre-Notice Site Cleanup Program (1989, revised 1996)—offers no further remediation letter, and attorney general can issue CNTSs; sites proposed for the NPL may be eligible if PRP can provide suitable assurances; 1996 revision shifted liability from joint, strict, and several to proportional share	•Environmental remediation tax credit—25% income tax credit for developers who restore contaminated sites, maximum of $40,000 per year and $150,000 per site; credits begin after the first $100,000 of development costs (this floor waived for sites in enterprise zones); credits may be transfered to new owners •Property tax credit in Cook County (Chicago area) for redevelopment and cleanup costs
Indiana	•Environmental Remediation RLF—$10 million over 3 years for "activities necessary or convenient to complete remediation activities on brownfields," conducted by "political subdivisions" through a priority ranking system, based on factors such as the ability of the community to contribute money to the project, the property's new use, the level of community support, and what the political subdivision has done to get that support; different rankings are used for communities of different size	•Voluntary cleanup program (1993)—offers COC, issued by Indiana Department of Environmental Management (IDEM), followed by a CNTS from governor's office; any contaminated site may be determined to be eligible	•Brownfield revitalization zone tax abatements—available in locally designated "brownfields zones"

Table 2-1. (*Continued*)

State	Financing program targeted to brownfield situations	Voluntary cleanup programs and assurances provided	Incentives to attract private investment to brownfields
Iowa	• Physical Infrastructure Assistance Program—offers loans, loan guarantees, or cost shares, adaptable to brownfields projects meeting development criteria	• Voluntary cleanup program (1997)—offers limited liability relief; rules still being written, expected October 1998	• TIF mechanism allows cities or counties to reimburse response action costs over a 6-year period
Kansas		• Voluntary Cleanup and Property Redevelopment Program (1997)—in process of writing regulations; anyone capable of gaining access to a contaminated property for assessment and/or cleanup activities, and adjacent property owners, can receive NFA determination	
Kentucky		• Voluntary cleanup program (1997)—offers NFR letter for property owned by local units of government; assurance transferable to future owners	
Louisiana		• Voluntary cleanup program (1996)—can offer COC and exemption from liability; eligible sites must qualify as "identifiable area of immovable property"	
Maine	• Pilot low-interest RLF for municipalities, for site assessments of tax delinquent properties	• Voluntary Response Action Program (1993)—offers COC for all pollutants identified in site assessment	

Maryland	•Brownfields Revitalization Incentive Program—low-interest loans/grants to persons conducting voluntary cleanups (pool is up to $1 million for fiscal year 1999)	•Brownfield Cleanup/Smart Growth program (1997)—offers COC or NFR letter; sites contaminated after 10/1/97 not eligible	•Brownfields Revitalization Incentive Program—5-year, 50% state tax credit (and optional 20% local, for a total tax credit of 70%) to offset increase in property tax due to remediation; tax credits may be extended to 10 years in designated EZs; incentive available in jurisdictions that agree to contribute 30% of the increase to the state's Brownfield Revitalization Incentive Fund
Massachusetts	•Redevelopment Access to Capital Program (pending in legislature)—$15 million for environmental insurance fund to pay for unanticipated costs associated with an approved cleanup or to guarantee private loans made for cleanup and redevelopment up to $500,000 (requires equal private investment) •Brownfields Redevelopment Fund—$30 million for low-interest loans to "eligible persons" for site assessment (up to $50,000) and cleanup (up to $500,000) in economically distressed areas; cities, redevelopment authorities, and community development corporations can seek similar amounts in the form of grants	•Brownfields Program (1998), supplements 1994 Clean Sites Initiative—offers CNTS liability protection to "eligible persons" (i.e., owners or operators) and "eligible tenants" who did not cause contamination or operate the site at the time of release, as well as site owners downgradient from an off-site source of contamination; site can be transferred before cleanup is complete and still get protection	•Massachusetts Economic Development Initiative—5% state investment tax credit, as well as 10% abandoned building tax deduction, both geared toward properties in economic opportunity areas; priority for state capital funding

Table 2-1. (*Continued*)

State	Financing program targeted to brownfield situations	Voluntary cleanup programs and assurances provided	Incentives to attract private investment to brownfields
Massachusetts (continued)	• Reclamation Pay Back Fund—up to $500,000 in loans for assessment/ cleanup, to cities or towns that certify that they will pay back the loan with half the property taxes generated by the redevelopment • Remediation tax credits ranging from 25% to 50% (depending on the level of cleanup) for innocent parties who "diligently" pursue site cleanups in economically distressed areas		
Michigan	• Site Reclamation/Site Assessment Grants—$45 million in bond proceeds; $35 million for assessment and cleanup at sites where a developer has been identified, and $10 million for assessment at sites "with redevelopment potential," available until funding exhausted • Cleanup and Redevelopment Fund —capitalized annually at $30 million per year (fiscal years 1997 and 1998, proposed fiscal year 1999); also supports Revitalization RLF program loans to local governments	• Natural Resources Environmental Protection Act (1994, amended in 1995)—completion of a baseline environmental assessment and submitting it to DEQ prior to or within 45 days of purchase provides an exemption to liability for existing contamination; nonliable new owners must use "due care" when redeveloping the property; cleanup standards are based on land use	• 10% single business tax credit ($1 million cap) for innocent party's development costs (not cleanup costs) on a property included in a brownfield plan of a Brownfield Redevelopment Authority; credit carries forward for 10 years

	•Revitalization RLF—$4 million for loans to cities for site assessment, demolition, and removal actions, with an interest rate of 2.25%, repayable over 15 years with 5-year deferral of repayment and interest to allow cities to repay loans from tax increments collected by a Brownfield Redevelopment Authority •Brownfield Redevelopment Authorities, which have TIF/bonding authority, can set up a site remediation revolving fund from tax increments captured after remedial actions are paid for •Clean Michigan Initiative—$675 million bond issue on November 1998 ballot, includes $335 million for cleanup activities; $20 million of this designated for grants and loans to local governments and Brownfield Redevelopment Authorities for cleanup of sites "with redevelopment potential"		
Minnesota	•Contamination Cleanup Grant Program—$7.8 million in grants to cities for cleanup administered by the Department of Trade and Economic Development for sites with development potential; up to 75% of project costs •Metropolitan Council in Twin Cities area offers brownfields project grants in a 7-county area	•Voluntary Investigation and Cleanup Program (1988)—offers 6 levels of assurance ranging from no action letters to COCs	•Hazardous waste subdistrict TIF, values brownfields at $0 to maximize increment/redevelopment finance resources

Table 2-1. (*Continued*)

State	Financing program targeted to brownfield situations	Voluntary cleanup programs and assurances provided	Incentives to attract private investment to brownfields
Mississippi		•Voluntary Cleanup and Redevelopment Act (1998)—regulations expected January 1999; NFA letter available, and relief transferable to future owners	
Missouri	•Brownfield Redevelopment Program—offers loans and loan guarantees geared toward capital improvements, for parties that have purchased properties abandoned or underutilized for at least 3 years •50% or $100,000 grant to investigate site feasibility •Public infrastructure grant up to $1 million	•Voluntary cleanup program (1994)—"clean letter" issued by state; PRFs may apply	•Brownfield Redevelopment Program—offers state tax credits for up to the entire remediation costs; tax credits of between $500 and $1300/year (for up to 10 years) for each new job created; capital investment tax credit of 2%; income exemption of 50%; tax abatement of up to 15 years for local property taxes
Montana	•Controlled Allocation of Liability Act and orphan share fund—offers reimbursement for expenditures beyond applicant's responsibility from an orphan share fund, depending on fund levels •Loans available through state Board of Investments program may apply to brownfield sites	•Voluntary Cleanup and Redevelopment Act (1995)—offers NFA letters; program can be used to address a portion of a site	

State		
Nebraska	• Nebraska's Remedial Action Plan Monitoring Act (1995)—offers NFA letter; any site is eligible	
Nevada	No program in place	
New Hampshire	• Brownfields Program (1996)—offers NFA letter, COC, and CNTS	• Brownfields sites exempt from state hazardous waste generator fees • Municipalities can abate taxes at brownfields sites • Site investigation and environmental consulting assistance to municipalities through state planning office
New Jersey	• Industrial Sites Recovery Act (1993; amended 1998), replaced Environmental Cleanup and Responsibility Act—offers NFA letter; PRPs can participate; 1998 amendment authorized CNTS, applicable to subsequent owners • Hazardous Discharge Site Remediation Fund—$75 million low-interest loan/grant program; loans and grants up to $1 million to private entities for remediation activities; $2 million in grants and loans available to local governments for orphan sites and sites obtained through tax sale certificates or by voluntary conveyance for redevelopment purposes	• Qualifying environmental opportunity zones (EOZs), designated by municipalities, offer low-interest loans from the Site Remediation Fund for cleanups within the EOZ, supported by incrementally increasing real property tax abatements (to offset cleanup costs) for up to 15 years as needed • 25% matching grant money to qualified persons for innovative technology cleanups and/or for nonrestricted or limited/restricted reuse • Tax rebates from the state of up to 75% of cleanup costs

Table 2-1. (*Continued*)

State	Financing program targeted to brownfield situations	Voluntary cleanup programs and assurances provided	Incentives to attract private investment to brownfields
New Mexico		• Voluntary Remediation Act (1997)—regulations being finalized; offers CNTSs to prospective purchasers; participants can get COCs; removes lender liability	
New York	• Clean Water/Clean Air Bond Act —$200 million earmarked for Environmental Restoration Project grants to investigate and/or remediate brownfields; non-responsible municipalities can use grants to cover up to 75% of investigation and remediation costs at sites which they own or co-own with not-for-profit organizations	• Voluntary cleanup program (1994) —offers NFA letter; covers any contaminated property for which the federal government does not have lead remedial responsibility; PRPs are excluded on class 1 or 2 registry sites; most TSDFs, and sites subject to other enforcement action requiring the PRP to clean them up	
North Carolina		• Voluntary cleanup program (1995)— offers NFA letter; PRPs are eligible	
North Dakota		No program in place	

| Ohio | •Urban Redevelopment Loan Program—makes loans up to $5 million to municipalities or nonprofit economic development organizations for real estate activities leading to developable parcels in distressed areas, including site remediation
•Water Pollution Control Loan Fund —low-interest loans issued for water-related brownfields activities, up to $3 million per project, for terms up to 20 years
•Ohio Water Development Authority —loans extended to public or private entities for "remediation of property"
•Competitive Economic Development Program—grants to small cities (less than 50,000) and small counties for business expansion and retention purposes; cities, in turn, may loan up to $500,000 to businesses for brownfield remediation, for projects that will create or retain jobs
•Urban and Rural Initiative Grant Program—grants up to $500,000 or $1 million (depending on population) to nonprofit and governmental organizations in "distressed areas" —requires a 25% match (program fund currently depleted) | •Real Estate Cleanup and Reuse Program (1995)—private licensed site professional, working with each site, develops NFA letter that is offered by the state; PRPs are eligible | •Brownfield Site Cleanup Tax Credit Program—state franchise or income tax credit of lesser of 10% or $500,000 for Phase I and II assessment and cleanup costs, or 15% or $750,000 in economically disadvantaged areas
•Tax abatements for up to 10 years for the increase in property tax due to remediation |

Table 2-1. (*Continued*)

State	Financing program targeted to brownfield situations	Voluntary cleanup programs and assurances provided	Incentives to attract private investment to brownfields
Oklahoma		• Oklahoma Brownfields Voluntary Redevelopment Act (1997)—offers risk-based certificates of no action necessary and COCs; any "abandoned, idled, or underused industrial or commercial facility complicated by environmental contamination" is eligible; PRPs are eligible	• Oklahoma Sales Tax Code exempts sales tax on machinery, fuel, chemicals, and equipment used in cleanup projects • Oklahoma Quality Jobs Act—provides quarterly incentive payments for 10 years to firms that locate on a minimum 10-acre site that qualifies as an NPL site, a Superfund site, or an official Superfund deferral site
Oregon	• Capital Access Program offers loan portfolio insurance for environmental evaluations and brownfields redevelopment projects • Credit Enhancement Fund offers loan guarantees for environmental evaluations and brownfields redevelopment projects as an allowable use, providing loan or credit guarantees to specific businesses • Brownfield Redevelopment Loan Fund can finance environmental evaluations (if certain criteria are met); feasibility studies or site remediation not eligible for BRLF support	• Voluntary cleanup program (1991)—offers NFA letter; liability release available through prospective purchaser agreement	

	• Special Public Works Fund available to small local and tribal governments for environmental evaluations on municipal property
	• Job Creation Tax Credit Program —tax credit of $1000 per new job created to firms that increase employment by 25 jobs or 20% within 3 years from start date (with program)
Pennsylvania	• Industrial Sites Reuse Program— provides loans and grants to municipalities and private entities for site assessment and remediation; maximum of $200,000 for site assessment, or $1 million for remediation per year; all require a 25% match; loans carry a 2% rate for terms up to 5 years (for assessments) or 15 years (for remediation)
	• Infrastructure Development Program —provides public and private developers with grants and loans for site remediation, clearance, and new construction, up to $1.25 million per project at 3% interest for 15 years
	• Land Recycling Program (1995)— offers CNTS, release from liability for approved cleanups; PRPs may participate
Rhode Island	• Industrial Property Remediation and Reuse Program (1995)—offers letters of compliance to responsible parts, and CNTSs (transferable) to volunteers, prospective purchasers, and lenders; PRPs are eligible
	• Rhode Island Mill Building and Economic Revitalization Act— offers a 10% tax credit on the cost of substantial rehabilitation for certified sites

Table 2-1. (*Continued*)

State	Financing program targeted to brownfield situations	Voluntary cleanup programs and assurances provided	Incentives to attract private investment to brownfields
South Carolina		• Informal voluntary cleanup program operated under state's Hazardous Waste Management Act (since 1988, expanded 1995)—offers completion letter; all sites are eligible	
South Dakota		No program in place	
Tennessee		• Voluntary Cleanup Oversight and Assistance Program (1994)—offers NFA letter indicating that obligations under consent orders have been completed	
Texas		• Voluntary cleanup program (1995)—offers COC that provides a liability release to all nonresponsible parties, including prospective purchasers and future lenders	• 4-year property tax abatements are available to certificate of completion recipients, on a sliding scale (100% the 1st year, then 75%, 50%, 25%)
Utah		• Voluntary Release Cleanup Program (1997)—awaiting program rules; offers COC	

State			
Vermont	• Tax incentives for rehabilitation of existing properties in designated "downtown" areas; not specific to brownfields, contaminated properties are eligible	• Contaminated Properties Redevelopment Program (1995)—offers COC covering contamination identified in site plan; 1998 amendment expanded liability protection to current owners	• New 5-site pilot program (1998) caps prospective purchaser's share of cleanup at 130% of the cost estimate in state-approved corrective action plan
Virginia		• Voluntary Remediation Program (1997)—offers "certification of satisfactory completion of remediation"	• Defines "environmental restoration sites" holding certificates of completion as a separate class of property and allows local governments to adopt an ordinance partially or fully exempting that class from taxation
Washington	• Remedial Action Grant Program—up to $1 million per project can be awarded from a $20 million remediation project pool	• Voluntary cleanup program (1997), replaced Independent Remedial Action Program—offers NFA letter for some sites, and CNTS for sites with heavy level of state oversight	
West Virginia	• Low-interest loans for site assessment and cleanup, about 50% of loan (rest comes from bank) at a 5% rate • RLF targeted for remediation (authorized but not yet funded)	• Brownfields Program (1997)—offers certificate of completion that provides liability relief	

Table 2-1. (*Continued*)

State	Financing program targeted to brownfield situations	Voluntary cleanup programs and assurances provided	Incentives to attract private investment to brownfields
Wisconsin	• Brownfield Grant Program—$10 million for public or private use, for investigation and cleanup; 20 to 50% match required. 7 grants must go to communities with less than 30,000 people • State Trust Fund Loan Program—loans local governments and other public districts up to $750,000 per year, at top rates of 7%, for projects that "benefit the public," including cleanups and redevelopment • Land Recycling Loan Program—$20 million for loans to municipalities, for site assessment and cleanup, with rates at 55% of market interest rates • Loan guarantees (up to $4 million) for brownfields redevelopment, expected to leverage $20 million in private lending • Brownfield Environmental Assessment Program ($750,000) conducts Phase I and II audits at city-nominated properties	• Land Recycling Act (1994)—offers certificate of exemption from liability that is transferable to new owners; voluntary party liability exemption exempts parties that did not recklessly or intentionally cause contamination, defines 5 situations in which lenders that acquire contaminated property are exempt, and exempts municipalities that acquire properties through means such as tax delinquency, foreclosure, or eminent domain	• Development Zone Tax Credits, 50% of remediation costs in designated zones • Business Improvement Districts (BIDs)—can impose taxes to raise revenues for Phase I and II assessments • Environmental Remediation Tax Increment Financing—TIF districts can be created to recoup remediation costs, with increment based on value-added worth of the clean site; municipality sets up the district and receives payment through increments generated once the property is privately owned • Cancellation of delinquent taxes allowed as part of agreement to clean up contaminated property

- Dry Cleaner Environmental Response Fund (available 9/98)—funded through industry tax, will reimburse up to $500,000 per facility to clean up solvent discharges
- Environmental Fund—$20 million for cash and bonding for state-funded cleanups at priority brownfields and landfills
- $2.5 million in state-administered CDBG funds for small cities (less than 50,000 people) earmarked for site assessment and cleanup
- Stewardship funds can be used to redevelop brownfields into parks or trails, or to restore riverfronts or rivers

Wyoming	No program in place

Cleanup standards:

- Typically, vary by intended use and applied on a case-by-case basis

Assurances provided:

- Covenants not to sue
- Liability releases
- Certificates of completion
- No further action letters

Financial assistance:

- Twenty-six states offer some type of financing programs (such as grants or loans)
- Twenty-three states provide incentives such as tax credits or abatements
- Nearly all of these programs have been in existence for two years or less

As the brownfield reuse issue continues to evolve, more and more states have begun to recognize the critical role that financial incentives must play if state voluntary cleanup programs are to be used more widely and effectively. Financing disparities and investors' fears of uncertainty continue to tip the economic development balance away from older industrial sites toward undeveloped greenfields locations. As indicated previously, essential financing to carry out site assessments and cleanup activities simply is not available to many prospective purchasers. Because brownfields redevelopment needs are so diverse, the key to effective financial assistance lies with a combination of existing and new sources.

New Uses for Old Tools.   States are especially well positioned to promote brownfields reuse projects by giving a new twist to their existing economic development finance programs. As with federal programs, many state efforts were designed and their rules defined long before brownfields concerns surfaced. States are beginning to enhance brownfields initiatives—as the Department of Housing and Urban Development (HUD) has tried to do with its community development block grant (CDBG) program—simply by recognizing site assessment and remediation needs as legitimate project development activities within the context of the common financial assistance initiatives noted in the following text.

Loan programs. Nearly every state offers economic development loans, either directly or through development agencies, authorities, or corporations. These programs are capitalized from a variety of sources—general appropriations, fee collections, or repayments from previous federal or state project loans. Offered for years in nearly every state, these efforts could be better targeted to the specific financing needs of brownfields.

Most jurisdictions require collateral before issuing a loan, so that if the business defaults the state does not lose its entire investment. The public or quasi-public agencies making the loans, therefore, are potentially subject to the same type of lender liability that private financiers face. If state programs are to more effectively promote brownfields cleanup and reuse, and to make capital available to the types of borrowers that private lenders avoid due to environmental concerns, then they will have to assume some of this liability. Because of public interest or community concerns, state lending agencies may be in a better position to work with new purchasers or existing owners of contaminated sites—for example, by offering more flexible loan terms—to encourage cleanups and stimulate new development activity.

Loan guarantees. Many states offer loan guarantees to minimize various risks that make financial institutions hesitant to lend. Small businesses, start-ups, and new technology ventures typically are viewed as especially risky and often are addressed in state programs; environmental risks are rarely addressed but could be the focus of a guarantee initiative.

A loan guarantee program makes commercial lenders more likely to offer loans to operations whose fiscal health would ordinarily make lending to them a questionable risk. Guarantees serve to lower what bank regulators term the *risk ratios;* the guarantee strengthens the performance of a bank's loan portfolio in the eyes of regulators because the guaranteed portion of the loan cannot be subject to default or become—in banking lingo—nonperforming. Loan guarantees provide banks with a sought-after backstop. Although loan guarantees do not solve the problems caused by concerns over liability, they do address the issue of diminished collateral value. Since the issue of collateral is much less important for a loan backed by a guarantee, the problem of a facility's lost market value due to contamination is reduced.

Business development corporations. An important source of investment capital, especially for small companies, are publicly chartered private development banks, usually called business development corporations (BDCs) or development credit corporations. Currently, these operate in about 30 states. BDCs are authorized by state law and operated under state rules, but are privately administered. Several states, especially those with constitutional restrictions on using state funds to help private business, have chartered BDCs as an alternative to direct loan and loan guarantee programs. To date, though, little BDC financial assistance has been directed to brownfields projects.

BDCs generate most of their capital from private sources such as banks, insurance companies, and similar institutions that purchase shares of stock, provide advantageous loans, or extend lines of credit to the corporation. Some of the more recently established BDCs have used state-granted tax incentives to attract individual and business investments. Often, participation in a BDC allows the financial institution to participate in less risky companion or shared loans as part of a resource package assembled by the BDC to finance a business project. Most financing is directed to small companies that use the funding for construction activities and working capital.

Enterprise zones. More than 30 states nationwide currently administer their own enterprise zone programs to spur investment and job creation in distressed areas; operating independent of the new federal initiative, most were launched in the mid-1980s prior to the emergence of the brownfield issue. States have designated more than 1400 zone areas. Although programs vary by provisions, eligibility requirements, and economic development "carrots," several common incentives can be found in most state programs, including:

- Tax credits, reductions, or abatements on sales, materials, inventory, and property
- Job training help or employer tax credits
- Loans, loan guarantees, and other types of capital assistance
- Management and technical assistance and related services earmarked to the zones

Many state zone programs could be better used to influence brownfields redevelopment. For example, loan and grant programs, as well as tax abatements, could be targeted to brownfields projects. Technical assistance services could be tailored to brownfields issues, such as site characterizations or liability, and brownfields users, such as manufacturers or developers.

State Financing Innovations—Trend-Setting Programs. Today, the more creative state assistance programs have been designed around the basic "Development 101" concept that resources are needed to make any project happen, brownfield or not. However, in the case of brownfields projects, lender concerns, investor expectations for return, and borrower creditworthiness issues must be addressed in the context of contamination, site preparation, and marketable reuse. At the same time, many states and cities have recognized that these public investments are often recoverable, either through sale of the site or from the new tax revenues that the project generates, and more and more they are willing to go forward with the projects.

As indicated, many of these initiatives are very new—just a year or two old—but already they are helping to make the numbers work for brownfields projects. They recognize that no specific type of public-private partnership—and no single approach—fits the financing needs of all brownfields projects. They are organized into two broad categories:

- Tax incentives and abatements applied to brownfields projects
- Financial assistance and tax incentive programs targeted directly at promoting brownfields reuse

Tax Incentives and Abatements Being Applied to Brownfields Projects. Minnesota has modified its tax increment financing (TIF) laws to recognize one of the realities of brownfield sites—stigma. By defining a hazardous waste subdistrict, cities can value brownfields at zero for TIF purposes. This boosts the increment and the potential to raise proceeds for cleanup and redevelopment.

Owners of brownfields sites in New Jersey's designated Environmental Opportunity Zones can negotiate with local communities and arrange to use some of their annual property tax levy to cover up to 75 percent of their site cleanup costs, instead of paying the money to their local tax collectors.

Sites in Connecticut can take advantage of a seven-year deferral of increased property taxes that could occur because a clean site becomes more valuable. Similarly, Idaho offers a seven-year, 50 percent tax break on the property's appreciation due to remediation. Texas takes a four-year sliding scale approach, starting with a 100 percent abatement the first year.

Maryland cities that agree to participate in the state's brownfields revitalization program can get a 50 percent property tax credit to offset the increased value of a cleaned-up site; to be eligible, sites must be in cities that agree to contribute 30 percent of the new tax revenues to the state's new brownfields incentive fund, which will offer grants and loans for voluntary cleanups.

Ohio is trying to level one aspect of the site selection playing field by offering a state franchise or income tax credit for Phase I and II assessment and cleanup costs. Site owners can claim the lesser of 10 percent or $500,000 for these purposes.

Illinois provides a 25 percent income tax credit of up to $150,000 per site—this is available to developers who spend at least $100,000 to restore contaminated sites, and these credits are transferable to new owners. Cook County, Illinois offers a companion 25 percent credit for properties there. Wisconsin offers a 50 percent credit for remediation spending in designated development zones.

Wisconsin has also addressed a tax issue that has proven problematic in cities all over the country, namely the issue of payment of back taxes on abandoned sites. Wisconsin allows cancellation of delinquent taxes for new purchasers as part of an agreement to clean up contaminated property.

Indiana permits localities to designate brownfields zones, and properties within these zones can get special tax abatements.

Financial Assistance and Tax Incentive Programs that Can Be Targeted Directly to Promote Brownfields Reuse.   Massachusetts is developing a reclamation payback fund. Under this program, a private bank would make assessment and cleanup loans of up to $500,000 at sites where cities certify that loan repayment will come from one-half of the property taxes generated by redevelopment of the site.

Several states target their brownfields tax incentives to job creation linked to site reuse. Pennsylvania gives $1000 per job to existing firms that increase their employment by 20 percent within three years. Florida offers what it terms a $2500 "bonus refund" for jobs created by brownfields reuse projects.

Michigan recently authorized the establishment of Brownfield Redevelopment Authorities, which have TIF and bonding authority. These authorities focus on brownfields projects, and they can set up a site remediation revolving fund from tax increments captured after cleanup is paid for.

Missouri offers a variety of property and job creation tax incentives, for up to 10 years, as part of its Brownfield Redevelopment Program. Site reusers pick from the menu according to their project needs, and package items together; the value of the incentives can total up to the entire cost of the remediation. The state also offers loan guarantees geared to properties abandoned for at least three years.

Oregon is starting a brownfield project pilot insurance initiative as part of the state's capital access program. This will bring the advantage of environmental insurance to small sites.

Connecticut has launched a cleanup program targeted at dry cleaners, who can get up to $50,000 to help with site remediation costs and pollution prevention measures. Funding comes from the state's 1 percent surcharge on dry cleaning services. Wisconsin is launching a similar initiative.

Finally, as indicated in the state program chart, some states are providing funds directly for brownfields purposes. For instance, a few states, such as Michigan, New York, and Connecticut, have set aside general obligation bond proceeds specifically for brownfields purposes. Nearly a dozen states have provided substantial general appropriations to capitalize loan and grant funds.

Local Brownfields Initiatives: Emerging Financing Tools

New Missions for Old Workhorses

Practically speaking, the benefits of bringing new business activity to established city locations has been outweighed by the risks accompanying the acquisition of brownfields sites. Environmental assessment and even small-scale cleanups remain significant costs that channel investment away from previously used facilities to greenfields sites. In many instances, local governments have begun to explore a variety of financial incentives to offset some of these risks. Many of these efforts will involve placing a new brownfield spin on longtime, tried-and-true financial assistance tools.

Tax increment financing. The TIF mechanism, available in nearly 40 states, has traditionally been used for numerous types of economic revitalization efforts, usually in economically distressed or abandoned areas—the typical brownfields locations. The TIF process uses the anticipated growth in property taxes generated by a development project to finance public sector investment in it. TIFs are built on the concept that new value will be created—an essential premise of most brownfields initiatives—and that the future value can be used to finance part of the activities needed now to create that new value. The key to TIF is the local commitment of incremental tax resources for the payment of redevelopment costs.

TIF bonds are issued for the specific purpose of redevelopment—acquiring and preparing the site; upgrading utilities, streets, or parking facilities; and carrying out other necessary site improvements. This makes them an ideal financing tool for brownfields projects; in fact, many cities with brownfields success stories helped bring them about with TIF financing. TIF programs are easily used with other types of funding, such as grants or loans.

However, many jursidictions have been hesitant to use TIF mechanisms for brownfields projects; if projected development fails to materialize or unanticipated complications arise, it can be difficult to retire the bonds. Some local economic development practitioners also cite the complexity of many TIF initiatives as a practical disadvantage; the initiatives can require a lot of time to put into place, and high levels of technical expertise and negotiating savvy to move a project from concept to implementation—especially a project made more difficult by environmental concerns.

Tax abatements. Tax abatements are commonly used to stimulate investments in building improvements or new construction in areas where property taxes or other conditions discourage private investment. States must usually grant local governments the authority to offer tax abatement programs, and most allow only certain areas to participate, such as economically distressed communities or deteriorating neighborhoods—typical brownfields locations.

Tax abatement programs must be carefully designed to target intended beneficiaries without offering unnecessary subsidies, a feat often difficult to accomplish. Because of this, tax abatement programs have numerous critics. Yet the key advantage of tax abatements is that they give local governments a workable, flexible incentive that helps influence private investment decisions. This can be important in efforts to promote brownfields reuse.

Community development block grant "float." Generally, CDBG recipients are unable to use their entire block grant allocations in the year received; long-term, larger projects (such as infrastructure construction) approved for funding take more than a year to plan and carry out. According to HUD rules, funds not needed to meet current project costs remain in the federal treasury until the city actually needs them; it is not unusual for CDBG funds awarded one year to be drawn down a few years later as big capital projects move toward completion.

When a city can show that previously awarded CDBG funds will not be needed in the near term, it may tap its block grant account on an interim basis—using what HUD calls a *CDBG float*—to finance short-term, low-interest construction financing for projects that create jobs. Any developer, not-for-profit agency, or private company that can obtain an irrevocable letter of credit from a lender is eligible to apply for such financing. (The letter of credit satisfies HUD's concern that the funding will be available for its originally planned purpose.)

Proceeds may be used to pay all costs for the purchase of land and buildings and for site and structural rehabilitation—including environmental remediation—or new construction. Float funds can also finance purchase of machinery and equipment. Maximum loan size is determined by the amount of funds in a jurisdiction's CDBG account available to cover the float. Float loans cannot be extended for more than 2 years; the interest rate is limited to 40 percent of the prevailing prime rate. A few municipalities, notably Chicago, have financed brownfields cleanup activities via the CDBG float mechanism.

General obligation bonds. Virtually all communities can issue general obligation (G.O.) bonds for (in the words of one city attorney) "any proper public purpose which pertains to its local government and affairs." Economic development practitioners can make a strong case that a bond pool to support brownfields cleanup and reuse projects could create jobs and enhance the local tax base, which are appropriate public purposes. Cities traditionally issue G.O. bonds for acquiring land, preparing sites, and making infrastructure improvements—key elements in a brownfield redevelopment strategy. Moreover, the city's ability to repay this bond debt would be enhanced by the growth in property tax revenues as more brownfields are brought back to productive uses.

Cities Step Up—Creative Local Financing Efforts

Chicago recently negotiated an agreement with Environmental Protection Agency (EPA) Region 5 as part of a fine levied on the Sherwin-Williams paint company. The city got $950,000 as part of EPA's settlement with Sherwin-Williams over environmental violations. The money will be used to clean an abandoned 130-acre industrial site on the city's southeast side.

Several federal programs have proven to be interesting brownfields financing tools when in local hands. For example, some communities have leveraged HUD block grant resources by using CDBG floats for brownfields purposes.

Community development block grant floats are local financing tools built on HUD's block grant foundation. They are not often used, but they have great potential to assist with smaller brownfields projects. The concept behind a float is this: a float is sort of like a city's advance on its block grant allowance.

Generally, CDBG recipients are not able to spend their entire annual allocation in the year they receive it from HUD, and unspent funds remain in the federal treasury until drawn down.

But when a city can show that previously awarded block grant funds will not be needed in the near term, it may tap its block grant account on an interim basis, using what HUD calls a CDBG float, to finance short-term projects that create jobs. Any developer, not-for-profit organization, or private company that can secure an irrevocable letter of credit from a lender may use float proceeds. Float loans can finance site and structural rehabilitation, including cleanup. Groups such as the Greater Southwest Development Corporation in Chicago have used CDBG floats to generate the $25,000 to $50,000 needed to investigate and clean up small sites in key areas. The floats are generally repaid from project development proceeds.

A second federal tool with good local potential is low-income housing tax credits. There is growing interest in reusing brownfields properties for residential purposes. And this interest will be further fueled as state voluntary cleanup programs become more established and the impacts of recent lender liability and cleanup expensing provisions are absorbed by the market.

Low-income housing tax credits can play an important role in attracting private investment to these projects. For example, the Circle F project in Trenton, New Jersey was developed on a manufacturing site that dated to

1886. Working with a local neighborhood organization, the city subdivided the site and targeted the older front half of the parcel for senior citizen housing. The back half remained light industrial. It was used by an existing manufacturer that needed more space.

Trenton officials selected a longtime local nonprofit developer to undertake the housing project. The developer fronted the $500,000 for site cleanup and preparation, and applied for and received an allocation of $8 million in federal low-income housing tax credits. These credits attracted a private lender, Nat West Bank. The bank helped finance the project, and assumed the role of a limited partner in the project in order to get the benefit of the tax credits.

The tax credits will translate into a 12 percent return on investment for Nat West. As bank officials noted, low-income housing tax credits are one form of Community Reinvestment Act activity that can be very profitable. In the case of Circle F, they were linked to brownfields considerations without undermining that profitability.

In the Twin Cities area of Minnesota, the six-county Metropolitan Planning Council has its own taxing authority, which the council is now using to raise revenues for a brownfield cleanup initiative. The 1-mill levy generates about $6.5 million each year for brownfields grants and loans.

And Cheektowaga, NY agreed to a tax abatement swap to promote brownfields cleanup. A local developer wanted a tax break on a hotel he was constructing, and the developer offered to clean up an old steel site in exchange for that tax abatement. This is a new twist on the concept of swapping development rights.

Refocusing Existing Local Development Programs

Every local government already uses a variety of financial assistance programs and incentives to promote economic and business development; like federal and state programs, local offerings can be more explicitly packaged and promoted for potential developers and lenders to use to clean and rehabilitate brownfields sites. A growing number of cities are examining ways to do this. Alternatives being considered in some places include:

- Earmarking water, sewer, and wastewater charges for brownfields cleanup activities

- Earmarking some portion of grant, loan, or loan guarantee program funds to applicants proposing site characterization or cleanup projects

- Developing a municipal "linked deposit" program targeted to brownfields borrowers

- Channeling some portion of loan repayments from existing city programs to brownfields projects

- Devoting monies raised from fines or fees to a brownfield financing pool

- Using small amounts of public funds to seed a private, shared-risk financing pool devoted to brownfields redevelopment

In addition, cities can explore other low- or no-cost techniques to stimulate the flow of capital to promising brownfields redevelopment undertakings. For example, Chicago and Cleveland are considering ways to more easily convey tax-delinquent properties to new owners with viable reuse plans. Other cities are contemplating modifications in their zoning requirements in specific cases to provide developers with the opportunity to earn a greater return on their investments and offset more site preparation costs.

New Types of Local Brownfields Finance Initiatives

Many brownfields sites have the potential to become economically viable, hosting new business activity and jobs. However, many of these sites require some level of public investment to achieve this viability. Federal and state resources will not be sufficient to address all the prospective site cleanup and reuse possibilities identified by jurisdictions across the country; the large number of applicants for the handful of EPA brownfields pilot sites designated to date is testimony to that. Existing local programs can meet some of this need, but clearly cannot meet all financing gaps in many areas. Therefore, communities must consider establishing new brownfields incentive programs of their own. These could help with site characterization and cleanup costs, development costs, or both types of activities.

Competing public needs and objectives, as well as limits to public resources, are facts of life in every community; recognizing this, local officials could consider two approaches to promoting brownfields finance. First, they should identify and set aside public sources that can be mostly self-sustaining, stable over time, and relatively isolated from changing political tides. Given the inherent limits of public funding, some type of cost recovery is essential to the sustainability of local public financing of brownfields projects. Against this backdrop, local programs can—as they evolve and become more established—enhance their own flexibility by offering forgivable loans, recoverable grants, lengthy repayment terms, recovery upon property transfer, and similar conditions.

Second, public resources should be marshaled in the context of an explicit, strategic brownfield approach. Generally, local officials should give priority to sites with greater development potential as they reach decisions on financial assistance. In many cities and towns, this may mean supporting several smaller sites in a declining area rather than the one big abandoned plant that has come to signify "brownfields" to the community. Momentum for brownfields cleanup and reuse—and justification for public sector involvement in it—can be created and maintained with visible successes, even at small sites. Moreover, smaller brownfields projects are more manageable and often more significant in terms of real benefits than a single large, more contaminated site.

Federal Tools: Working to Fit Existing Programs to Brownfields Needs

State and local governments, as indicated here, can find creative ways to help companies and investors overcome the difficulties contamination can bring to the site reuse process. However, the federal government—whose programs, policies, and regulations form the foundation on which many state, local, and private development finance initiatives are built—must play a stronger, more visible role if brownfields financing is to become more widely available.

Existing programs such as those offered by HUD can play a critical role in local economic development, and such efforts need to incorporate brownfields project circumstances into their guidelines. Cities and towns across the country use HUD resources to support a wide variety of financial assistance programs—loans and loan guarantees, grants, and technical assistance—to help spur economic revitalization and growth. New HUD initiatives, as part of the agency's Economic Development Initiative (EDI) program, will play an important role in state and local strategies to encourage the renovation and reuse of older industrial facilities.

Community Development Block Grants

The CDBG program is one of the most useful federal initiatives remaining to provide direct funding for activities that support the reuse of industrial sites. Distributed by HUD according to formula, CDBG resources can be used to finance the rehabilitation of privately owned buildings and sites, covering specific costs related to labor, materials, construction, or renovation. They also can pay for services such as entrepreneurial counseling, preparation of work specifications, loan processing, and site inspections.

Block grant funds are particularly well suited to the new generation of industrial site reuse projects, which bring a much stronger focus on environmental concerns. Large and small cities can use CDBG funds for grants, loans, loan guarantees, and technical assistance activities. This makes the program a highly versatile tool to stimulate private investments in targeted distressed areas, such as those with a concentration of largely abandoned, obsolete industrial facilities.

Section 108 Loan Guarantees

A related HUD program, known as Section 108 loan guarantees, enables local governments to finance physical and economic development projects too large for front-end financing with single-year CDBG grants. Under Section 108, localities issue debentures to cover the cost of such projects, pledging their annual CDBG grants as collateral. The debentures are underwritten and sold though public offering by a consortium of private investment banking firms assembled by HUD, which guarantees each obligation to ensure a favorable interest rate. Local governments can use their annual CDBG allocations to pay off these obligations, although most use income generated from the development project for some or all of the payments.

Activities undertaken with money from loans guaranteed under Section 108 must meet the basic requirements of the CDBG program. Among the eligible activities are property acquisition, clearance or rehabilitation of obsolete structures, construction of public improvements such as water and sewer facilities, and site improvements. Chicago was one of the first cities to tap the Section 108 program for resources to use in preparing identified brownfields sites for redevelopment.

Empowerment Zones and Enterprise Communities

Empowerment zones (EZs) and enterprise communities (ECs) are geographic areas targeted to receive special federal treatment and incentives in order that private investment and other economic activity might be attracted to them. Depending on the plan developed for each area, benefits can include financial, regulatory, and technical assistance.

In December 1994, HUD and the Department of Agriculture named 95 enterprise communities (65 urban and 30 rural), as well as 9 empowerment zones (6 urban and 3 rural). Designation brings several benefits to the selected areas, including $100 million in social service grants for each of the urban EZs, $40 million for each rural EZ, and $3 million for each EC. In addition, designated communities can compete for as much as $2.5 billion in new tax incentives to induce investment in the targeted distressed areas.

Applicant jurisdictions were required to specify how they would use these resources to confront economic distress and unemployment. Many applicants identified the problem of brownfields and stated that overcoming associated barriers was a critical element of their local economic revitalization strategy. Detroit and Chicago, two zone designees, have placed special emphasis on brownfields activities.

In August 1997, Congress authorized designation of a new round of zones, and HUD is currently developing the selection process.

Tax Incentives that Could Influence Brownfield Activities

Several existing federal tax incentives could contribute to brownfields redevelopment activities. Industrial development bonds (IDBs), targeted to manufacturing projects, can play an important part in a brownfield reuse or business retention strategy, especially for efforts seeking new industrial uses for old industrial sites. The popularity of IDBs stems from their versatility as a development finance tool, a versatility that fits well with brownfields activities.

Rehabilitation tax credits were devised by Congress in the 1970s to discourage the unnecessary demolition of sound older buildings and to slow the loss or relocation of businesses from older urban areas. Across the country, the credits have helped attract redevelopment capital into all types of projects in blighted and ignored areas not ordinarily considered for investment. A number of brownfields success stories involve renovation and reuse of old industrial structures. The Parke-Davis lab project in Detroit is a prominent example of a situation in which rehab tax credits played a key role in the project's economic viability. The rehab tax credit is well suited for packaging with other economic development grant and loan programs; it can be an ideal complement to a brownfield redevelopment initiative in an older industrial area. Congress originally intended for rehab credits to help level the economic playing field and balance the development costs between older established (and often declining) areas and new sites; this concept naturally extends to making brownfields sites economically competitive with greenfields.

What Have We Learned from Brownfields Successes, and What Does It Mean for the Future?

Encouraging the cleanup and redevelopment of brownfields requires a comprehensive package of solutions—some direct, some indirect—that become relevant at different points in the process. Sometimes, financial incentives and liability relief are critical in terms of persuading individuals to undertake projects in the first place. However, other factors—such as interagency coordination and project leadership at the local level—assume greater importance once a project is under way by helping to slash time frames and save costs.

The Northeast-Midwest Institute has examined a wide range of brownfields reuse success stories—nearly 50 from around the country over the past 5 years. This analysis has reinforced the notion that no single recipe for success exists, and it has also revealed that some common ingredients of success are divided into five groupings: (1) players and institutional capacity; (2) community involvement; (3) regulatory and legal issues; (4) costs and financing; and (5) risk management and cleanup. Finally, broader brownfields policy considerations based on this analysis are offered.

Players and Institutional Capacity

Presence of a Proactive Local Government Entity or Redevelopment Authority.   One of the most critical ingredients for success is the presence of a strong local government entity. For half the projects examined, the city acted as a "brownfield broker," essentially helping interested buyers acquire contaminated and/or abandoned properties. For several other projects, the city played a key role in helping a company to clean up and redevelop its contaminated site, as well as to locate potential buyers or end users. Local officials also were invaluable in terms of facilitating community involvement and helping parties navigate difficult regulatory requirements.

In addition to playing a key logistical role, cities often provided financing to make a brownfield project economically viable. In Bridgeport, Connecticut, Westinghouse has spent over $1 million to clean up its Bryant Electric facility, and the city contributed $700,000 toward demolition and site preparation to make way for a new end user. Without the city's involvement, the site might only have been remediated and not actually redeveloped.

In Louisville, Kentucky, the city has been working with an expanding business to acquire an abandoned, contaminated property. This project has been complicated by many factors, including environmental contamination at the site and uncertain remediation requirements. The city oversaw relations between the Kentucky Department of Environmental Protection, the Landbank Authority, and the prospective purchaser, Louisville Dryer Company. In addition, the city dedicated funds to this project (for personnel and site assessment) from grant money provided under USEPA's Brownfields Pilot Site Program. Because of the city's involvement, Louisville Dryer has been able to remain involved in a real estate transaction it might otherwise have abandoned long ago.

Appropriate Institutional Capacity at the Local Level—Consolidating Brownfields Project Management Teams Under One Roof. While cities clearly play an essential role in brownfields redevelopment, often they are not set up to effectively manage such projects. The problem is that brownfields initiatives require involvement by personnel from a range of departments (e.g., planning, law, economic development, and environmental protection), which can create administrative snafus. In addition, efforts can be complicated by the fact that these departments often have conflicting missions and mandates. Most local officials agree that establishing a single entity for oversight of brownfields initiatives is key. The Worcester Redevelopment Authority (WRA) in Worcester, Massachusetts, is an example of such an entity. The WRA acquires properties, coordinates remediation, and facilitates site redevelopment work. Similarly, the Port of Seattle assembled under one lead manager a group of staff members who were dedicated to the Southwest Harbor redevelopment project. This team, which worked out of one office location, included members of the port's marine facilities as well as staff from legal, engineering, environmental, and finance offices.

Strong Public-Private Partnerships.   Public-private partnerships—usually between private parties, the city, and the state—are essential. In particular, those locales that have forged alliances between business interests and public sector objectives have seen significant results. In Wyandotte, Michigan, BASF, the city of Wyandotte, and the Michigan Department of Environmental Quality teamed up to redevelop BASF's South Works into a public recreation area. In St. Paul, Minnesota, the St. Paul Port Authority and the Minneapolis Pollution Control Agency worked with Texaco to transform an old petroleum tank farm into a new light industrial business park.

Project Leadership—Individuals Make a Difference.   Many projects have been successful because of certain key individuals who possess strong leadership, persistence, and creativity. For example, in Sacramento, California, the federal courthouse development has been spearheaded by the city's Wendy Saunders. When Saunders took maternity leave over the summer of 1996, many participants indicated that the project was virtually on hold until she returned. The same has been true of Kevin Geaney with the Lawrence Gateway Project in Lawrence, Massachusetts. For years, redevelopment of the city's Oxford Paper site was at a standstill; when Lawrence hired Geaney, however, the process finally began to move forward. Several people interviewed indicate that Geaney was the catalyst for launching the Lawrence Gateway Project.

Coordination Between Local, State, and Federal Government Entities. Many brownfields projects are burdened by high assessment and remediation costs and by long, drawn-out time frames—a situation that is only exacerbated when multiple government agencies (i.e., local, state, and federal) are involved. For many of the projects examined, streamlining interagency coordination was critical in terms of resolving overlaps in administrative jurisdictions and oversight. Lawrence, Massachusetts, tackled this problem by establishing two interagency task forces—teams composed of local, state, and federal representatives—that ironed out key issues, facilitated decision making, and coordinated the multiple regulatory issues connected with the project. In Sacramento, California, redevelopment of Southern Pacific's rail yard was made possible by an innovative memorandum of understanding (MOU) between the city, the state, and Southern Pacific. The MOU articulated roles and responsibilities for each party and established a third-party oversight entity, the Environmental Oversight Authority, which was tasked with overseeing all site assessment and cleanup in lieu of the state environmental agency.

Community Involvement

Strong Community Participation. In almost every case analyzed, carefully orchestrated public outreach and involvement plans were implemented from the outset. Without this critical community buy-in, many project participants note, their efforts could easily have fallen apart. In Minneapolis, community participation was central to the redevelopment of the Johnson Street Quarry into a discount shopping center. The Minneapolis Community Development Agency assembled a neighborhood task force, which met monthly in a televised public forum to discuss project plans. In a written report, the group expressed numerous concerns about traffic, noise, and public safety and called on the city to implement a series of traffic control measures and infrastructure improvements before it would support the initiative. The city and developers unanimously agreed to meet the task force's demands, and the project moved forward with strong public support.

Capitalizing on a Community's Vision. Most local officials agree that brownfields initiatives should dovetail with a community's vision for growth. For example, where brownfields redevelopment is part of a concerted downtown revitalization program, it stands a better chance of securing public and private investment as well as gaining political and community support. In Chattanooga, Tennessee, cleanup and reuse of riverfront property dovetailed with the city's broader Vision 2000 initiative, which sought to revitalize neighborhoods, remediate the environment, and attract new businesses throughout the city.

Appropriate Job Training. Many communities are eager to ensure that brownfields redevelopment and the presence of new business translate into job opportunities for area residents. Job training often is necessary to ensure that residents acquire the appropriate skills. In St. Paul, Minnesota, the St. Paul Port Authority launched an innovative job training program, the Employment Connection, which helps link brownfields redevelopment with neighborhood wealth creation. The port determines the specific employment needs of local businesses, connects with various neighborhood groups, and creates a customized training package for companies. Businesses pay 10 to 15 percent of the costs for the training package; the balance is provided by the state and private corporations/foundations. This program will help ensure that area residents are properly trained for job openings at the Crosby Lake Business Park.

Regulatory and Legal Issues

State Voluntary Cleanup Programs—Availability of Liability Relief. As indicated in Table 2-1, nearly 40 states have established programs encouraging voluntary cleanup of contaminated sites. These state initiatives—many of which offer financial incentives, liability relief, and simplified cleanup standards—have significantly encouraged brownfields cleanup and reuse. For example, the contaminated Holden-Leonard Mill, in Bennington, Vermont, was in a holding pattern for years despite the fact that a party was interested in buying the site. With the enactment of Vermont's 1994 Contaminated Properties Program, however, site assessment and cleanup are now proceeding. One key reason is that both the seller and prospective purchaser now can obtain some liability closure once a state-approved cleanup plan has been completed.

Clarity Between the State and USEPA in Terms of Liability Relief. Although many states are offering some form of liability relief, participants interviewed still consider fear of EPA involvement a barrier to redevelopment, particularly in terms of securing financial backing. The agency has responded by stating publicly its intention not to interfere at sites that are participating in state voluntary cleanup programs. Some regional EPA offices have put this commitment in writing by including "comfort language" in their Superfund memoranda of agreement with states. These moves to bless state voluntary cleanup programs have been greeted favorably by brownfields practitioners.

Costs and Financing

Property Location and Market Conditions. The old real estate adage, "location, location, location," applies to brownfields as it would to any other property. If a brownfield is situated in a desirable location—near a bustling downtown, along a scenic waterfront, or by a busy highway interchange—redevelopment is more likely to occur than not, despite potential environmental contamination. In other words, the economics of the project may make sense even when costs for cleanup are factored into the equation. For example, Southern Pacific decided to remediate its Sacramento rail yard facility as well as to pursue redevelopment because the company saw potentially huge profits (due to the site's prominent downtown location). Similarly, in Chattanooga, Tennessee, RiverValley Partners decided to purchase and redevelop old industrial property along the Tennessee River, wagering that any contamination discovered would be offset by huge returns on the investment.

However, if a brownfield is situated in an economically depressed area— on the outskirts of town or in a blighted neighborhood—redevelopment is a more difficult proposition. In these cases, the role of the public sector to encourage brownfields reuse becomes critical. The threat of environmental contamination is a major deterrent, but there may be other concerns: inadequate neighborhood safety for employees, blighted conditions surrounding the site, lack of access to skilled labor, or the potential for property devaluation. In these cases, the city or the state may offer financial incentives to companies willing to locate in economically disadvantaged areas. Alternatively, the city may assist in redeveloping a portion of a neighborhood—as was the case in Bridgeport's West End, surrounding the Bryant Electric facility—in order to trigger a domino effect of revitalization.

Possibility of "Piggybacking" onto Public Works Projects. Brownfields projects may be coordinated with public works initiatives—including transportation projects, historic preservation efforts, and green corridor planning—in order to access innovative funding sources. In Lawrence, Massachusetts, for example, cleanup and redevelopment of the old Oxford Paper plant seemed financially infeasible until city personnel came up with the idea of piggybacking the project with a nearby highway expansion, allowing Lawrence to draw on much-needed state highway funds for demolition and remediation.

Benefit of Being a Large Development Company. Brownfields projects present less of a financial risk to large development companies, which can essentially distribute risk among different projects. These firms also may be able to finance initiatives themselves, without the involvement of banks, enabling them to take on projects of greater risk. In recent years, developers have become increasingly interested in contaminated sites. According to *Crain's New York Business* (November 1996), "Once seen only as a blight . . . [brownfields] are now becoming an opportunity for a number of savvy real estate developers." This sentiment is echoed by an attorney for Benderson Development Company (developer of the Ernst Steel site in Cheektowaga, New York), who notes, "Benderson views brownfields as the last frontier on which to make money."

Availability of Public Sector Financing. Private parties frequently are not able or willing to act on their own to ensure that a brownfield site is redeveloped to its full potential. With assistance from the public sector, however, numerous projects are able to move forward. Public sector funds typically support front-end activities such as environmental assessment and remediation, demolition, and site preparation, whereas private sector funds more often support redevelopment and construction of new facilities. The projects examined by the Northeast-Midwest Institute used a variety of public funding incentives and programs, typically one or more of those identified earlier in this section.

Availability of Private Sector Financing. A majority of the cases examined benefited from some form of private sector financial involvement. Following are key sources of private sector support.

Responsible parties. Prior to redevelopment, some sites were cleaned up by the parties responsible for environmental contamination. Remediation of the Bryant Electric plant in Bridgeport, Connecticut, for example, is being carried out and financed by Westinghouse, which agreed to turn the clean property over to the city for $1. Bridgeport officials, in turn, are financing demolition of the 500,000-square-foot building to make room for a new manufacturing facility at the site. Similarly, the Texaco tank farm in St. Paul, Minnesota, was cleaned by Texaco and then sold to the St. Paul Port Authority, which provided site preparation and infrastructure improvements before selling off parcels to interested buyers. In several instances, the responsible parties not only financed the cleanup but also chose to redevelop the property (i.e., opted to retain ownership of the land rather than transfer it to a local government entity). In Sacramento, for example, Southern Pacific is financing remediation at its 244-acre rail yard facility and redeveloping the site for a mixture of commercial, residential, and recreational activities.

Purchasers. Some case study projects were cleaned up by the new property owners when a responsible party could not be identified or held accountable, or where the party was financially insolvent. At the Ernst Steel site in Cheektowaga, New York, the Benderson Development Company purchased the property and assumed responsibility for the cleanup. Likewise in Detroit, Michigan, the Acetex Corporation purchased neighboring H&H Wheel's property to accommodate its business expansion and assumed responsibility for remediation in the process. Even where purchasers do not actually shoulder remediation costs, they usually invest money in redeveloping the site (i.e., by renovating or building new facilities).

Commercial banks. Many banks are reluctant to loan money on brownfields projects until remedial work at the site has been certified as complete, either with a no further action letter or a covenant not to sue. In Detroit, Michigan, Comerica bank loaned the Acetex Corporation $2 million to finance redevelopment at H&H Wheel once remedial work had been completed. At the Circle F factory in Trenton, New Jersey, Nat West Bank loaned Lutheran Social Ministries $4 million to finance construction of a senior citizen housing complex.

Foundations. Several projects analyzed in this book received funding from private foundations. For example, the Lyndhurst Foundation provided $10 million for construction of the Tennessee Aquarium and Ross's Landing in Chattanooga, Tennessee. In St. Paul, Minnesota, job training for new businesses at the redeveloped Texaco tank farm site was financed in part by several area foundations.

In-kind work. In an effort to facilitate redevelopment of mill sites in Oregon, one electric utility and several other private sector entities teamed up and contributed over $100,000 in legal, financial, and administrative services. In St. Paul, Minnesota, at the former Texaco tank farm (now the Crosby Lake Business Park), Northern States Power Company installed utility lines at its own expense, and U.S. West strung fiber-optic lines at no charge to the developer.

Risk Management and Cleanup

Risk Management (in Lieu of Risk Elimination) and Cleanup Standards Tailored to End Use

Contaminated industrial sites traditionally have been cleaned according to the most stringent residential standards, such that children could ingest remediated soils. Increasingly, however, states are recognizing that such cleanups are prohibitively expensive and sometimes unnecessary, especially if a site's end use is commercial or industrial. As such, many states now are allowing certain contaminants to be left on site, provided that the potential for human exposure or environmental harm is eliminated (i.e., by an impervious surface such as an asphalt covering or a building). Less stringent cleanup levels for commercial and industrial settings usually are accompanied by engineering or institutional controls, such as deed restrictions, to ensure that inappropriate uses (i.e., residential housing) never occur at the site. The Port of Seattle worked with the Washington Department of Ecology to derive site-specific cleanup action levels for soils based on future industrial land use. Redevelopment of the Carol Cable plant in Warren, Rhode Island, moved forward because of a new law that allowed cleanups to be tailored to end use and that permitted the use of engineering and institutional controls where contaminants would be left in place.

Use of Innovative Remedial Technologies

Many cities are realizing that immense cost savings can be achieved by implementing creative cleanup technologies. In Lawrence, Massachusetts, for example, a soil vapor extraction system was utilized to treat contaminated soils on site, rather than sending them off site for incineration. In Worcester, Massachusetts, officials used ground-penetrating radar to identify the locations of underground storage tanks. Remedial costs at the Ernst Steel site in Cheektowaga, New York, were offset by the use of an experimental hydrogen sulfide liquid treatment that immobilized lead in soils—a process that saved the company upward of $300,000.

Broader Policy Conclusions

Public Dollars Leverage Private Investment

Minimal public sector investment often leverages private sector dollars. For approximately $370,000, Chicago was able to demolish an eyesore, clean up environmental contamination, and provide a clean, secure lot for Scott Peterson Meats—a strong neighborhood company—to use for employee parking. The city's commitment gave Scott Peterson Meats the impetus (and its lenders the willingness) to invest $5 million in the project, which in turn led to the hiring of 100 additional employees. Without that critical public funding, local officials feel that private investment in Scott Peterson's project would never have materialized. EPA Brownfields Pilot Site grants also have been instrumental in leveraging both public and private funds.

Brownfields Projects Often Trigger
a Ripple Effect of Revitalization

Local officials stress the importance of choosing brownfields projects that lead to further development, so that a ripple effect of economic revitalization may occur. This was the case in downtown Worcester, Massachusetts, with development of the Medical City project. In conjunction with other downtown redevelopment initiatives, Medical City has created a domino effect of economic growth over the past five years. "You can't address one isolated brownfield and expect it to survive alone," says the city's Dave Dunham.

Brownfields Projects Can Be Pilot
Sites for Shaping Broader Policies

Many projects identified by the Institute actually were pilot sites used to craft policies and strategies for a broader area. For example, in Louisville, Kentucky, lessons learned during cleanup and redevelopment of the Ni-Chro Plating site helped establish policies for Louisville's citywide brownfields program. Similarly, construction of the federal courthouse in Sacramento, California, has served as a template for policies affecting the remaining 240 acres of Southern Pacific's Sacramento rail yard, and the first two years of the Oregon Mill Sites Conversion Project will guide future redevelopment at rural mill sites. At both the Oregon and Sacramento projects, generic remedies are being formulated for common environmental contaminants—a step that should save time and lower costs for cleanups at similar properties.

The Worcester Redevelopment Authority (WRA) served as an institutional pilot for the creation of a regional brownfield redevelopment authority in Massachusetts. Because the WRA so successfully handled the many challenges associated with the Medical City project, the Commonwealth of Massachusetts in 1995 passed legislation establishing a regional body modeled on the WRA—the Central Massachusetts Economic Development Authority (CMEDA)—to oversee brownfields initiatives.

Experience on the Ground
May Be Used to Guide Future
Brownfields Policies

To effectively shape brownfields policy, it is critical to learn from experiences on the ground—to understand what is needed, what is lacking, and how public funds may best be spent. This requires a constant exchange of information between practitioners in the field and officials in government, a dialog that demands some kind of forum or facilitating body.

The Brownfields Working Group in Louisville, Kentucky, is a consortium of public and private entities that teamed up to identify and overcome barriers associated with contaminated site reuse. The group is creating a database of brownfields sites, facilitating community involvement in decision making, and working to streamline a process by which sites effectively may be brought back to use. Similarly, Chicago's Brownfields Forum is a broad-based interdisciplinary task force, launched in 1995, that seeks to identify policies to encourage brownfields reuse in the city. The Forum incorporates lessons learned through the city's hands-on experience in cleaning up and redeveloping contaminated sites under its Brownfields Pilot Site Program. At regular roundtable discussions, Forum members offer policy suggestions as well as identify critical information gaps that warrant further research. In this way, Chicago has been able to address immediate needs on the ground while also charting sensible brownfields policies for the future. Many other localities across the country have established similar brownfields policy working groups, including Buffalo, New York; Detroit, Michigan; and Cuyahoga County, Ohio.

The Challenge: Confronting the Next Generation of Environmental and Economic Issues Affecting Site Redevelopment

Underused or abandoned industrial facilities are a national concern. Confronting the environmental and economic issues affecting site reuse requires a deliberate, multidimensional approach that often does not neatly fit with the rules and procedures of federal, state, or local economic development or environmental programs. Financing has emerged as a key barrier to brownfields reuse. Site assessment and cleanup requires financial resources that many firms lack and find difficult to secure. And, without financing, private reuse projects cannot go forward, even if their proponents want them to. This further undermines efforts to revitalize the distressed areas that are home to so many abandoned, contaminated sites.

Yet despite the barriers, brownfields reuse opportunities are real. Dozens of diverse projects have been documented, ranging from an old Uniroyal tire factory in Sacramento that was cleaned and converted into an office/retail complex to a Soo Line rail yard in Minneapolis that is being redeveloped as a light industry park. These projects have been carried out in a way that makes economic sense, and that builds on the competitive advantages boasted by specific sites. Such success stories suggest that liabilities can be worked out, that financing can be secured, and that cleanup can be accomplished—in short, that brownfields redevelopment can be achieved.

The challenge that local governments, the states, and the federal government face now is to provide the tools that make the economics of redevelopment projects work. At the same time, it is important to emphasize that incentives can make a site economically viable but that the public sector alone cannot carry the brownfields reuse load. Redevelopment on a wider scale can only be achieved if public policies and programs foster a climate that invites private investment in these projects.

Charles Bartsch is a senior policy analyst at the Northeast-Midwest Institute, specializing in economic development issues, notably industrial site reuse, federal and state technical and financial assistance, tax incentives, technology transfer, and manufacturing modernization. He has authored and co-authored many publications and reports.

From 1981 to 1984, Mr. Bartsch served as a legislative and federal programs analyst for the National Council for Urban Economic Development. From 1979 to 1981, he served as a presidential management intern on the urban policy staff at the Department of Housing and Urban Development, where he contributed to the president's 1980 urban policy report on business development and neighborhood revitalization issues. He received a master's degree in urban policy and planning from the University of Illinois at Chicago, and a B.A. from North Central College in Naperville, Illinois.

Mr. Bartsch often testifies before congressional committees on issues of economic development and recovery. His writings on economic development issues have been published in many books and journals. He also has advised several local and state economic development programs, including Chicago's Brownfields Task Force and Ohio's Edison Technology Centers program. In addition, he has taught college-level courses on manufacturing competitiveness.

3

The Industrial Company

Gary Ballesteros of Rockwell International openly discusses the owners' viewpoints concerning remediation of brownfields properties. In the introduction to this book, we listed four categories of brownfields. The corporate discussion in this chapter is of the fourth category (currently operating sites in danger of becoming brownfields due to historical contamination), which potentially represents the largest number of brownfields. The issues discussed must be addressed adequately, or corporate brownfields redevelopment will be delayed. It is important to understand the sellers' point of view of brownfields regeneration.

3.1 Brownfields Redevelopment of Current Industrial Sites

Gary W. Ballesteros

Introduction

Like any transaction, a brownfield redevelopment project requires two parties: a willing buyer and a willing seller. What has often been ignored in discussions about brownfields redevelopment (and what will be discussed in this chapter) is the viewpoint of an industrial seller. In particular, this chapter will examine the impediments that would stop an industrial owner of a brownfield property from redeveloping that property into a new useful asset.

There is no question that brownfields redevelopment is a good thing. Obviously, converting an underutilized industrial parcel into a community asset is both economically and socially beneficial. But if it is such a good thing, why aren't industrial companies selling their land at a record pace all around the country?

While it is true that there is a growing trend toward developing more and more brownfields properties, few of the projects completed to date have involved properties owned by ongoing industrial companies. Instead, many brownfields redevelopment projects involve vacant, abandoned, condemned, or bankrupt sites whose owners are eager to see them turned into productive assets. But that is because the owners are cities that have taken over abandoned parcels due to tax delinquency, banks that hold a security interest in an abandoned parcel, or bankruptcy trustees wrapping up the affairs of a defunct company. In addition to these types of brownfields, there are a number of desirable parcels that are held by current ongoing industrial companies either as excess real estate or perhaps as underutilized plants. Unless the owners of these facilities can be convinced to be willing sellers, many potential brownfields projects will die in their infancy, and an important inventory of brownfields properties will remain unavailable.

What Stops the Industrial Seller from Selling?

Fear of Liability

Why wouldn't a company committed to a profitable bottom line be interested in turning a nonproductive or underutilized asset into cash? The primary reason is fear of liability. For many of the parties in a brownfield transaction, the fear of liability is the single strongest impediment to completing the deal. Indeed, one of the only reasons an industrial company would not jump at the chance to unload excess real estate is because it fears doing so may expand or accelerate its liability exposure and therefore end up costing the company far more than could be gained by the real estate transaction.

The industrial company's fear of liability is not unreasonable, because selling the brownfield property could expand liability exposure in at least four potential ways.

Expanding the Class of Potential Plaintiffs. Perhaps the best explanation of the company's fears can be illustrated through a hypothetical example. Let us assume that BigCo owns contaminated industrial property that it does not need. It sells this property to a brownfield developer who puts in a shopping mall, a day care center, and luxury apartment lofts. So long as the property and the contamination were lying dormant, BigCo was relatively safe from the risk that anyone alleging to be harmed by the contamination could bring suit. But now there are large vulnerable classes of people who may actually or hypothetically be exposed to the historical contamination caused by BigCo and who may see the deep pockets of BigCo as a plaintiff's treasure trove. Even if the contamination is cleaned up as a result of the deal, residual amounts may be left behind or new exposure pathways may be created. This is a real concern to companies, and one that by itself may kill the deal. Without further incentive, many companies prefer to sit on the property and pass up opportunities to sell it in order to minimize the risk of future toxic tort or environmental litigation.

Kicking the Sleeping Dog. Developing a parcel of property will undoubtedly bring new scrutiny to the environmental condition of the property. Many industrial parcels have a long manufacturing history, and the owners are convinced, even in the absence of any confirmatory evidence, that if you test you will find contamination of some sort. Therefore, many owners would rather not know what is below the surface of a plant. But once a property is targeted for development, then the buyer, the lender, the investors, and the new tenants will all have questions about the environmental condition of the property. Owners fear that opening up that can of worms is not worth the marginal benefit of selling the property, particularly if the holding cost of continuing to maintain the property is fairly low.

The Sticky Liability. The perception is fairly widespread that once stuck with an environmental liability, a company will always have that environmental liability. Due to the rigorous liability scheme of CERCLA, owners of contaminated property know that they will forever be one of the four categories of persons identified by CERCLA as PRPs. (See CERCLA Sec. 107(a)(1) - (4)). Moreover, as an ongoing industrial concern, the industrial owner knows that it is a "deep pocket" and an easy target for an EPA collection action. Thus, owners of contaminated properties appropriately question what is to be gained by selling the property when they will still be liable for the contamination regardless of whether or not they own the property.

The Time Value of Money. Even if the transaction does not expand the company's existing liability, it will certainly accelerate the timing of when that liability is due. As mentioned earlier, once the sleeping dog has been kicked, any of the problems caused by that sleeping dog must now be addressed. For some companies, this presents an issue because business managers are trained to delay expenditures where possible in order to take advantage of the time value of money. Thus, when given the choice of pay now or pay later, many companies will opt to pay later and let the problem, and therefore the property, sit idle.

Cash Flow Problems

Beyond the fear of liability, there are other secondary issues that also add to a seller's trepidation. One of these secondary issues is the fact that selling a brownfield property will actually result in negative cash flow. In other words, rather than "selling" properties, owners will be required to pay someone to take properties off their hands. This is because the costs of cleanup and redevelopment often exceed the fair market value of the property in its current contaminated condition. In contrast, continuing to hold on to the property may result only in minimal ongoing tax and maintenance obligations. Thus, if the decision to sell is already fraught with liability concerns, that decision becomes even more unpopular when the industrial owner realizes it is not going to net any money out of the deal. Instead, it will have to come up with a large cash outlay.

If It Isn't Broke, Don't Fix It

One of the other secondary impediments to sale is inertia. Many industrial sellers are not willing to devote the time and energy needed to closing a brownfield deal unless there is a compelling incentive to do so. If property is dormant and has only minimal carrying costs, then why do anything about it? Why not just continue to carry it along at low cost? Sometimes the costs of carrying the property are not insignificant and therefore the company will take the incentive to move the property off the books. But often there is not a great deal of enthusiasm for seeking out a brownfield deal until there is a real problem that needs to be solved.

What Can Be Done to Convince the Seller?

In order to convince an industrial company to part with its prime brownfield property, the buyer must overcome these concerns. Because the fear of liability is the seller's primary concern, that is the most important obstacle to overcome. In this author's view, the only way to eliminate (or at least minimize) legal liability is through protective legal mechanisms—in other words, through a binding, legally enforceable document such as a contract containing sufficient indemnities and covenants not to sue. A vague assurance from an environmental agency that there is not a problem at the property (such as is often offered in no further action letters) is probably insufficient. A sophisticated seller such as an industrial company will, and should, insist on a legally enforceable document that either eliminates or caps its liability. As is discussed elsewhere in this book, a prospective purchaser agreement often accomplishes this goal quite well for a buyer. But state and federal environmental agencies have yet to introduce similar legal documents that can provide similar comfort to sellers.

Thus, in order to alleviate the seller's fear, the liability protections must be built into the private transaction. This requires, at a minimum, two things. First, the contract for sale must include adequately broad assumptions of existing environmental liabilities by the buyer and indemnification protections for the seller. Second, the seller must be convinced that these protections are backed up by significant financial strength. Anyone can promise to indemnify you for future liability, but unless that person has the money to back it up when the time comes, the promise is relatively meaningless.

In order to demonstrate that their money is where their mouths are, many brownfields developers have formed alliances with insurance companies. However, experience shows that many industrial companies are very skeptical of the sincerity of insurance company commitments to cover environmental liabilities. For the past 20 years a pitched battle about whether comprehensive general liability (CGL) policies cover environmental liabilities has been going on in the courts between most major insurance companies on one hand and corporate industrial America on the other. The insurance companies have fought tooth and nail to avoid the determination that their CGL policies cover environmental liabilities. The result is that many industrial companies believe that an insurance company will make good-sounding promises when it's time to sell you a policy, but will back away from those promises when it comes time to pay up. Whether this perception is in fact right or wrong is almost meaningless. The fact is that many companies believe it, and therefore it is a bias or perception that must be dealt with.

To overcome the industrial skepticism regarding insurance companies, the financial backing for a brownfield deal must be created through insurance mechanisms broader than the traditional CGL policies. Environmental impairment liability insurance, or "gap" coverage, is attractive and is most often written in clearer language that leaves little doubt as to the coverage it provides. If the promises are backed by this type of insurance from a large and reputable insurance company, then a seller should be assured that the money will be there if the need arises.

In summary, a seller's legitimate fear of liability should be overcome if a brownfield buyer (1) agrees to assume the existing environmental liabilities at the property, (2) promises to indemnify the seller for future liabilities, and (3) demonstrates that its promises are backed by solid insurance commitments (or other secure financial mechanisms such as escrow accounts, letters of credit, etc.).

If an industrial company seller receives all of these concessions, then the primary deal obstacle (fear of liability) has been overcome. But what about the secondary obstacles? How can you overcome the inertia, the fear of kicking the sleeping dog, or the related argument that the time value of money means that the payment of cleanup costs should not be accelerated?

One persuasive tool for getting around these obstacles is the belief held by many industrial environmental managers that it is better (i.e., less expensive) to confront environmental issues earlier rather than later. This school of thought believes as follows: time generally does not make contamination better. While some natural attenuation can occur, there is also great risk that the longer an environmental problem is allowed to fester, the more likely it becomes that the contamination will migrate, or will potentially degrade into more harmful chemical breakdown products (such as vinyl chloride). Thus, it may be less expensive to voluntarily acknowledge and manage an environmental issue early rather than wait for it to develop into an intractable problem that you are ordered to clean up. Moreover, if contamination exists on property you own, it is far better to address that contamination when you still own the property, have jobs in the area, and can therefore influence and control the remediation process. This school of thought has obvious merit, and frankly is the way most responsible corporations handle their environmental issues. Thus, presenting this position to a reluctant seller who is otherwise receiving adequate liability protections may cause the secondary issues to fade in importance.

Another argument around the natural resistance toward kicking a sleeping dog is the fact that the dog will not always be asleep. Other external events may require the company to acknowledge the existence of the problem. For example, publicly held companies are required by Securities and Exchange Commission (SEC) regulations to disclose the "material effects that compliance with Federal, state and local provisions . . . relating to the protection of the environment may have upon" the business (17 CFR Sec. 229.101(c)(xii)). Other SEC requirements, such as the mandatory "Management's Discussion and Analysis" provision, require disclosure of any known trends, demands, events, or uncertainties that are reasonably likely to have a material impact on earnings, specifically including environmental issues (17 CFR Sec. 229.303). Under the SEC regulations, companies may be required to divulge latent contamination anyway, thus kicking the sleeping dog.

Conclusion

In order to reach the sizable stockpile of potential brownfields properties that exists throughout corporate America, the savvy brownfields purchaser must speak the language of business. Although redeveloping brownfields properties has unquestionable benefits for society, even the most altruistic and benevolent corporation must be concerned with its long-term liability and profitability. A cautious businessperson will understandably be skittish about the possibility of expanding a company's liability exposure in a deal promising small financial gain. However, the hurdles can be overcome with appropriate legal safeguards and a sales pitch designed to accommodate the seller's concerns.

Gary W. Ballesteros is currently the assistant general counsel for environmental affairs at Rockwell International Corp, a $7 billion high-tech electronics company headquartered in Costa Mesa, CA. Prior to joining Rockwell, Mr. Ballesteros was a partner in the Environmental Law Department of the Chicago law firm of Jenner & Block.

While at Jenner & Block, Mr. Ballesteros' practice focused solely on environmental matters and involved primarily environmental litigation, as well as corporate transactions and client counseling. In Chicago, Mr. Ballesteros was active in the effort to develop contaminated brownfields properties. Mr. Ballesteros successfully negotiated the first prospective purchaser agreement ever entered into by the Illinois Attorney General's Office, which led to the successful brownfields redevelopment of a 31-acre parcel of industrial property in the heart of Rockford, Illinois. He has also written extensively on brownfields issues before, including articles published in Chicago Lawyer, *the* ABA Business Law Section Annual Proceedings, *and* The Brownfields Book *(Jenner & Block and Roy F. Weston, Inc., 1997).*

Mr. Ballesteros is a 1988 cum laude graduate of the University of Michigan Law School. He received a B.S., also cum laude, from the College of Commerce and Business Administration at the University of Illinois in 1985.

4

Illinois, Cook County, and Chicago Activities

This chapter looks at the efforts of Chicago (and Illinois and Cook County) to face brownfields issues. Chicago is an excellent example because of the many brownfields properties and the proactive efforts of the city, and the positive efforts of the state to modernize remediation regulations.

In Sec. 4.1, Mary Culler is very clear about the directions the Chicago redevelopment effort will take. The city is moving aggressively, and at the same time recognizes the uncertainties involved and the need for flexibility. Culler summarizes many aspects of the Chicago program, and the incentives to companies and private developers.

In Secs. 4.2 and 4.3, William Trumbull provides a view of the beginnings of the programs, of the pilot programs, and of the state and county issues.

In Sec. 4.4, Jessica Rio presents a case study of a successful pilot test that provided many lessons for the administrations.

4.1 Furthering Industrial Development

Mary Culler

Overview

The history of Chicago is closely tied to our country's pattern of industrial development and evolution from its agricultural base to the current information and distribution economy. This linkage between Chicago and industry continues today. According to a recent economic assessment of the Midwest prepared by the Federal Reserve Bank of Chicago, there is abundant evidence that over the last 10 years the region has been economically recovering from its painful adjustment period in the 1970s and 1980s, when it was labeled part of the Rust Belt. A resurgent strength in the region's mainstays, agriculture and manufacturing, has brought renewed interest in redeveloping urban industrial property. One of the main shifts is the understanding that industry is changing and that there is a need to adapt to alternative industry technology and locational preferences.

Industrial activity is clearly a critical segment of Chicago's economic base, providing well-paying jobs for Chicago residents, assuring a diversified economic base, and stimulating growth and employment of the economy. The Midwest, with Chicago at its helm, is experiencing a significant upswing in local industrial and manufacturing investment. It is critical that the city protect its current industrial base, as well as prepare itself to provide an environment suitable for and advantageous to industrial growth and development. In order to benefit from this current regional trend toward manufacturing growth, Chicago needs to be poised and ready to accommodate industrial expansion and development. Assuring continued industrial investment in viable manufacturing districts is critical to Chicago's long-term fiscal health.

It is therefore critical that suitable land be made available over the long term for new industrial investment, and to accommodate expansions of existing businesses. An uncertain future not only restricts expansion and in-migration of industrial users, but also drives out existing industries that require the land use stability needed to justify long-term capital investments. Chicago's goals are therefore to assemble and clean up new properties to ensure new business opportunities, and to maintain stable industrial districts so that firms will not be forced out by residential development. Failure to do this will force industry to seek suburban locations where land is available and the land-use patterns are relatively stable.

The city of Chicago has already taken actions to encourage, attract, and retain a significant amount of industrial growth and development. An active industrial expansion and protection policy allows the city to continue to add significant amounts of higher-value industrial space to the tax rolls and retain an equally significant number of jobs that might otherwise be lost. Building on some of the city's unique attributes and pursuing growth opportunities through thoughtful and proactive planning efforts can slow the flow of industry and jobs to the suburbs. Thus the revitalization of the city's industrial areas can be achieved.

Developing the Ideal Industrial Development Environment

In 1995, the city developed the Model Industrial Corridor Initiative to help foster the continuing investment in the city's industrial base and the maintenance and growth of job opportunities found there. This program provided support to selected not-for-profit neighborhood organizations and industrial councils for strategic development planning in the city's industrial corridors. Not only do Chicago's industrial corridors constitute the city's primary resource of space for industrial development, they are already home to a large portion of the city's industrial base.

The city's industrial land use policy sets forth a strategy for industrial development that is uniquely urban. It recognizes that the growth of Chicago's economic base will take place in built environments, and will be driven largely by the expansion and modernization of existing companies. The primary goal of the industrial land use policy is therefore to foster the expansion and modernization of Chicago's industrial companies by enhancing the physical environments in which they operate. After all, Chicago is a city of industrial neighborhoods. Just as residential communities need appropriate public investment to make them pleasant places in which to live, industrial areas require specific types of improvements to meet constantly evolving needs. This policy is the backbone of a broad economic development agenda that includes technology assistance, workforce training, and investment incentives.

The model industrial corridor planning process resulted in three studies for the North, West, and South sides of Chicago. Entitled "Corridors of Industrial Opportunity," the documents clearly identified overall planning goals that the city should follow in order to encourage industrial development. The plans specified that industrial corridors must be served by a well-maintained infrastructure that accommodates modern production and transportation. Corridors must have well-defined boundaries and separation from incompatible activities so that operations are not in conflict with neighboring uses. They must also be safe places where employees feel secure and companies do not hesitate to entertain customers. Industrial corridors must be provided with physical amenities, such as attractive streetscapes, trees, signage, and gateways—items too often overlooked in industrial development strategies.

The city implemented a number of initiatives to reinvent its industrial areas. Industrial capital improvement dollars are being targeted to corridors to ensure that bridges can accommodate industrial loads, viaducts have adequate clearances, and streets are rebuilt to industrial standards.

These improvements are critical steps in attracting modern industrial businesses. In addition, the Industrial Street Vacation Program closes old streets and turns outdated street patterns into more useful and attractive spaces for companies throughout the city.

The city's industrial land use policy, which encourages the redevelopment of brownfields, is supported by the following strategies that incorporate existing programs and new initiatives in a corridor-targeted approach.

Create Accessible and Attractive Environments Throughout the City's Industrial Corridors

Corridors are target areas for coordinated public and private investments in the physical environment. The city has three goals in implementing these investments.

1. Provide efficient access to major transportation links and smooth internal traffic circulation. Most industrial capital improvement dollars are being targeted to corridors to ensure that the infrastructure meets the needs of today's industrial business.

2. Strengthen the physical identity of corridors and provide the range of amenities companies expect to find in a contemporary industrial park environment.

3. Facilitate the development of brownfields or underutilized and contaminated industrial corridor properties.

Assure Stable Land Use Within the Corridors Through Improved Zoning and Land Use Regulation

The city is committed to assuring that appropriate land use is maintained in industrial corridors and the following actions are being followed:

1. Nonindustrial developments in industrial corridors undergo full review through the planned development process.

   The rezoning of land in a corridor to a nonindustrial zoning classification is avoided. It is recognized, however, that in particular cases such rezoning may be appropriate. All proposals for rezoning of parcels in industrial corridors to other uses require approval by the Chicago Plan Commission.

2. Full consideration is given to the operational needs of existing industries when reviewing proposals to rezone property near industrial corridors.

The development of land outside of corridors should be consistent with the long-term needs of the area. Where a proposed rezoning adversely impacts an industrial corridor, the rezoning is modified or discouraged. These actions are meeting the needs of modern industry and encouraging the attraction and expansion of industrial businesses in Chicago.

3. Planned Manufacturing Districts (PMDs) are established in areas where conflicting land uses are especially a problem.

This action serves as a means to assure companies that locate in industrial corridors, but are concerned about zoning changes or land use conversions that may ultimately be incompatible with the character of the manufacturing district, that the zoning will not change. A PMD designation protects industry by limiting future land use and development to those uses that are compatible with industry. Chicago has successfully designated three PMDs to serve large industrial users such as Federal Express, Republic Windows and Doors, and Finkel Steel.

Other Necessary Components

Revitalizing the inner city requires an economic strategy to build viable businesses that can provide attractive employment opportunities. Economic development in inner cities will only come from recognizing the potential advantages of an inner-city location and building on the base of existing companies, while dealing frontally with the present disadvantages of inner cities as business locations.[1]

Manufacturing is critically important to the city of Chicago because of its strong employment base. Today, manufacturing accounts for an estimated 165,000 jobs in the city, and wholesaling provides an additional 59,000. According to a study completed by the Boston Consulting Group, companies cite Chicago as a prime location because of the availability of labor and the proximity to transportation, suppliers, and customers. Despite these advantages, the city is experiencing a decline in manufacturing. One of the primary reasons Chicago is losing many of its strongest manufacturing and wholesaling companies is because the city lacks available sites for immediate development.

Although some manufacturers located in Chicago report dissatisfaction with some social issues, most ultimately leave the city because sites for expansion are not available quickly enough. Once expansion needs force companies to move, these companies begin to consider a variety of factors in selecting a new location. Most want a site that provides reasonable safety and proximity to workers. It is only once a site is identified that costs become a deciding factor.

The city understands that the most important action it can take is to increase site availability. What is notable is that there is plenty of underutilized land in Chicago. Brownfields abound! The shortage arises because most available sites are small and scattered and do not meet the needs of industrial users. The city must aggressively assemble sites large enough to attract modern manufacturing. This requires substantial time and investment with no guarantee of success.

In addition, brownfields sites often require environmental remediation, which is very costly. In some areas, speculators drive up the price of land, hoping it will be rezoned and purchased for residential or commercial use. These higher prices, along with site preparation costs, make redevelopment of the parcel impossible. In addition, where existing industrial buildings are available at the right price, they are often functionally obsolete, with multiple floors and low ceilings. Given the time, investment, and risks involved, private developers and interested companies typically find industrial development in the city and the redevelopment of brownfields unattractive and economically skewed.

Public sector intervention is therefore necessary to reinvigorate private industrial development of brownfields. The city is taking a lead role to leverage its unique resources in industrial site development by:

- Ensuring zoning consistency through Planned Manufacturing Districts to preserve land for industrial use.

- Using the multiple legal tools and established programs to assist in the process of assembling land. For example, the city pursues sites that are tax delinquent, ultimately taking title through the Tax Reactivation Program. This program waives all back taxes and allows the city to take title for very little cost. While this program is especially useful for brownfields, the process can take 18 months, which can be too long for a company to wait. This is why it is so important that the city start now to pursue this approach. The city also pursues a similar process for abandoned buildings in major disrepair. A demolition lien allows the city to take title to those buildings in severe need of demolition or repair for a fraction of a possible acquisition price.

- Identifying the financial assistance necessary to leverage private investment. This is critical because the costs of brownfields redevelopment can be high.

At this time, the public sector is aggressively pursuing the redevelopment of brownfields on three fronts:

1. Creating large industrial parks capable of accommodating tenants of all sizes

2. Developing individual, scattered sites in established industrial areas

3. Providing targeted assistance to specific companies interested in expanding their current locations

Chicago knows that brownfields will only be redeveloped if the city can make them attractive to a developer or company. This requires that the city invest heavily, financially and strategically, in pursuing the often difficult real estate puzzles that must be solved before a large property can be assembled. Once the real estate issues have been taken care of, there are often environmental and other site preparation needs that are essential to the successful marketing of the site. More often than not, the costs associated with this process for a developer are greater than any returns seen by sale or lease. For interested companies, the process is often just too time consuming and they cannot wait.

Another important factor is that many of these sites are in areas that are in need of major infrastructure and other area improvements. The public sector does not often hold the most lucrative properties in its portfolio, but those that need the most help. Infrastructure improvements and other surrounding improvements must often be addressed in order to see the successful redevelopment of the brownfield site. A company must be confident, for example, that the road not currently built will be there in the near future. Subsequently, the public sector must act as a private developer and follow through on the longer-term plans for the area. This confidence in the public sector is essential. The city provides proof of its commitment through the programs provided by the city to redevelop brownfields. The benefit is clear—the city can retain manufacturing/wholesaling companies that are growing, and potentially attract firms that are relocating from other areas.

If We Assemble It, Will They Come?

In 1998, Arthur Anderson's Real Estate Services Group completed an objective analysis of the industrial development market in Chicago. The study provided the city with a framework for making strategic decisions about industrial retention and development opportunities. The full report provided extensive data about the major industrial corridors in the city, and served as a resource for supporting and strengthening industrial land use and development decisions.

The study identified demand opportunities that could add 1.8 million square feet of new industrial space per year during 1998–2005. The space would have positive economic benefits for the city, including potential property tax benefits of almost $220 million and an estimated 31,000 new or retained jobs. These benefits support a continued strategic city industrial policy to accommodate and stimulate industrial growth.

Although the analysis identified more than 2200 acres of potential industrial property, the majority of the sites are currently unavailable. This is because there are numerous barriers to development, including issues related to site preparation, infrastructure, and ownership. The city must take advantage of industrial development potentials that current trends appear to support. Bringing these properties to market will require using the city's eminent domain authority, funding capabilities, environmental and planning capacities, and zoning powers to assemble enough acreage to meet demand.

The city must also recognize that the overall composition of the industrial sector has shifted. While Chicago has traditionally been identified as a Rust Belt city dominated by older and dying smokestack industries, the reality is that the smokestack manufacturing sector has decreased to an estimated 30 percent of occupied space. The remaining occupied space consists of 48 percent distribution space and 21 percent flex/services space. Further, 96 percent of future industrial growth is projected to be in the distribution and services sectors. This is an important factor to those looking to redevelop brownfields. Two major recommendations came out of the study supporting the city's need to assemble and clean up brownfields.[2]

1. The first priority of the city's industrial policy is to continue to retain existing industries and accommodate their replacement and expansion needs. This calls for a need to look at brownfield redevelopment even more closely as existing clean assembled land is limited.

2. To accommodate emerging demand, the city must expand its land acquisition, assembly, zoning, cleanup, and marketing efforts to build an inventory of available sites. Land assembly and improvements remain time consuming. This intensive process must continue to be managed in order to assure a future supply for emerging demand conditions.

Sweetening the Deal

In order to attract interested companies and developers, the city is aggressively pursuing the tools necessary to redevelop key industrial areas. Within several of the city's key industrial corridors, Chicago has established eight industrial tax increment finance (TIF) districts (see Fig. 4-1). TIF is a powerful tool that enables cities to self-finance their economic development programs. TIF funds can pay for public improvements and other economic development incentives using the increased property tax revenue the

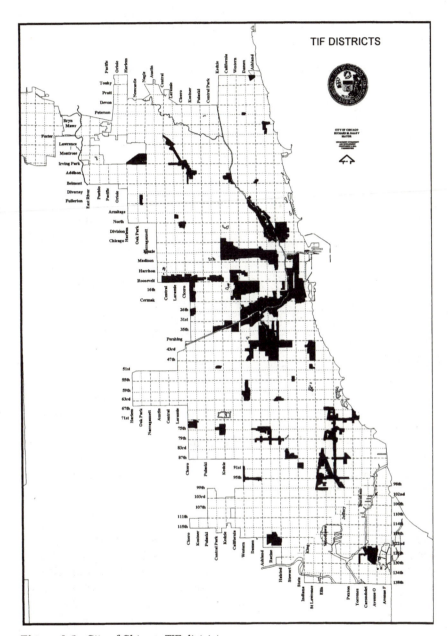

Figure 4-1. City of Chicago TIF districts.

improvements help generate. The TIF districts will help Chicago attract and retain industry by financing measures such as:

- Improvements to streets, viaducts, and other infrastructure elements
- Land assembly and site preparation
- Environmental cleanup of brownfields
- Rehabilitation of deteriorating or obsolescent buildings
- Incentives to attract or retain employers
- Job training, education, and other workforce readiness programs

Over the years, TIF has become an indispensable part of Chicago's industrial development strategy and a key method for redeveloping brownfields. TIF assistance has played an important role in the retention and attraction of several large industrial employers. The city has also been successful in using TIF to attract and retain employers in key industrial areas. As of December 1997, TIF projects in Chicago have created approximately 8,300 new jobs and saved approximately 24,300 others.

Most importantly, TIF provides the needed tools and authority to acquire, clear, and otherwise prepare land for redevelopment. Without this authority the city is unable to acquire and clean up sites. This designation allows the city to address the urgent need to assemble and prepare sites for redevelopment. TIF revenue is also the only way Chicago can pay back loans for environmental remediation borrowed from the federal government. The city has borrowed $50 million from HUD's 108 program for four key sites in the city that are located in TIFs. This federal loan will allow the city to get a head start and begin assembling sites for development prior to TIF increment being generated. Revenue generated by new development in these TIFs will pay back the HUD loan. Extensive increment generation models are created for each site showing the estimated dollar amounts each TIF will create over its 23-year life.

There are an estimated 209 vacant industrial sites in the designated industrial TIFs totaling approximately 631 acres. These sites, many of which are brownfields, represent opportunity areas for business expansion or new development. The vacant opportunity sites could accommodate an estimated 12.5 million to 15.3 million square feet of new or expanded industrial space with opportunities for 14,500 to 18,800 additional workers.

TIF Explained

Tax increment finance (TIF) districts provide assistance to businesses. TIF provides assistance with business expansions, facility rehabilitation, or worker training plans. Tax increment financing is the city's way of retaining and creating jobs and stimulating investment in targeted areas.

How TIF Works for Business

When previously underutilized buildings or land are redeveloped and put to more productive use, the city is able to capture the increased tax revenue from that project to pay for incentives. For example, if there is vacant land next to your building, the city could help you assemble the land and address any environmental problems that would allow you to expand your business. If that expansion necessitated more employee training, the city could help with that, too. So the program does not raise your taxes! It captures the increased value of surrounding properties and reinvests it in the immediate area.

Eligible Business Costs

Here are some business costs eligible for TIF reimbursement:

- Land acquisition
- Site preparation
- Environmental remediation
- Building rehab or repair
- Job training/retraining
- Public improvements (street, sewer, water, viaduct, etc.)
- Studies and surveys
- Relocation costs
- Certain financing costs

Although new private construction and purchase of machinery and equipment are not eligible costs under TIF guidelines, the city of Chicago offers other programs to assist businesses with these types of costs. If your business needs to expand and you can demonstrate a proven financial gap, TIF may be the answer.

Tax increment financing is one of many tools the city uses to help businesses grow, compete, create jobs, and provide products and services to Chicagoans.

How Do TIF Districts Work?

First, the city works with local officials, community groups, businesses, and developers in identifying an area not living up to its potential. The city examines the land to see whether it is eligible to be a TIF district. Properties and buildings within potential TIF areas must display some of the following characteristics:

- Age/obsolescence
- Illegal uses of individual structures
- Violation of minimum code standards
- Excessive vacancies
- Overcrowding of facilities
- Lack of ventilation, light, and sanitary facilities
- Inadequate utilities
- Excessive land coverage
- Deleterious land use of layout
- Depreciation of physical maintenance
- Lack of community planning
- Dilapidation
- Deterioration

Because of these problems, the prospective TIF area contributes less to the city's tax base than similar areas. In order to eliminate some of these blighting characteristics from a neighborhood, the city creates a TIF redevelopment plan to eliminate those conditions and help revitalize the neighborhood. Public hearings are held to provide input to the redevelopment plan. Once the redevelopment plan has been approved, the city council formally creates a TIF district.

Equalized assessed value (EAV) is the county assessor's way of assigning similar taxes to similar structures and spreading the property tax burden equitably among properties. EAV can also be a useful tool for determining the relative strengths and weaknesses of similar properties in different areas. TIF-eligible areas have lower EAVs than comparable areas in the city because of all the qualifying physical characteristics. Therefore, they are contributing less to the city's tax base than similar land uses.

Once an area is declared a TIF, the amount of tax the TIF area generates is set as a baseline, and the seven taxing entities within the city that generally receive portions of property tax revenue generated in the area continue to receive that set amount during the life of the TIF. Schools, parks, and other agencies continue to receive the same amount of revenue, so there is no loss of revenue to those local taxing bodies.

As vacant and dilapidated properties developed with TIF assistance return to productive and appropriate uses, the value (EAV) of those properties increases, creating an incremental increase in the revenue generated within the TIF district. The increment created between the baseline and the new EAV is captured and used solely for improvements and redevelopment within the TIF district. This increment can be used as a source of revenue to pay back bonds issued to pay up-front costs, or can be used on a pay-as-you-go basis for individual projects.

The maximum life of a TIF district is 23 years. As the TIF expires and the city's investments in the redevelopment projects within the designated TIF are paid back, property tax revenues are again shared by all the different taxing entities. Those taxing entities realize a budget windfall, receiving the higher revenues based on much higher EAVs that would not have been possible without the TIF.

Other Financial Tools

The city is committed to offering incentives to attract and retain employers. While TIF is a strong tool that is proving successful, it is limited in one key area: TIF funding cannot be used for new construction. A new development may therefore not have significant TIF-eligible costs. The city therefore offers other incentives that work together to provide an appealing package to a company or developer.

Established businesses, new enterprises, communities, and real estate developers and brokers have all benefitted from the wide range of business assistance provided by the Department of Planning and Development (DPD). All of the business assistance programs are designed to encourage development and redevelopment of commercial and industrial properties and to promote the retention and expansion of businesses throughout Chicago.

Loans

DPD's loan programs are designed to fill gaps where conventional financing falls short of a company's needs in projects that create jobs and/or help revitalize a targeted area.

Micro loan:

Available for—small businesses that will create or retain jobs for low- to moderate-income Chicago residents.

Terms—low interest rate; up to $20,000.

CD float loan:

Available for—any private developer, not-for-profit organization, or individual business that can obtain an irrevocable letter of credit from a qualified bank. Funds can be used for new construction, demolition, rehabilitation, land/building acquisition, machinery, equipment, and related costs.

Terms—short-term, low-interest interim financing; minimum loan is $1 million.

Bank participation loan:

Available for—commercial or industrial businesses expanding in or moving to Chicago who apply through a participating bank.

Terms—the city provides low-interest financing for 50 percent of eligible project costs. The city's participation will not exceed $250,000 for commercial businesses and $500,000 for industrial businesses.

Revenue bonds:

Available for—industrial and not-for-profit companies wishing to expand or modernize their facilities and/or operations.

Terms—tax-exempt bond issued by the city on behalf of the company; offers long-term financing at lower rates than conventional financing.

Enterprise Zone Incentives

In addition to all the other city assistance and incentive programs available to all businesses, those located within one of Chicago's six state-designated enterprise zones (see Fig. 4-2) are eligible for:

Tax credits:

- *Investment tax credit*—0.5 percent for investment in machinery, equipment, and buildings (in addition to the 0.5 percent statewide investment tax credit). If the investment results in a 1 percent increase in employment, an additional 0.5 percent may be available.
- *Jobs tax credit*—a $500 income tax credit for each job created for dislocated or disadvantaged workers. (A minimum of five jobs must be created during the year in which a jobs credit is taken.)

Tax exemptions:

- *Sales tax exemption*—building materials used for remodeling, rehabilitation, or new construction are exempt from both city and state sales taxes.
- *Machinery and equipment sales tax exemption*—6.25 percent sales tax exemption on machinery and equipment used in the process of manufacturing or assembly, or in the operation of a pollution control facility.

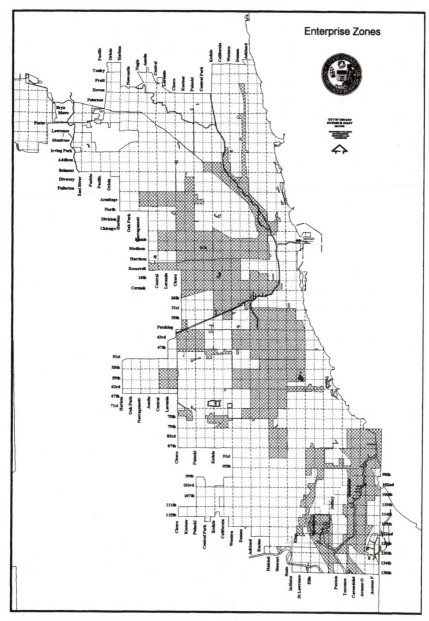

Figure 4-2. City of Chicago enterprise zones.

- *Utility tax exemption*—an exemption from the state tax on gas and electricity, as well as the Illinois Commerce Commission's administrative charge for businesses certified by the Department of Commerce and Community Affairs.

- *Real estate transaction tax exemption*—exemption for taxes on the transfer of title on commercial and industrial properties.

Tax deductions:

- *Income tax deduction for financial institutions*—financial institutions may deduct an amount equal to the interest received from a loan for development in an enterprise zone from their taxable income.

- *Corporate contribution tax deduction*—corporations that make donations to a designated zone organization may claim in their state returns an income tax deduction at double the value of the contribution.

Empowerment Zones

EZ incentives are available for business, not-for-profit organizations, individuals, governmental entities, and community groups located in the federally designated empowerment zone (see Fig. 4-3).

- *Employer wage tax credit*—20 percent of the first $15,000 of qualified wages paid to each employee who meets certain criteria.

- *Increased Section 179 deduction*—up to $37,500 on the cost of property in the year in which it is placed in service.

- *Tax-exempt bond financing*—for purchase and acquisition of qualified empowerment zone property.

Property Tax Reductions

Available for—commercial and industrial businesses; new construction, substantial rehabilitation, or reutilization or reoccupation of buildings that have been vacant for at least two years.

Terms—assessment levels are reduced from 36 percent on industrial buildings and 38 percent on commercial buildings to 16 percent for 10 to 12 years.

These incentives all work together encourage the attraction of industry and the redevelopment of brownfields.

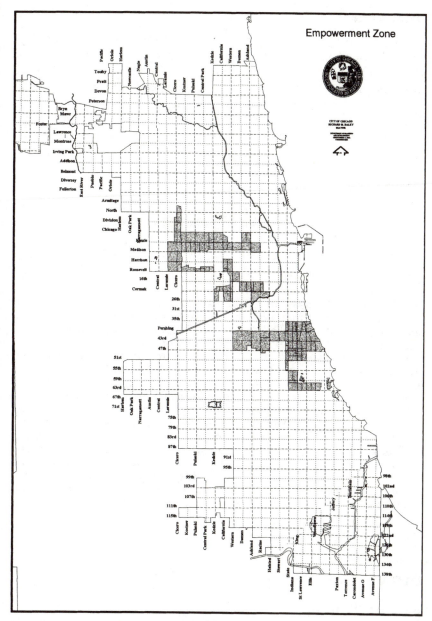

Figure 4-3. City of Chicago empowerment zones.

Summary

The city is leveraging a wide variety of tools to help with the redevelopment of brownfields. Without aggressive public sector initiatives such as TIF and other legislative tools available to the city, the private sector alone would not take on the challenge. The real estate and environmental obstacles are just too daunting. It is therefore critical that the city continue its policy of acquiring and assembling land so that it is poised to deliver sites to companies or interested developers in a timely manner.

References

1. Porter, Michael E., "New Strategies for Economic Development." The Boston Consulting Group, in partnership with The Initiative for a Competitive Inner City. (June 1998)
2. Arthur Andersen LLP, "City of Chicago Industrial Market and Strategic Analysis." (March 1998)

Mary Culler is the assistant commissioner for industrial development with Chicago's Department of Planning and Development. She is responsible for coordinating the city's industrial development projects. Current projects include, among others, a 200-acre business park development, a 62-acre development on a "silver shovel" brownfields site, and expansion of Chicago's historic Stockyards industrial park. Ms. Culler identifies and promotes incentives for retaining existing companies and attracting new industry to the city. She negotiates development deals with developers, brokers and businesses. Previously, Ms. Culler was with the Environmental Protection Agency (EPA) in Washington, DC, where she was the associate director of the National Brownfields Office. She has led efforts on other environmental policies regarding hazardous and solid waste issues and air regulations. She has also worked in the U.S. Senate on natural resource and energy legislation. Ms. Culler graduated from Indiana University, majoring in journalism and French, and she holds a masters degree from Harvard University.

4.2   The Chicago Brownfields Initiative

William Trumbull

Background

When Chicago was incorporated as a city in 1837, its city seal was inscribed with the motto *urbs in horto*, which means "city in the garden." With that vision, the Chicago Department of Environment works to maintain a balance between environmental health and economic development, which are often viewed as distinct and competing goals. Our city is our environment, and continued economic prosperity requires a healthy environment, just as a sustainable environment depends on a healthy economy. In dense urban areas, residential communities, neighborhood parks, commercial areas, and industrial facilities relate to and depend upon each other. Our garden is not just a place for flowers; it is our place to work, educate and recreate.

The intrinsic relationship between the economy and the environment is best demonstrated by the brownfields movement, which focuses on environmental cleanup for economic development. In the past five years, Chicago and many other municipalities have demonstrated that environmental contamination is not an insurmountable liability, but may be used as a tool to spur economic development in former industrial areas. The Chicago Brownfields Initiative has grown from a small pilot program into an interdepartmental city program that cleans sites, creates jobs, and improves city neighborhoods. Chicago is committed to applying public resources to distressed properties that otherwise would not be returned to productivity by market forces alone. In addition to the city program, a number of private sector brownfields transactions have recently occurred in Chicago, many without the assistance of public subsidies.

From a regional land use perspective, efforts to preserve rural communities and productive farmlands, wildlife habitats, and biodiversity have reinforced the need to limit urban sprawl by encouraging brownfields and infill development. The unfolding story of brownfields initiatives across the country will tell us much about how we value urban and rural landscapes, clean land, clean air, and our ability to access them.

Charles Bartsch of the Northeast-Midwest Institute reportedly coined the term *brownfield* at a 1992 conference called "New Uses for Old Buildings" to describe abandoned industrial properties in which environmental liabilities have led to disinvestment. Out of that meeting came the idea of modifying the federal tax law to allow cleanup costs to be treated as a deductible expense for redevelopment projects. A similar bill, with limited geographic applicability, was finally enacted in 1997. In the intervening years, several policy initiatives, state and local laws, and economic incentives have been developed to encourage private, public, and nonprofit brownfields activities.

An operating definition of a brownfield is a vacant or underutilized property passed over for development due to real or perceived contamination. A brownfield can be the several hundred acres of a shuttered steel mill, a small abandoned factory, or a vacant service station. This is not just an urban issue, since such a property could probably be found in just about any community. At the most basic level, a brownfield is a real estate and economic development issue, complicated by environmental costs and liabilities and acquisition and development problems. The General

Accounting Office (GAO)[1] estimates that there are over 500,000 brown-fields across the country. Based on a compilation of several computer data bases of land use and contamination parameters, the Army Corps of Engineers[2] identified 711 potential brownfields sites in Chicago.

It is difficult to actually count the number of brownfields in a community. The current definition is not restrictive enough, and sufficient information to count sites is often not available or is held independently by multiple sources. There are many reasons a site may be vacant or under-utilized. For example, the extent of environmental contamination is often an unknown, or a site may be vacant because of location and access problems having nothing to do with the condition of the property. Current tax status, applicable mortgages, or liens all affect the ability of a property to be marketed. Unless a more restrictive definition concerning environmental conditions, ownership, and land use is applied, brownfields are difficult to count. Consequently, a real inventory is not feasible.

The Chicago Brownfields Initiative comprises three basic strategies:

1. The Chicago Brownfields Forum to define the issues and recommend solutions

2. The Pilot Sites program for municipal cleanup and redevelopment

3. Incentives to encourage private sector investments in brownfields

Tax incentives are discussed in Sec. 4.3.

The Chicago Brownfields Forum

With support from the John D. and Catherine T. McArthur Foundation, more than 100 representatives from government, business, finance, environmental, community, and civic organizations gathered for 2 days in December of 1994 and outlined the barriers to brownfield redevelopment in the city. The forum was conceived as a broad-based, interdisciplinary task force to inform public policy on brownfields issues. For several months, smaller work groups met to draft recommendations. In May of 1995, when the Forum concluded, the groups had produced 63 recommendations, which were then consolidated into 9 project teams, each team led by a public, private, or nonprofit entity. The Brownfields Forum Final Report and Action Plan[3] was published in November of 1995.

While the Forum was in session, the city undertook the redevelopment of five pilot sites, which will be discussed in the following text. The pilot projects and Forum were mutually reinforcing; the sites informed the policy of real-life complications, while the expertise of the Forum participants often served as a resource for moving the projects forward.

Many barriers have been addressed or eliminated, either by direct actions of Forum participants or through the indirect influence of the recommendations. For example, while the Forum was in progress, the Illinois legislature was debating amendments to the Environmental Protection Act to create the Illinois Site Remediation Program, which provides for the voluntary cleanup of sites based on end use standards. Completing the program leads to a letter of no further remediation, which provides liability protection from the state. The Forum did not, as a group, participate in the legislative process, but the parallel activities kept a broad group informed of the issues, and led to improved legislation (Sec. 4.3).

To address concerns of lending institutions that were reluctant to make loans on contaminated property, Forum participants produced a Model Lending Package, which outlines a process for making loans on brownfields properties.[3]

To increase community capacity, the Brownfields Institute was established by the Chicago Association of Neighborhood Development Organizations. The Institute developed an eight-course curriculum to educate community groups on the issues of cleanup, marketing, and financing redevelopment projects.[3]

These are just a few of the many activities triggered by recommendations from the Forum. In June of 1998, with support from USEPA Region 5, the progress of the Brownfields Forum was formally evaluated. Just as the Brownfields Forum Final Report outlined the barriers to redevelopment, this evaluation indicated success in overcoming barriers.

The Brownfields Pilot Program

While the Brownfields Forum was identifying barriers, an interdepartmental work group from the City Departments of Environment, Planning and Development, and Law was demonstrating that environmental cleanup can lead to economic development. With $2.0 million of general obligation bond money, the group selected five sites based on:

- Site access and control
- Information on contamination
- Redevelopment potential

The thought was that the five sites could be investigated and, with any remaining money, one or two sites could be remediated.

Table 4-1 summarizes key information about the five pilot projects. All five sites have been cleaned up for just over $1 million (excluding demolition costs, which were allocated from another city program). One site has

Table 4-1. Chicago Brownfields Pilot Sites Projects (1995–1997)

Project name	Size (acres)	Demolition costs	Cleanup costs	Estimated annual tax revenues	Jobs created	Jobs retained
Verson Steel	7.0	$350,000	$211,000	$209,000 (wh)	125	500
Scott Peterson	1.8	$400,000	$303,000	$23,500 (pk)	100	200
Blackstone Manufacturing	3.3	—	$592,000	$43,000 (pk)	7	200
Madison Equipment	2.1	—	$2,800	$62,000 (wh)	7	25
14th & Union	2.5	—	$4,000	Univ. of IL expansion	NA	NA
Total	—	$750,000	$1,112,800	$337,500	239	975

Tax revenue estimates based on a coverage ratio of 0.3 and assessments of $2.25 per square foot for warehouse (wh) and $1 per square foot for parking (pk).

been transferred to the University of Illinois at Chicago to provide space for expansion; the other 4 are now in productive reuse, having created 332 new jobs and retained another 950 jobs in the city. One important lesson was learned in the process from the Madison Equipment site, which had been vacant for years because of perceived contamination. With an investment of $2800 for an environmental assessment, the site was found to be clean! Formerly tax delinquent, the site may generate $353,000 in real estate taxes annually.

In addition to the original pilot sites, the Department of Environment has completed eight other remediation projects that are in various stages of redevelopment. Cleanup costs average $2.15 per square foot, which is within the range of the annual tax revenues received from the property after development. These costs have been minimized by carefully screening projects, remediating to industrial use standards, using engineered barriers and institutional controls, and competitively bidding projects to environmental consultants under term agreement contracts with the city.

The Chicago Brownfields Program routinely conducts Phase I and Phase II environmental site assessments to screen sites for acquisition, cleanup, and development. Chicago has negotiated term agreements with 19 environmental consultants, which allows the city to solicit competitive proposals for site-specific projects. With the consultant term agreements in phase, Chicago has been able to utilize the expertise of each firm for a wide range of projects from standard underground storage tank removals to complex remediation and removal projects.

All investigations are undertaken with the intent of entering the site into the Illinois Site Remediation Program. If Tier 1 cleanup objectives cannot be achieved, engineered barriers may be used as part of the corrective action. In such cases, it is always helpful to have an end user with site development plans in hand to present to the Illinois EPA.

Since the success of the program is measured in terms of tax revenues generated and jobs created, we often compare cleanup costs against those factors. As a result of careful site selection, competitively bid cleanup costs, and a market demand for sites, remediation costs are down to $2.15 per square foot, which is on a par with the annual tax revenues generated from industrial properties in the city However, it is necessary to have a final user to achieve the goals of increased jobs and expanded tax revenues.

Summary

The Chicago Brownfields Initiative has taken an aggressive position to turn environmental liabilities into economic assets, to clean up properties, to create jobs, and to improve neighborhoods. Based on successful projects, sound policies, and relevant research, the initiative will continue to break new ground in urban redevelopment. Our city is our garden, and we need to maintain it as a place to live, work, and play. Over the last few years we have learned a few lessons. All brownfields are local, and the true cost has to be measured in terms of the economic degradation and the lost opportunities to communities blighted by brownfields properties. Brownfields are as much a real estate transaction and economic development issue as they are an environmental problem. Consequently, successful projects require multiple resources from a broad range of participants. With adequate tools, other municipalities can use land acquisition authorities, enforcement actions, environmental litigation, and condemnation to effectively spur redevelopment.

References

1. U.S. General Accounting Office (GAO), *Report to the Chair, Committee on Small Business, House of Representatives.* (1995)

2. Army Corps of Engineers, unpublished report. (1996)

3. City of Chicago, *The Brownfields Forum.* (1995)

William C. Trumbull is the deputy commissioner in the Chicago Department of Environment responsible for the city's Brownfields Initiative, which links environmental remediation with economic development.

Mr. Trumbull is on the board of directors of the National Association of Local Government Environmental Professionals and the Chicago Brownfields Institute Advisory Council, and is active in several other environmental and economic development organizations.

Prior to working for the city, Mr. Trumbull was a regulatory issues manager in the Environmental, Health and Safety Department of Amoco Corp., focusing on petroleum transportation issues. He began a career in the petroleum industry as an exploration geologist. Mr. Trumbull has a B.S. in geological sciences from the University of North Carolina at Chapel Hill.

4.3   State and County Programs

William Trumbull

Illinois Site Remediation Program

The Illinois Site Remediation Program, like many other state voluntary cleanup programs, encourages private sector remediation projects by providing cleanup objectives based on the future use of the property; it also provides for a no further remediation (NFR) letter from the state at the completion of the cleanup. In Illinois this method is known as the Tiered Approach to Corrective Action Objectives (TACO)[1] (see App. 1).

The NFR letter provides liability protection for developers who have met specified cleanup objectives. In Illinois, the NFR letter is further supported by a memorandum of understanding with USEPA Region 5, which essentially states that the federal government will have no interest in sites cleaned up under the state program.[2]

The statute defines three tiers that can be applied to determine remediation objectives. The tiers can be obtained directly from a table, or can be site-specific and determined from risk-based calculations that take into account the contaminants, exposure routes, and possible receptors. This provides for flexibility in developing cleanup objectives, allows for the exclusion of unlikely exposure routes, and considers area background concentrations.

Tier 1 remediation objectives are listed in lookup tables based on conservative risk models. To complete a Tier 1 evaluation, you must know the extent and concentration of contamination, the groundwater classification, and the intended land use (residential or commercial/industrial).

Tier 2 evaluations can be conducted on a site when Tier 1 objectives are not met and when institutional controls or engineered barriers can be used on the site. Tier 2 information can allow for less stringent but equally protective remediation objectives.

Tier 3 objectives can be applied to sites that cannot be handled under the first two methods. These situations may involve simple sites where physical barriers limit remediation, or complex sites where full-scale risk assessments and alternative modeling methods can be applied.

For the applicant entering the Illinois Site Remediation Program, the goal is the issuance by the state of a no further remediation (NFR) letter. A site qualifies for an NFR letter when all program requirements and applicable remediation objectives have been met. The NFR letter must be filed with the county recorder of deeds to ensure that current and future users of the property will be informed of the environmental condition of the property.

In the year since the program rules were finalized, 70 sites in Chicago have been voluntarily remediated under the state program, which demonstrates that, given the opportunity for certainty in the cleanup process, responsible landowners will undertake environmental projects that will result in economic development.

Institutional controls are a legal mechanism for imposing restrictions and conditions on land use. If applied to a site, these restrictions are contained in an Illinois EPA NFR letter. They may include restrictive covenants, deed restrictions, negative easements, or ordinances adopted by local governments.

For example, Lake Michigan is the potable water supply for Chicago and the surrounding metropolitan area. Much of the shallow groundwater in the city is contaminated, some of it by the residue from the Great Chicago Fire of 1871 and subsequent uses. Establishing a prohibition of the drilling of potable water wells in the city creates an institutional control that eliminates the need to consider ingestion of groundwater as a pathway in the risk-based calculations of TACO. As a result, a remediation project in the city only needs to demonstrate protection of human health and the environment, but does not have to remediate groundwater to the most stringent drinking water standards. In addition, if contamination extends under a city roadway, the Department of Environment can enter a right-of-way agreement that allows contamination to remain in place, but tags the location as part of the city permitting process so the presence of contaminants will be communicated to anyone requesting a permit to dig in the right-of-way.

Engineered barriers limit exposure and/or control the migration of contaminants on a site. Such barriers may be fabricated or natural; their purpose is to cut off the exposure route of a contaminant, and they may be used in developing remediation objectives for Tier 2 and Tier 3 evaluations. An engineered barrier must also be accompanied by an institutional control that assures the maintenance of the barrier. It must be considered a permanent part of a final corrective action and is, therefore, transferrable with the property and must also provide for procedures if intrusive work is necessary in the future.

Examples of acceptable barriers include caps made of clay, asphalt, or concrete, or a building. A clean soil cover is acceptable for eliminating the soil ingestion pathway. In redeveloping a site, it may be possible to place a parking lot over the contaminated area to serve as the engineered barrier.

Municipal Acquisition Tools

Municipalities undertaking a brownfield redevelopment may wish to acquire properties for several reasons. If funding is available, it may be easier for a city to clean a site than to find a private source of cleanup funds. A city may also be able to eliminate back taxes or other liens on a property through ownership. Site acquisition may be part of a larger land assembly strategy that can only be implemented by public actions. In Chicago, brownfield land acquisition has been accomplished through negotiated purchases, building demolition, lien foreclosure, tax reactivation for back taxes, and eminent domain. Each process has its own internal time frame, risks, and associated costs. For example, Chicago's tax reactivation program takes a minimum of 18 months and allows the delinquent taxpayer the opportunity to redeem. In any case, a city will not want to acquire a site it cannot afford to remediate, and it cannot afford to remediate a site that cannot be developed. A policy question arises when there is a site that is suitable for development, but for which there is no current user.

In 1997, two statutes were enacted that improve a municipality's ability to acquire brownfields properties.

- The eminent domain statute was revised so that environmental costs could be deducted from the determination of fair value in condemnation proceedings. Without this provision, municipalities faced paying a price based on a clean parcel of land, and then in addition having to pay again for the cleanup.

- The environmental lien was created so that abandoned properties more than two years tax delinquent could be investigated and remediated by the city, with the cost of remediation creating a foreclosable lien on the property. This statute follows the same principles as the demolition lien provisions used to tear down unsafe structures within the city.

Tax Incentives

Tax incentives are considered a valuable tool for encouraging private sector redevelopment projects. The major incentives now available to the private sector are the release from liability through state voluntary cleanup programs that should allow access to easier financing, and the availability of federal, state, and local tax incentive programs. Recent legislation at the federal, state, and local levels has created incentives that can be used to offset the cleanup costs a developer incurs as part of a brownfield redevelopment.

In Illinois, the state and Cook County (the taxing authority for Chicago and its inner ring suburbs) have created incentives that can be used in conjunction with the federal incentive.

- The federal tax incentive is a deduction against taxable income.
- The state incentive is a transferable tax credit.
- The county incentive is a transferable property tax classification that can be extended for up to 18 years.

Because the incentives can be layered, the resulting tax advantages can be significant (up to 50 percent of the remediation costs).

As an example, consider an industrial developer that has purchased a contaminated site for $100,000 and spends $500,000 on remediation and a further $1 million for redevelopment. In addition, assume that the developer has a taxable income of $5 million in the year in which remediation costs were incurred. Depending on where the site is located, different incentives will apply. The following scenarios show which incentives are applicable and the resulting tax benefits (see Table 4-2).

Scenario 1—the brownfield site lies within a federal empowerment zone or other targeted area within Chicago. In addition, the project meets the requirements of the Illinois and Cook County incentive programs.

Scenario 2—the brownfield site lies within Cook County, but outside a federal empowerment zone or targeted area, and therefore only state and county incentives apply.

Scenario 3—the brownfield site lies outside Cook County but within a federal target area and is, therefore, only eligible for the federal and state incentives.

Federal Tax Savings

We assume that the developer generates a taxable income of $5 million in the year the remediation costs ($500,000) are incurred on the site. Assuming a federal income tax rate of 34 percent, the developer would pay $1.7 million ($5 million × 0.34) in taxes. With the current federal deduction against taxable income, the developer would pay $1.53 million ($5 million – $500,000) × 0.34). This is a tax savings of $170,000. This calculation omits the capitalization of costs over a certain time period under prior law, which would have resulted in a much smaller tax savings.

Table 4-2. Tax Incentive Example Case Summary

	Federal deduction	State tax credit	County property tax savings	Total first-year tax savings	Percent of remediation costs
Scenario 1: federal/ state/county	$170,000	$40,000	$41,500	$251,500	50%
Scenario 2: state/ county	Ineligible	$40,000	$41,500	$81,500	16%
Scenario 3: federal/ state	$170,000	$40,000	Ineligible	$210,000	42%

State Tax Savings

Again, assuming a taxable income of $5 million in the year in which remediation costs are incurred on the site, and assuming a state corporate income tax of 4.8 percent, the developer would pay $240,000 ($5 million × 0.048) in taxes. Under the new Illinois tax credit, the following calculation determines the new amount due:

Total remediation costs = $500,000

Total costs applied to credit = $400,000 (includes $100,000 deductible)

Tax credit rate = 25 percent

Amount deducted from taxable liability = $400,000 × 0.25 = $100,000

New amount due: $240,000 (income tax) − $100,000 (tax credit) = $140,000

If taken in one year, the credit must comply with the $40,000 limit. Therefore, a total of $200,000 is due in the first year. The applicant can take another $40,000 in year 2 and the remaining $20,000 in year 3 to fulfill the total $100,000 credit due. This is a tax savings of $40,000 in the first year, with a total savings of $100,000.

County Tax Savings

For our scenario, we need to assume the following:

Acquisition costs = $100,000

Remediation costs = $500,000

Value of the redeveloped facility and parking lot = $1 million

Cook County aggregate tax rate (1996 level) = 9.647 percent

Multiplier (1996 level) = 2.1517

Given the 6(c) classification rate of 16%, the tax bill would be $1 million × 0.36 = $360,000

Equalized assessed value = 360,000 × 2.1517 = $774.612

Tax = 774,612 × 0.09647 = $74,727

Given the 6(c) classification rate of 36 percent, the tax bill would be $1 million × 0.16 = $160,000

Equalized assessed value = 160,000 × 2.1517 = $344,277

Tax: $344,272 × 0.09547 = $33,212

This is a tax savings of $41,515 in the first year.

If the Class 6(b) were factored into this calculation and compounded over the following 13 years that the full incentive is eligible to accrue, there would be a tax savings of up to 55 percent per year for up to 16 years (or 18 years if the two 1-year extensions were applied). However, this calculation has been omitted for the purpose of this scenario because the three variables that adjust the tax bill—multiplier, property value, and tax rate—will change throughout the term of the incentive.

References

1. Tiered Approach to Corrective Action Objectives (TACO): 35 ILL. Adm. Code Part 742. (1997)

2. Memorandum of Understanding between the Illinois Environmental Protection Agency and the U.S. Environmental Protection Agency Region 5. (July 1, 1997)

For biographical information on William Trumbull, see Sec. 4.2.

4.4   Burnside South/ Verson Steel—Case Study

Jessica Rio

In 1979 the Burnside Steel Foundry closed following an explosion on this site. The company filed for bankruptcy and abandoned the property. In the years following, scavengers cleared the site of all valuable materials, leaving dilapidated structures and debris that posed a health and safety threat to neighboring residents and businesses.

The city of Chicago chose the site as a Chicago Brownfields Pilot Site on the recommendation of the Southeast Chicago Development Commission. The city acquired the site through the Tax Reactivation Program, then cleared the site, demolished structures, and removed contaminated soil at a cost of approximately $760,000.

Verson Steel, a company that makes large steel presses, is located directly east of the site. Verson had contemplated leaving Chicago, taking along 500 jobs, in part because of the condition of the Burnside site and in part because of space constraints. When the city remediated the Burnside site, Verson became interested in acquiring it for expansion. The company has since begun a $31 million expansion, adding 125 jobs to be filled by community residents. Verson will use the site for storage and truck staging and in the future for manufacturing space.

Review of the Project

Chicago selected the site as one of five for the Brownfields Pilot Sites program, which began at the same time as the Chicago Brownfields Forum. The pilot was funded with $2 million in general obligation bonds. The city's idea was to get hands-on experience with the challenges of redeveloping brownfields at the same time it was convening public and private partners to break through these barriers. The issues identified in the site cleanup program were among those addressed by the Brownfields Forum. Likewise, the projects that resulted from the Brownfields Forum would facilitate both public and private redevelopment projects.

The city selected sites where the outlook for development was good because a neighboring business had expressed interest in the property.

The Burnside Steel/Verson project was four years from selection as a Brownfields Pilot to reuse. Below is a timeline of the major acquisition, cleanup, and development activities.

1994

January	Private environmental consultants submit proposals to perform a Phase I site assessment. The city of Chicago can send requests for services to a preapproved list of qualified consultants who maintain term agreements with the city.
February	Phase I site assessment starts. When the Chicago Department of Environment conducts a site visit as part of the Phase I investigation, the site contains part of a dilapidated corrugated metal building, two concrete silos, open pits formerly used as quench tanks, several transformers, a structurally unsound chimney, construction debris, other illegally dumped materials such as cars and tires, and overgrown vegetation and small trees.
March	Phase I site assessment completed. The Phase I report reflects the potentially hazardous site conditions in terms of the surface debris. It also shows the potential for soil contamination from heavy metals and petroleum hydrocarbons in areas of the site that contained certain steel-making processes or that were exposed to certain hazards after the property was abandoned.
May	The city begins to characterize the surface debris and have it removed by private contractors.
June–July	More than 200 truckloads of debris are removed from the site at a cost of about $500,000.
August	Site clearance completed. The city secures the cleared property so illegal dumping cannot continue. Through this and other Pilot Sites, the city has learned early in the site rede-

velopment program that preventing further site decline is key. The city now temporarily secures many sites as part of cleanup, recognizing that the ultimate deterrent of environmental crimes on sites is productive reuse of the site.

1995

February   The city submits a work plan for a Phase II environmental site assessment to the Illinois Environmental Protection Agency (IEPA) as part of Illinois' voluntary cleanup program. Illinois' voluntary cleanup program is, coincident with this cleanup and the Chicago Brownfields Forum, moving toward a risk-based, site-specific approach. The program now is called the Illinois Site Remediation Program. The testing and cleanup guidelines for the program are known as the Tiered Approach to Corrective Action Objectives (TACO).

March   IEPA approves the Phase II work plan.

April   The city begins acquiring the property through tax reactivation. The Tax Reactivation Program allows the city and private parties to bid on tax-delinquent property. The owner of record is given an opportunity to redeem the taxes and retain ownership. The program takes about 18 months for the city to complete if the owner chooses to relinquish the property. Aggressive use of this program continues to be a primary tool for the site control needed to direct further cleanup activities and redevelop. The city's other options for acquiring sites include lien foreclosure and condemnation, where the respective procedures are appropriate.

July   The city hires a consultant to conduct a Phase II environmental site assessment.

October   Phase II environmental site assessment completed. The Phase II report shows elevated levels of benzo(a)pyrene on one part of the site and elevated levels of lead in another area. The city remediates both conditions.

1996

May   The City begins the subsurface cleanup phase of the project.

July   Remediation is complete.

November   Property acquisition is complete.

1997

January   IEPA issues a no further remediation (NFR) letter to the city. The city's cost for site investigation, cleanup, and the NFR review totals approximately $211,000.

1998

Verson begins a $31 million plant expansion, adding warehouse capacity on the site. Once complete, the expansion will mean 125 new jobs.

Jessica Rio is the public information officer for the City of Chicago Department of Environment.

5

The Private
Developer—
Getting Started

This chapter represents discussions intended to be of assistance to private developers or companies considering brownfields redevelopment. This long discussion has been divided into the various steps in the process of redevelopment. The process begins with the developer doing preliminary investigations of the selected site.

In Sec. 5.1, Robert Rafson gives an in-depth view of what to look for at the site, and first assessment of costs.

In Sec. 5.2, Carey S. Rosemarin and Steven M. Siros discuss the facets of liability.

In Sec. 5.3, Andrew Warren discusses the Federal Brownfield Incentives, which affect private developers (as well as anyone else working in the field).

In Sec. 5.4, John Oharenko discusses the lender's viewpoint. Financing is something the developer has to obtain immediately. Oharenko writes succinctly and as an expert in the field. Most developers who have attempted to finance brownfields projects through traditional lending sources have met with resistance. Traditional loans are available, but often require special terms and conditions. For example, a lender may not make a 75 percent loan, but will make a 25 percent loan-to-value loan. The lender may insist that the borrower post additional loan collateral beyond the pledged brownfield project. Arranging financing for brownfields sites is possible, but the terms may or may not be helpful in making the overall economics feasible.

Financing is the central issue, and the subject of ways to find the money in loans, incentives, TIF programs, and so on, is discussed in many chapters in the book.

In Sec. 5.5, Donna Ducharme discusses the community and nonprofit institutions viewpoint. Ducharme has a very constructive attitude, and is in the unique situation of having worked with a privately financed institution that encourages brownfields redevelopment with financial support. This positive approach may not be available in other cities.

In Sec. 5.6, Harold Rafson begins the environmental engineering discussions, and deals with Phase I assessments.

5.1 Getting Started

Robert Rafson

Assessing a project's viability is most important in the planning of a project, and viability is determined by all the parts of a project; purchase price, redevelopment costs, sale/lease income, and financing. This section reviews the issues of concern to the developer in determining the viability of a brownfield project and is organized by the different aspects of the project and how each is affected by real or perceived contamination:

- Site

- Potential use

- Complexity of redevelopment

- Project financial evaluation

- Contractor financing

It is our hope that brownfields redevelopment projects will proceed and that this discussion will help developers or municipalities decide whether a project is viable. Since any redevelopment is a time-consuming, complex, expensive process, it is our hope that developers who embark on brownfields projects will understand some of the issues they will encounter and how the different aspects of such projects are affected by the special issues related to brownfields.

Site

The project selection begins with the first walkthough of the potential property. The developer will make an initial review of the property and its location. The developer will have a subjective impression as to whether the project has a chance of getting done with private investment, and, if not, then whether the project must either in part or in whole be assumed by the public sector.

Many projects have common issues: location, condition of the site and buildings, health of the neighborhood, and potential use after redevelopment. Many of these factors are affected by the extent of the real or perceived site contamination.

Location is a most important consideration in a real estate decision. Every real estate broker will tell you that property that has a good location has a much better chance of being sold or leased after redevelopment.

Fortunately, many brownfields are located in older and often more desirable locations where industrial properties were previously concentrated, and near multiple modes of transportation (highways, rail, and public transportation). Newer industrial parks are often located further from these amenities. If site problems can be overcome, the older industrial locations can be preferable.

One problem with contaminated properties, especially those that have been abandoned for long periods of time, is that the area can become blighted and deter other redevelopment. Potential owners buying neighboring properties would be cautious due to the threat of contamination. If such contaminated property is not redeveloped, it might continue to depress neighborhood property values.

Older industrial park locations often have infrastructure problems: low bridges, deteriorating streets, and antiquated sewers and other services. These problems can cause significant barriers to some potential users of industrial facilities.

However, older industrial parks have several significant environmental advantages over modern parks. One is that they are often in well-established and accepted industrial areas. The residents in the area expect that the businesses will cause some nuisances such as traffic, noise, and pollution. There is a limit to everyone's tolerance, but these areas tend to be more lenient since it's been that way for so many years. Many manufacturers will prefer to be in these locations.

Zoning indicates the limits of use for a property and that of neighbors. Although zoning of a property or area can change, the site must conform to allowed uses. Zoning laws also lend some level of stability to the site and neighboring uses.

Often zoning will also affect other aspects of the project redevelopment. In Chicago, there are 22 industrial corridors (see Fig. 5-1). Many of these corridors have restrictive zoning. Area 12 on the map is one of the Planned Manufacturing Districts (PMDs), which keep manufacturing jobs in the city but severely restrict other uses and limit property value. Any reduction in property value greatly impairs the redevelopment of brownfields properties that have many other barriers to overcome as well. These PMDs are critical to maintaining jobs within the city. The city should recognize that this larger benefit comes at a cost to the maximum potential

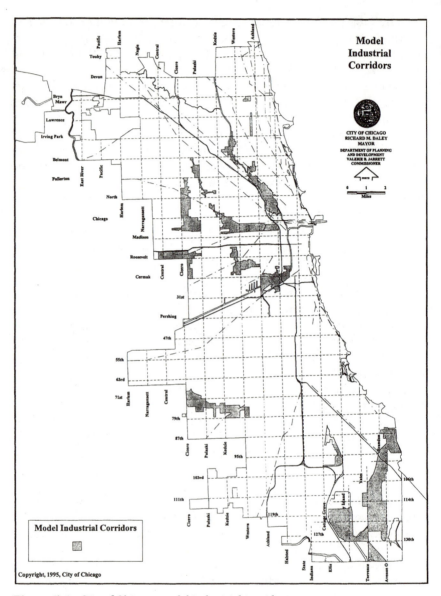

Figure 5-1. City of Chicago model industrial corridors.

for site redevelopment. Often the city has provided tax credits and other forms of assistance to foster the redevelopment process. Many of these industrial corridors are empowerment zones (providing incentives for labor and education) and enterprise zones (providing economic development incentives and tax breaks).

Major factors in the remediation and redevelopment of properties are the condition and state or repair of land, buildings, and other improvements, including the presence, quantity, location, and type of contaminants. A lack of care for properties results in multiple problems on contaminated sites, which hampers redevelopment efforts.

Owners of contaminated properties can be a significant barrier if they are uncooperative. Brownfields properties are by their nature problem properties. The owner may cling to the idea that the property has high real estate value and may feel entitled to a good price for the property. Once the owner overcomes the self-deception that the contaminated property has full fair market value, then negotiations on the property can begin in earnest.

Other barriers to redeveloping brownfields sites are the proximity of neighbors, previous uses of the sites, condition of the sites, existing structures, public sector view of use, aging infrastructure, high property taxes, and inner city traffic. Any one of these may prove to be insurmountable. The city of Chicago has been working to reduce the impact of some of these barriers and create an atmosphere conducive to redevelopment of all older sites, whether contaminated or not.

Older neighborhoods may have problems that are brought on merely by the existence of brownfields. Deteriorating buildings and contaminated properties are blights to their neighborhoods. They lower property values in the area and thus discourage redevelopment in the local area. The redevelopment of brownfields is important to the redevelopment of the neighborhoods. When the site is cleaned up and put back to productive use, the neighborhood as a whole may be improved.

It is ironic that often many of the neighbors who will gain benefits from the redevelopment of dilapidated and abandoned properties are the first to complain. An abandoned property, although an eyesore, does not make any noise, smoke, or traffic. In some older industrial areas surrounded by housing, residents have become sensitized to the emissions and noise of prior industrial uses. This stigma may cause added problems for some potential uses. Often the neighbors withhold comment until the redevelopment is complete and the business is in operation. This is a problem for all industrial users, but more severe where neighbors have been adversely affected in the past.

Therefore it is often very important to get the support of the local officials and neighbors prior to redeveloping one of these projects. I am aware of more than one project that was killed by last-minute requirements put on the subject site by city officials and neighborhood groups, causing the prospective site user to back out.

Potential Use

Potential use determines the site redevelopment plan and process. Each use may have specific site requirements. For example, if the use produces either excessive noise or noxious emissions, there should be a significant buffer between the site and the nearest residence. Use determines site requirements such as acreage, truck loading, parking, and building size and shape. Older facilities may have multistory buildings on the site that are unsuitable for many modern manufacturing uses. Older facilities may also have buildings in poor condition and with difficult access. There are many issues that might compromise a project due to the incompatibility of the use on a specific site. Unfortunately, modern facilities incorporate significant changes from the designs and systems of older industrial building and areas. However, often the end user would like to be located in a specific neighborhood and can open possible options for properties that may have been overlooked.

Modern industrial facilities are often planned to feature flexible forklift access and efficient material and process flow; single-story buildings with high clear heights are preferred. Very few brownfields properties fall into this category. Multistory or multiaddition buildings often do not provide good access for efficient forklift operation or process flow and thus limit potential use. If the product manufactured is transported through the production and warehousing process by other means such as conveyors, these building limitations may cause no barrier.

Most manufactured products are shipped to and from the plant by truck. Good configuration of loading docks and truck access is critical. Often additional truck loading space must be added to older buildings. The city of Chicago requires all new loading docks to be located off street. This may require redesign of a structure to accommodate off-street truck loading. Heavy trucking applications may require costly and major modifications of older industrial brownfields.

On the other hand, improvements and additions installed by previous owners may benefit new owners. The best potential function of the site may be for a similar use. Heavy power previously brought to the site may have significant benefit to a new user. The site may have interior cranes, racks, or other building improvements that may be a great boon. There could be large paved areas for truck storage or parking. But all of these benefits of locating a similar use in the same location may be overshadowed by the environmental liability brought on by the same use on the same site.

Presently, environmental laws are swinging back from owner liability to contributor liability. If a new user of the site processes, uses, or stores a contaminant that the previous owner used, the new user could become attached to the liability for existing contamination. New users must protect themselves against this risk and must consider the site, though clean to present regulations or laws, as possessing unknown contamination that could affect future ownership and liability.

It is important that developers help potential users to understand that they may want to look for facilities that do not contain the same chemicals used or stored by the potential user, especially if those are the chemicals that contaminated the site. Attachment of liability happens when an operator uses a chemical that has been used in the past. At that moment, it is impossible for the environmental engineer or regulator to determine whether the contamination was caused by the previous or the present user. For this reason, many attorneys and brownfields developers recommend against same-use redevelopment. This avoids the issues of contribution to the site's existing contamination.

Complexity of Redevelopment

35ILL Adm. Code Part 742 and 743 (Tiered Approach to Corrective Action Objectives, commonly known as TACO) has, since 1997, defined the cleanup standards in Illinois. When the potential use of a site and the decreased risks of some uses are considered, this definition can be significantly more relaxed than previous cleanup objectives. TACO thereby reduces the cleanup and reduces uncertainty, which is one of the most important benefits for the developer. Reducing uncertainty means that project costs can more accurately be estimated and planned.

Further, the city of Chicago's ordinance prohibiting the consumption of groundwater within the limits of Chicago eliminates the exposure pathway to drinking water. This has two important impacts on any redevelopment. First, it reduces or eliminates the cleanup costs of groundwater. Second, the cleanup can be much quicker, more accurately estimated, and less expensive. Groundwater cleanups are lengthy, uncertain, and expensive. The need to clean up groundwater and the uncertainty in cleanup time and expense may doom a project.

TACO and Chicago's groundwater ordinance have one further important effect on the redevelopment of contaminated properties: increasing the comfort level of the subsequent owner that once the site is clean, no further action will need be taken on the site. If a site is cleaned up with the oversight of IEPA through the voluntary cleanup program, at the conclusion of the approved remediation plan the state will produce a no further remediation (NFR) letter stating that the site is clean for the intended use. Technically, there is not much protection provided, but the subsequent site user feels much more comfortable and this makes the site more marketable and valuable. Discussed later is the importance of improvement in valuation on any redevelopment project.

Few sites have an obvious use such as the expansion of a neighboring business. For those few companies for which expansion is feasible, the possibility of expanding their present facilities without having to move often makes the project favorable. These companies are often willing to wait, in some cases for years, until the cleanup and site preparations are complete. Unfortunately, many companies do not have the luxury of waiting that long, and the cleanup and site preparation must be completed as quickly as possible. If the cleanup plan takes longer than a year, the project may not go forward because speculative redevelopment of these projects is very difficult.

Project Financial Evaluation

Any project has a list of income and expense items that need to be evaluated to determine whether a project is feasible. Remediation costs are only one part of the redevelopment costs, but contamination on the site adds difficulties, time, and costs. Financing is probably the most challenging hurdle to overcome, because most traditional lenders will not lend on a contaminated property but may lend to some companies or developers with a successful track record.

The following is a checklist and discussion of potential costs that must be reviewed for any project:

- Land cost
- Hard costs
- Soft costs
- Reserves
- Incentives

Land Cost

The land cost is often known, but can be variable when environmental contamination is present. The site may have a negative value due to the environmental conditions. In these cases, the owner may only have one option: to sell the site for little or no money. This low acquisition price may not be the whole acquisition cost and may be affected by other site factors.

Buildings and other structures may add or detract from the value, depending on the potential end use.

Land cost is not usually the most expensive component of a project. However, due diligence may require environmental or geological testing and could add $5,000 to $10,000 per acre or $0.15 to $0.25 per square foot to land acquisition. This cost may be required up front before negotiations can start in earnest. This due diligence is costly and is completely at risk, and therefore it is the most difficult funding to come by.

Off-site development costs and additional infrastructure may be required for larger projects to make sites feasible for the intended use. In some cases these improvements will be completed by the city but may require some matching funding from the beneficiaries of the improvements.

Hard Costs

The term *hard costs* refers to rehabilitation or construction costs related to materials and physical conditions. These costs include:

Building rehabilitation

Demolition and site work

Roofing repair or replacement

Structural repairs

Masonry repair or replacement

Concrete work

Tenant improvements

Construction contingency

Building Rehabilitation.   There are also components, listed here, that may be needed to make a building or facility complete. Only those items affected by environmental issues are discussed.

Demolition and site work

Redevelopment

Concrete work

Remaining construction items

Reuse of materials

Demolition and Site Work.   Often you must demolish portions of existing structures before you can build the intended facility. The demolition and site work can range from small interior or partition demolition to complete site demolition. Following is a list of some types of the demolition work and the environmental concerns associated with each.

Interior demolition concerns:

Asbestos
Interior frames and doors
  Lead-based paints
Electrical
Plumbing

Exterior demolition concerns:

Contaminated building materials
Debris on site
Contaminated land
Underground storage tanks
Unknown contamination and other buried materials

Hazardous materials concerns:

Storage in the building

Interior Demolition. Building interiors of brownfields structures have many hazards to deal with. In the following paragraphs are listed some of the environmental hazards to be addressed during the interior demolition. These same issues are necessary to address when demolishing an entire structure because the hazardous materials need to be removed and properly disposed of prior to demolition.

Asbestos. Asbestos-containing materials include floor tiles, fireproofing, insulation, roofing materials, and wallboard.

After World War II, asbestos was used as a binding and fireproofing material in many building materials. The previous list includes many of the materials commonly used in buildings built in the 1950s and 1960s. These materials must be encapsulated or removed in redevelopment of the property and, if friable (easily crumbled), must be removed properly prior to demolition.

In the case of vinyl floor tile containing asbestos, the tiles should be removed prior to redevelopment of a space. The removal is quick and inexpensive and removes an obvious hazard. If the redevelopment is for residential or some commercial use that would expose children (such as a day care center), the mastic used to apply the tiles must be removed from the subfloor because it often contains asbestos as well. In commercial spaces the new flooring will provide adequate protection from the small amount of asbestos contained in the mastic.

Lead-based paint. Lead-based paint is potentially an extremely expensive and difficult problem for residential redevelopment. The areas most affected are window sashes and doors. These are areas where the friction between windows and doors pulverizes the paint, rendering the lead free to be ingested. If part of the redevelopment is to replace windows and doors, this problem can be overcome and the remaining lead-containing paint could be encapsulated. Otherwise the removal is costly and time consuming and extends the redevelopment process timeline, which can make the project not viable.

Electrical demolition. Polychlorinated biphenyls (PCBs) can be found in the building transformers and fluorescent ballasts of older buildings. It is worth the effort to properly remove and dispose of these hazards. The total hazard to health and the environment is small, but the impact of non-removal on financing and sale of the property is significant. There is a perception that these transformers and ballasts represent a great liability; therefore, it is worth certifying that they do not contain PCBs or that PCBs have been removed.

Mercury vapor lights are commonly used in industrial facilities. Removal of these lightbulbs prior to demolition may reduce the hazard of debris.

Other than this issue, electrical demolition usually proceeds only to the point where the removal facilitates the relighting of the facility. In some cases complete removal is preferable to partial demolition; in others, the bulk of the electrical system can be salvaged. Each project is different and should be evaluated separately.

Plumbing demolition. Plumbing demolition has one of the greatest potentials for cost overruns. It is often more desirable and more effective to assume the complete demolition and replumbing of a building than to try to reuse existing piping. The older the piping, the more difficult it is to match and bring up to building codes. Americans with Disabilities Act (ADA)-approved bathrooms are now required in most facilities, and rehabilitation of existing bathrooms may be prohibitively expensive. Therefore, careful consideration of the plumbing redevelopment is necessary.

There is the potential that products used in prior manufacturing procedures will be trapped in process piping. Extra care should be taken in the removal of drain piping from areas previously used for plating operations, since cyanide, heavy metals, and strong acids and bases may be present in the piping and adjacent soils. Plugged drains can also contain other hazards, including explosion risks from trapped solvents or piping attached to undiscovered underground storage tanks. These areas that contain contaminants may need to remain on site as part of the remediation process. Capping may be an acceptable method of containing these wastes and preventing them from transport to groundwater or accessibility by ingestion or inhalation. All paving and concrete work, including all related excavation work, must be coordinated with the site remediation and health and safety plans.

Exterior Demolition. Some of the interior and exterior demolition issues overlap, but as a whole the exterior issues can be much more costly than the interior issues. In the redevelopment of a site, the complete demolition and removal of the slab may be required to complete a ground-up redevelopment. The following exterior issues to be addressed can occur in one or more stages of the demolition. Additionally, the complete demolition of a site will require the interior issues to be addressed prior to the exterior demolition.

Concrete and masonry. Concrete repair or replacement work includes consideration of the following items to estimate cost: patching, concrete removal, excavating, floor patching, saw cutting, concrete footings, foundation walls, floor sealing and vapor barriers, dock pilings, retaining walls, slab on grade or existing concrete, scissors lift pits, and stair footings.

Each of these areas requires some excavation or cutting into the existing concrete. Since contaminated sites may have contaminants in the concrete or under the flooring, it is imperative that the concrete contractor be aware of the hazards and that the contaminants be removed from any excavated areas, tested, treated, and disposed of appropriately.

Excavation also has the potential to uncover additional hazards other than just contaminants that have leached into the concrete pad and soils below. Underground storage tanks and retaining basins can contain hazardous and explosive materials.

As an example, in one building a worker was preparing an area for a concrete patch and used a saw to shorten a partially removed pipe. He was unaware that the pipe was a fill pipe to an underground gasoline storage tank. The sparks from the saw ignited the flammable vapors in the tank and blew up approximately 20 feet of the exterior wall and the roof in that section and created a crater approximately 15 feet diameter and depth. The worker was killed in the explosion. This is an example of how, without proper review of site environmental conditions and safety requirements, a simple repair can end in calamity.

There are many industrial processes that can cause leaching into the concrete. Most notably, plating operations involve both the hazardous chemicals that would be retained in the concrete (heavy metals, arsenic, chromium, lead, nickel, and copper) and the method used to transport these chemicals into the concrete (strong acids). Saw cutting concrete can release sufficient quantities to cause severe injury or death to the exposed workers. Hazardous materials handling training may be required for those working directly with contaminated concrete and excavated materials from below the slab.

New construction walls may reuse existing footings or may require new footings. The cleanup plans may require special handling of the materials excavated or may necessitate additional barriers when extending or capping the existing footings. Coordination between site remediation plans and construction plans is imperative.

Usually, once the concrete footings and slab are in place, the redevelopment can proceed as a normal redevelopment of the site.

Debris on site. Many abandoned sites become dumping grounds for other people's waste. This waste may be benign, but it may also contain unknown hazardous materials. Debris piles need to be carefully inspected prior to taking on the project. These materials could also contain liquid wastes that can contaminate the soil below and debris itself. Unfortunately, the real extent of the problem will not be known until all of the materials have been dug up and removed from site. These unknowns can create the potential cost overruns that make brownfields redevelopment risky.

Underground storage tanks. Many commercial and industrial buildings have underground storage tanks for many uses. Most commonly, fuel oil has been stored in the building and the tank is either underground or enclosed within a room in the basement. There are many rules that pertain to the requirements for removal, abandonment in place, or use of these tanks. Non–fuel oil tanks have different rules due to the different hazards they present to health and the environment.

Buildings with underground storage tanks are harder to sell than buildings without such tanks. For this reason, it is a generally accepted practice to remove these tanks regardless of the regulatory requirements that could allow abandonment in place. When the removal cost is high or when damage to the structure of the building is likely, abandonment in place is considered.

Once the storage tank is removed, the environmental engineer takes samples of the soil under and around the tank to determine whether the soil is contaminated and whether removal or treatment is required. After removal or treatment, it is desirable to obtain a no further remediation (NFR) letter. This comfort letter provides an added level of security that the conditions at the site are as the environmental engineer states them to be. It also provides protection by reducing the likelihood that EPA will require additional cleanup sometime in the future. This letter makes the subsequent owner and the lender feel that the property cleanup is complete and meets with regulatory approval.

Hazardous Materials. Buildings prime for brownfields redevelopment have often been neglected or abandoned. These buildings also often have hazardous materials stored within them. Paints, solvents, and other materials from the previous use must be handled properly and their disposal costs must be budgeted for in the planning of the project. If over-looked, this could affect the bottom line of the project, though the time or expense are not usually large enough to kill a project. This item is often uncovered in due diligence inspections, and it may be a requirement in negotiations that the existing owner dispose of these materials prior to closing. It is also a good idea to create a separation of liability from those contaminants and their ultimate disposition.

Redevelopment. Many items on this list are unaffected by contamina-tion. But if cleanup is necessary, it will affect the logistics of the project, including time required and costs. If the cleanup work occurs within the building, then structural, access, plumbing, concrete, and interior finish work may be affected. In some cases it is cheaper to remove the building, clean up the contaminant, and build a new, modern facility. Unfortunately, if this is necessary it is less likely that the project will be economically fea-sible. If the contamination is outside the building, then the cleanup work can usually be easily done, which may make the remediation project more achievable.

Each step of the redevelopment must be looked at carefully. In many cases, the redevelopment will reuse existing structures. Each of the con-struction budget line items must consider the cleanup and the require-ments set out in the remediation plan. For instance, there may need to be vapor barriers included in the concrete floor repairs or replacement to pro-vide an effective engineered barrier to protect the building's occupants from residual contaminants in the soil below the concrete floor.

Reuse of Materials. Reuse of any materials presently on site must be reviewed for its potential for short-term and long-term environmental impact (see Table 5-1).

All of these issues collectively affect the overall complexity of the rede-velopment of the site. Many contractors avoid this type of redevelopment due to the complexity and possibly due to their lack of experience dealing with these environmental issues.

Soft Costs

In addition to the hard costs of construction just outlined, the price of a redevelopment is often heavily impacted by the soft costs, including:

Environmental legal work

Environmental liens

Unpaid real estate taxes and liens

Insurance

Construction and property management

Site security

Legal services

Environmental engineering

Architectural/engineering services

Carrying costs

Marketing/broker commissions

Table 5-1. Potential Environmental Impact of Materials Reuse

Material/system	Potential environmental problem
Reused doors and windows	Lead-based paint
Reused wall covering and paint	Lead-based paint
Reused plumbing	Contaminants contained in and around existing piping and asbestos pipe insulation
HVAC	Asbestos on boiler, pipe, and duct insulation
Electrical	PCBs in transformers
Landscaping	Contaminants in soils
General requirements	Additional insurance may be required to cover remediation cost overruns
Environmental requirements	Environmental testing, planning, cleanup, purchase contracts, sale contracts, and indemnifications

Environmental Liabilities, Legal, and Liens.   Superfund and other environmental liens are relatively new impediments to the redevelopment of properties. They can easily change a project from profitable to an absolute financial disaster. These environmental liabilities come in four types: direct liens, residual contamination, lawsuits, and the impact of environmental liabilities on the remediation. We will now focus on environmental liens.

Direct environmental liens can now be placed on a property by USEPA, a state EPA, and/or a city for the cost of cleanup.

The federal government conducts remedial activities under CERCLA, commonly known as the Superfund law, and has the right to lien the property for the cost of the cleanup and to pursue the owners and operators, past, present, and future, as well as the creator of the contamination for these costs. These cleanups are triggered by the imminent danger to health and the environment and proceed immediately upon funding by the federal government.

There are two paths that eliminate the Superfund lien but may not eliminate the financial responsibility. If another governmental body gains title, the federal government will not typically pursue that city or department of government for the cost recovery, but may choose to pursue the subsequent owner. So, if the ownership of the property will be retained by the city or governmental department, these organizations may not have to worry about paying for the cleanup. The other option is for an individual or organization to enter into a prospective purchaser agreement (PPA). These agreements are covenants not to sue for the past remediation, and protect the prospective purchases from third-party lawsuits and against remediation expenses for existing contamination. More details on the PPAs are contained in Sec. 7.3.

State and city environmental liens are negotiated much like USEPA PPAs, based on the economic viability of the project. Unfortunately, these agreements are strictly financial and do not protect the subsequent owners from past liabilities (such as toxic tort).

Unpaid Real Estate Taxes and Liens.   Many brownfields properties will have many more encumbrances than just environmental contamination. Often the reason the business occupying the site did not remediate its own environmental problems is that it has had some financial trouble. This is evidenced by delinquency on the business's other obligations and often creates additional liens on the property. These liens can come as mechanical liens (liens for work completed for the business or on the property), tax liens (for unpaid real estate taxes), and so on.

Insurance.   Insurance is a standard risk management tool for every project. Coverage typically includes standard liability insurance, builder's risk insurance, and performance bonding. General contractors often provide insurance coverage and can additionally insure the developer or property owner during the construction process. Environmental insurance, discussed in Sec. 7.5, also can provide capping of cleanup costs and protection against third-party lawsuits and/or unknown cleanup liabilities (including migration of contaminants off site).

Construction and Property Management.   Fees are typically included for the general contractor to manage the subcontractors as well as the other aspects of the project. This allows the contractor to include all the out-of-pocket and labor fees incurred during the project. Property management fees may be required if the project has occupants at the time of redevelopment.

No developer would forget to include construction management as a normal element of the redevelopment costs. Most construction management costs are estimated by taking a percentage of total project costs. This must be higher than the percentage for normal construction due to the complexities, hazards, and potential for overruns in time and costs. Therefore the standard percentage for the construction management number usually will justifiably need to be higher for a brownfield redevelopment.

Site Security. Security during construction is necessary on many projects, and more so on brownfields projects because they are often located in depressed areas. Comprehensive site security is a cost that is not usually required for greenfields projects because the development of farmland or other undeveloped land is often far enough away from population that the site is relatively safer from normal theft and vandalism.

Legal Services. Legal work is a part of every project. In brownfields redevelopment, legal costs are significant, especially if long negotiations with governmental agencies, such as prospective purchaser agreements with Superfund, are required. Environmental lawyers provide guidance on the implications of embarking on a project and help in understanding the risks involved. Additional discussion of the environmental legal role can be found in Sec. 7.2.

Legal contracts and their costs are standard for all developments. Some reserves are incorporated into this estimate for disputes that arise during construction. We hope there will be no problems with any of the contracts with general contractors or subcontractors, but it would be rash to think that such difficulties will not happen. The best scenario would be one in which the disputes can be resolved between the parties. Often the disputes need to be resolved by legal means and some reserves should be budgeted.

In addition, legal work related to purchase contracts, leases, sale contracts, and removal of liens can be extensive, and should be considered at a higher budget figure than usually provided.

Environmental Engineering. Environmental engineering (Phase I, II, and III) work is now a standard part of all projects. Banks require at least a Phase I assessment for every loan secured by real estate. Brownfields projects require the expertise of the environmental engineer to determine the extent of contamination, recommend a course of action, estimate the costs of remediation, apply for appropriate permitting, negotiate with EPA, and oversee the remediation work.

Architectural/Engineering Services. Architectural costs are a common part of any development or redevelopment. If the site reuses structures or buildings, there will need to be a structural engineer to work with the architect to verify that the building is sound to redevelop. These design costs for redevelopment can approach the expense of building a new structure. Remediation work may need to be included in the construction plans, and coordination between the environmental engineer and architect will be necessary.

Civil engineering is often a straightforward part of the redevelopment plans. Environmental work often entails work on the soil and subsurface structures and heavily affects the ground condition, drainage, and other factors that impact the landscaping and connections with other services (sewer, drinking water, stormwater, etc.). Significant changes in the landscape of the property might affect groundwater flow and the design, implementation, and cost of long-term ground and water treatment.

Carrying Costs. Loan and equity interest and other carrying costs are standard on all projects. Marketing time will be much longer than for a standard redevelopment, and therefore interest and other carrying costs will be higher. It is very difficult to estimate the amount of additional time these projects will require for development, marketing, and conclusion of a sale or lease.

Marketing/Broker Commissions. Marketing of a brownfield site costs more due to the significantly increased time and effort required, and finding brokers to handle a brownfield property may be difficult. Brokers see environmental problems, residual risk, and regulatory involvement as deal breakers, and many will not list properties with these problems. Since the property will likely sell below market value, the broker may ask for a higher than standard commission. Additional costs of time and marketing materials, including copies of all environmental data, need to be factored into the project cost.

Reserves

Reserves for cost overruns are difficult to estimate, and yet these must be considered while keeping the project costs competitive. The three main reserves that need to be set aside would be environmental, construction, and carrying costs. Environmental reserves for unforeseen problems and cleanup overruns have been minimized by well-defined regulations and improved testing and estimation procedures. Many environmental engineers and developers reserve as much as 50 percent over anticipated costs.

Rehabilitations of existing structures have much higher risks of cost overruns than new buildings. Depending on the redevelopment plan, the appropriate amount of reserves will vary. Carrying costs may be greater as these projects take more time than expected.

Both the developer and the lender may require larger contingency reserves on these projects due to the added uncertainties posed by real and perceived environmental issues. These reserves add to the capital requirements and, on some types of cleanups (e.g., pump and treat), may be as high as 50 percent of the anticipated costs to protect against cost overruns. Environmental cleanup cost cap insurance can eliminate the need for these reserves, but this insurance is costly. Additional reserves should be set aside to cover potential overruns in cleanup, redevelopment, marketing, and legal costs.

Tenant improvements. There may be tenant-specific improvements included in the costs to make the lease or sale attractive to a subsequent user. In some contracts these improvements are stipulated and the costs are borne by the developer. In these cases the costs must be accounted for in the anticipated redevelopment or construction cost.

Incentives

Up to this point, we have only discussed costs. However, there are also benefits available in the form of grants and tax relief. This section discusses some of the benefits available for rehabilitation and brownfields redevelopment. Some are direct payments for work completed, but most are reductions of future costs, such as reductions in property taxes or income tax. These incentives are also discussed in Secs. 4.1 and 4.3.

Mayor Daley has said that industrial development in Chicago is brownfields redevelopment. With the goal of industrial development and the desire to maintain jobs in the city, a series of incentives have become available for the redevelopment of industrial properties and specifically brownfields.

The property tax incentives, as discussed in Sec. 4.1, help the resale value of the property but do not get the property cleaned up or the reconstruction completed. These tax incentives do, however, affect the value of the property and make the project more viable.

There are three groups of redevelopment funds that come in the form of grant money. These are especially important for the most difficult projects, which, without the assistance of city, state, or federal money, would not go forward. These funds come in the form of:

Tax incremental financing (TIF) and other special arrangements

Community development block grants (CDBGs)

Industrial revenue bonds (IRBs)

TIF funding is a relatively new tool, and is provided on a local basis by the city. These funds are specifically for the redevelopment of TIF districts (see Chap. 4, Fig. 4-1). The city pays for a portion of the redevelopment costs of an area (perhaps including streets, bridges, other infrastructure, and building redevelopment), since that redevelopment increases the value of the properties within the district, thus creating additional property tax revenue to repay the expenditures prepaid by the TIF. This type of financing is critical for infrastructure in older industrial areas that have low viaducts and aging water mains, sewers, and streets.

CDBGs can provide initial funding for TIFs but can also provide funding for direct grants and financing. Construction loans and loans for equipment purchasing at low rates improve project economics and can help tip the balance in marginal redevelopments (such as brownfields).

IRBs are bonds created by a city or lender to fund a project. These loans generate lower interest rates for financing for the project. They can fund relocation, site purchase, redevelopment, and equipment purchase. IRBs can also tip the balance in favor of a project that may not otherwise be viable.

Financing

Financing is heavily affected by contamination. There are many reasons a lender may not want to get involved in a contaminated site. A contaminated site has lower value because of cleanup costs, it is harder to sell, and there are potential liabilities to being connected with the site, especially if the note is unpaid and foreclosure may be necessary.

These days it is easier to determine the costs of remediation. There are better techniques to test a site and, because the environmental cleanup business is much more mature, estimated cleanup costs are much more accurate than in the past. Therefore, the cleanup costs could theoretically be deducted from appraised value to arrive at the present value of the property, and a percentage of that amount could be borrowed and paid upon completion of the cleanup.

Unfortunately, the determination of value of a property is affected by many other factors and thus greatly reduces the willingness of lenders to lend. It is easier to borrow on a property in a hot sales area that could be sold very quickly than a contaminated property in a rough industrial area.

Fortunately, many banks have expanded their mission statements to include community redevelopment and have included brownfields redevelopment as part of that mission. Each developer will have to pass all of the normal issues of acquiring a construction loan and/or mortgage and will have to reserve much higher contingencies due to the additional real and perceived risks of this type of real estate development. The economic determination of each project is be discussed in Sec. 6.2.

Borrowing on one these projects may also require a lower loan-to-equity ratio. This means that a developer's investment does not go nearly as far on a brownfield project as on clean property. The overall profit of the project must be higher because the equity investment requires a higher return than a bank loan, especially considering the risks.

All of these factors affect the lending on a contaminated property and whether the project will go forward. Ultimately, getting funding on a project is the most important portion of brownfields redevelopment. If you can't get financing, you often cannot get the project completed. If you can fund the project fully by private funding, then you typically pay more interest and a greater return to the investors. You also have less money for cleanup and other construction items.

The trends of banking are beginning to support the businesses that enter into brownfields redevelopment. In the past lenders may have considered lending to brownfields redevelopers akin to lending money to a gambler. This required even the most experienced redevelopment organization to be funded by all equity, making the brownfields developments few and very far between. The only cleanups taken on by developers were the simplest and quickest ones. The bulk of redevelopment was being done by the organizations that created the contamination.

Redevelopment of these more difficult sites is becoming a more standard part of industrial redevelopment. Project development and management will require additional attention to detail to protect against all of the soft and hard redevelopment costs and will necessitate handling many of these issues regardless of the existence or absence of contamination.

In the following sections of the book, we will take the elements of redevelopment and issues of concern and reduce them to actual costs to determine whether projects are viable. Finally, after due diligence we will have our final chance to decide whether to go forward with a project or not. Environmental, health, and safety liability issues make the decision more worrisome. The decision-making team makes the call of whether to take on the risks and liability relative to the potential benefits of the project.

Robert Rafson, P.E., has achieved recognition in several fields, including brownfields redevelopment, innovative soil remediation technologies, modern insulation technology, and gaseous emission controls. He serves as CEO of three technology-based companies: Greenfield Partners (brownfields development), Rafson Engineering, Inc. (environmental engineering), and Service Insulation (modern industrial process insulation). Over the past five years he has been involved as a private brownfields developer of seven sites and has done environmental consulting on many additional sites. Over the past 17 years, he has been involved in engineering and design of insulation applications. Mr. Rafson is author of several papers, and has presented technical papers and participated in panel discussions on environmental and brown-fields issues.

5.2 The Current Law of Liability of Owners Under the Superfund Statute

Carey S. Rosemarin and Steven M. Siros

The purpose of CERCLA is to force responsible parties to remediate contaminated properties. Accordingly, the statute's definition of the classes of persons who may be held liable to pay for environmental remediation—potentially responsible parties (PRPs)—is exceedingly broad. Section 107(a) of CERCLA, 42 U.S.C. § 9607(a), defines four classes of persons who may be held liable, and describes what they may be held liable for.[1] The first category of such persons is primarily at issue in this section, and indeed in this book. It includes simply "the *owner and operator* of a vessel or facility . . . from which there is a release, or a threatened release, which causes the incurrence of response costs, of a hazardous substance." The statute also says that PRPs may be held liable for all costs of environmental remediation and damages to natural resources.

The composition of the "owner and operator" category of PRPs is more complex than its description suggests. For example, cases have been litigated over the issue of whether a landlord is the owner in a case in which the tenants caused a release of hazardous substances. Also, in the 1980s and 1990s, a major debate raged over whether a bank that foreclosed on a loan secured by contaminated property became the owner, and thus liable under the statute. This chapter reviews these issues and describes various ways in which persons connected with contaminated property can become liable as owners.

Owners in General

A current owner who causes a release of a hazardous pollutant may be held liable for remediation of the contaminated property.[2] This aspect of owner liability under CERCLA is straightforward. However, current owners of contaminated property may be held liable even when the release of hazardous substances occurred prior to the current owner's acquisition of the property.[3]

Landlords as Owners

Landlords can be held liable as owners of contaminated property even though the actions of the tenants, and not the landlord, caused the contamination. In *Weyerhaeuser Corp. v. Koppers Co.*,[4] a landlord was deemed a PRP solely due to the landlord's status as the owner of the property. In *Koppers* it was undisputed that the landlord's tenant, and not the landlord, contaminated the property.[5]

Lenders and Trustees as Owners

The term *owner* could include any person who holds title to real estate. Thus, absent qualification, the language of the statute could have qualified numerous lenders as PRPs because lenders often hold the legal title to real estate, including contaminated real estate, that secures a mortgage. The same is true of trustees and other fiduciaries who, in their official capacities, hold title to property that may contain hazardous substances. Therefore, the statute exempts from the definition of owner "any person, who, without participating in the management of a vessel or facility, holds indicia of ownership primarily to protect his security interest in the vessel or facility."[6]

As might be detected from the complexity of this language, a great deal of controversy has arisen about whether lenders and fiduciaries could become PRPs.[7] There are numerous situations in which lenders may require borrowers to take certain actions to improve profitability. Issuing such instructions or advice was thought to be a perilous exercise in light of the terms of the exemption, and in the mid-1980s a number of court cases addressed the issue of whether actions by lenders rendered them liable under CERCLA. If a lender or trustee holds indicia of ownership for reasons other than to protect a security interest, the lender or trustee may be subject to CERCLA liability. Where a lender's oversight and control turns into "participation in management," that lender may become a PRP. This issue is discussed further in Chap. 7.2.

The remainder of this section deals with the potential liability of fiduciaries, which involves analyses similar to those applied to determine whether lenders may be held liable. Like lenders, fiduciaries may be held liable where the fiduciary acts in a manner other than primarily to protect its ownership interest in the property. In *City of Phoenix v. Garbage Services Co.*,[8] the court noted that a trustee could be held personally liable under CERCLA where the trustee had the power to control the use of trust property and knowingly allowed the property to be used for the disposal of hazardous wastes.

The Asset Conservation, Lender Liability, and Deposit Insurance Protection Act of 1996 ("the Act") was enacted to provide lenders and fiduciaries additional protections from CERCLA liability. Specifically, the Act provided that a fiduciary's liability is limited to the assets held in the fiduciary capacity. It also set up a "safe harbor" by which a fiduciary cannot be held liable in its personal capacity for undertaking (or directing another person to undertake) the following activities:[9]

1. Performing response actions under the direction of state or federal on-site officials

2. Otherwise addressing a release of hazardous substances

3. Restructuring, renegotiating, or terminating a fiduciary relationship

4. Including in the fiduciary agreement a covenant relating to environmental compliance

5. Monitoring or inspecting a vessel or facility

6. Providing financial advice to the settlor or beneficiary of the fiduciary relationship

7. Administering a vessel or facility that was contaminated prior to the fiduciary relationship

However, the Act does not provide complete protection for fiduciaries. First, the Act defines a fiduciary as someone acting for the benefit of another person.[10] Thus, the fiduciary is not protected from CERCLA liability if it benefits from the fiduciary relationship beyond that which would represent customary or reasonable compensation. Second, a fiduciary is not protected from CERCLA liability where the fiduciary's own negligence causes or contributes to the release or threatened release of hazardous substances.[11]

Operator Liability

In addition to owner liability, CERCLA also imposes liability on operators of contaminated facilities. There is often a close relationship between owners and operators of contaminated sites, and the purchaser of a brownfield should be cognizant of the boundary between the two.

An *operator* has been defined as "one who has the authority to control the site and abate damage caused by the disposal of hazardous substances regardless of whether that authority is actually exercised."[13] Some courts have held that ability to control is not enough, ruling instead that to incur liability as an operator, the defendant must have actually exercised such control. The court so held in *United States v. Kayser-Roth Corp.*[14] It ruled that operator liability attached where Kayser-Roth exerted practically total influence over the subsidiary's operations, including control over the subsidiary's monetary accounts and personnel decisions.[15]

Corporate CERCLA Liability

To the extent that a corporation owns the property at issue, analysis of whether the corporation will be held liable as an owner is similar to determination of whether an individual or partnership owns the property. In certain situations, CERCLA liability can be imposed on individual shareholders or parent corporations. Historically, corporate shareholders (individual shareholders and parent corporations) have been safeguarded by the concept of limited liability.[16] Corporate shareholders are generally viewed as entities separate from the corporation itself, and are therefore protected from the corporation's liabilities by the corporate veil "unless specific, unusual circumstances" dictate that the veil be pierced and that the corporate form be ignored.[17] These common law principles of corporate law are generally determined by reference to state rather than federal law.

Traditionally, proof of three elements is required to successfully disregard the corporate form, or "pierce the corporate veil": (1) the corporate entity must be a "mere instrumentality" of another entity or individual (i.e., its "alter-ego"); (2) the corporate entity must be used to commit a fraud or wrong; and (3) there must have been an unjust loss or injury to the plaintiff.[18]

In the CERCLA context, the issue has arisen whether a parent corporation should be held liable when the subsidiary is the ostensible owner or operator. Some CERCLA cases held that the corporate veil may only be pierced where the corporate form was used to commit a fraud or wrong.[19] However, other courts adopted a more liberal test to determine when the corporate veil should be pierced. Recognizing that the statute authorized the imposition of liability on both owners and operators, these courts determined that parent corporations could be liable as operators of their subsidiary corporations even though the traditional elements necessary to pierce the corporate veil were not present.[20] In determining whether the corporate veil should be pierced under this more liberal standard, courts considered whether the parent corporation or shareholder had a substantial financial or ownership interest in the subsidiary and whether the parent corporation controlled the management and operations of the subsidiary.[21]

A 1998 Supreme Court decision shed some light on whether the traditional common law approach or the more liberal approach to piercing the corporate veil was appropriate. In *United States v. Bestfoods*,[22] the Court noted that CERCLA owner or operator liability can be imposed on a shareholder or parent corporation only if the corporate veil can be pierced under traditional corporate law principles. The court explained that a parent's influencing general management decisions of the subsidiary would not be sufficient to render the parent liable directly as an owner or operator under CERCLA. But the *Bestfoods* decision went on to note that operator liability could be imposed on a parent corporation where the parent corporation manages, directs, or conducts operations specifically related to the contamination—that is, operations having to do with the release or disposal of hazardous waste, or decisions about compliance with environmental regulations. In other words, the court held that if the parent engaged in such activities on its own behalf,[23] rather than on behalf of the subsidiary, the parent could incur CERCLA operator liability directly and not on the basis of piercing the corporate veil.

In sum, under CERCLA, a corporate shareholder can be held liable for remediation of contaminated properties owned by a corporation in two circumstances. First, a shareholder can be liable where the elements necessary to pierce the corporate veil are present, that is, where the corporation was formed for fraudulent purposes or is the alter ego of the shareholder. Second, a shareholder may be held liable where the shareholder manages or controls polluting operations at a contaminated site on its own behalf.

Conclusion

CERCLA liability is generally based on a person's status with regard to contaminated property. CERCLA owner or operator liability can attach to a variety of different persons in a variety of different circumstances. Section 7.2 describes in further detail the various ways a prospective purchaser can seek to avoid, or at least minimize, potential CERCLA liability exposure.

References and Notes

1. Section 107(a) of CERCLA reads as follows:

   Notwithstanding any other provision or rule of law, and subject only to the defenses set forth in subsection (b) of this section—

(1) the owner and operator of a vessel or facility,

(2) any person who at the time of disposal of any hazardous substance owned or operated any facility at which such hazardous substances were disposed of,

(3) any person who by contract, agreement, or otherwise arranged for disposal or treatment, or arranged with a transporter for transport for disposal or treatment, of hazardous substances owned or possessed by such person, by any other party or entity, at any facility or incineration vessel owned or operated by another party or entity and containing such hazardous substances, and

(4) any person who accepts or accepted any hazardous substances for transport to disposal or treatment facilities, incineration vessels, or is selected by such person,

from which there is a release, or a threatened release, which causes the incurrence of response costs, of a hazardous substance, shall be liable for—

(A) all costs of removal or remedial action incurred by the United States Government or an Indian tribe not inconsistent with the national contingency plan;

(B) any other necessary costs of response incurred by any other person consistent with the national contingency plan;

(C) damages for injury to, destruction of, or loss of natural resources, including the reasonable costs of assessing such injury, destruction, or loss resulting from such a release. . . .

42 U.S.C. § 9607(a)

2. 42 U.S.C. § 9607(a). *In re CMC Heartland Partners,* 966 F.2d 1143 (7th Cir. 1992) (current owner of contaminated property liable under CERCLA)

3. *New York v. Shore Realty,* 759 F.2d 1032, 1052 (2d 1985) (holding current operator of site personally liable under CERCLA for contamination caused by prior owner); *Mathis v. Velsicol Chemicals,* 786 F. Supp. 971, 974 (N.D. Ga. 1991) (noting that CERCLA liability can be imposed on the current owner, regardless of who caused the contamination)

4. *Weyerhauser Corp. v. Koppers, Co.,* 771 F. Supp. 1406 (D. Mar. 1991)

5. *See also Folino v. Hamden Color and Chem. Co.,* 832 F. Supp. 757 (D. Ver. 1993) (landlord's status as owner of property rendered it a PRP even though landlord's tenants and other third parties contributed to the contamination)

6. 42 U.S.C. § 9601(20)(A)

7. *See United States v. Fleet Factors Corp.,* 901 F.2d 1550 (11th Cir. 1990) (lender could be liable under CERCLA if lender has the capacity to participate in management of facility); *United States v. Nicolett, Inc.,* 712 F. Supp. 1193 (E.D. Pa. 1989) (CERCLA liability exists where lender participated in management and operation of facility); *United States v. Maryland Bank and Trust Co.,* 632 F. Supp.

573 (D.Md. 1986) (CERCLA liability attaches where bank purchased property after foreclosure and held property for four years)

8. *City of Phoenix v. Garbage Services Co.*, 827 F. Supp. 600 (D. Ariz. 1993)

9. 42 U.S.C. § 9607(n)(4)

10. 42 U.S.C. § 9607(n)(5)

11. 42 U.S.C. § 9607(n)(3)

12. 42 U.S.C. § 9601(20)(E)(ii).

13. *See HWR Systems, Inc. v. Washington Gas Light Co.*, 823 F. Supp. 318 (D. Maryland 1993) (CERCLA operator liability based on ability to control operators at the site including ability to control decisions regarding disposal of hazardous waste); *Nurad Inc. v. William E. Hooper & Sons, Co.*, 966 F.2d 837, 842 (4th Cir. 1992) (where the defendants lacked the ability to control decisions involving disposal of hazardous substances at the site CERCLA operator liability did not attach); *see also Clear Lake Properties v. Rockwell Int'l Corp.*, 959 F. Supp. 763 (S.D. Tex. 1997) (holding a tenant liable as a current operator of contaminated facility because the tenant had ability to exercise control over the site, regardless of fact that contamination was caused by prior owners and operators)

14. *United States v. Kayser-Roth Corp.*, 910 F.2d 24, 26–27 (1st Cir. 1990)

15. The holding of this case was affected by the Supreme Court's decision in *United States v. Bestfoods*, 1998 WL 292076 (Mar. 24, 1998)

16. *Carte Blanche v. Diners Club Int'l Inc.*, 758 F. Supp. 908 (S.D. N.Y. 1991) (noting the general rule that shareholders are not responsible for corporate debts and liabilities)

17. *RCS Engineered Products Co. Inc. v. Spartan Tube and Steel Inc.*, 102 F.3d 223 (6th Cir. 1996) (corporate shareholders are generally not responsible for corporate debts and obligations absent elements necessary to pierce the corporate veil)

18. *Bodenhamer Building Corp. v. Architectural Research Corp.*, 873 F.2d 109, 112 (6th Cir. 1989) (setting forth the traditional elements necessary to pierce the corporate veil)

19. *See United States v. Cordovo Chemical Co.*, 113 F.3d 572, 582 (6th Cir. 1997), reviewed on other grounds, 1998 WL 92076 (Mar. 24, 1998) (despite involvement of the parent corporation in the operations of its subsidiary, there was no evidence of fraud or improper use of the corporate form that would justify piercing the corporate veil); *Josyln Manufacturing Co. v. T. L. James & Co.*, 893 F.2d 80 (5th Cir. 1990) (corporate veil cannot be pierced where there is no evidence that the corporate form was used as a sham to perpetrate fraud or to avoid personal liability)

20. *Schiavone v. Pierce*, 79 F.3d 248, 254 (2d Cir. 1996)

21. *United States v. Nicolet*, 712 F. Supp. 1193 (E.D. Pa. 1989). *See also United States v. Kayser-Roth Corp.*, 910 F.2d 24 (1st Cir. 1990) (parent corporation liable for environmental contamination caused by subsidiary corporation due to the parent corporation's exercise of control over financial and management affairs of the subsidiary corporation); *In re Acushnet River & New Bedford Harbor*, 675 F. Supp. 22 (D. Mass. 1987) (in determining whether to pierce the corporate veil, the court looked for evidence of pervasive control by the parent over its subsidiary)

22. *United States v. Bestfoods*, 1998 WL 292076 (Mar. 24, 1998)

23. *Bestfoods*, 1998 WL 292076 at *7

For biographical information on Carey S. Rosemarin and Steven M. Siros, see Sec. 1.2.

5.3 Lender Liability and Federal Brownfield Initiatives

Andrew Warren

Lender Liability

Lenders face potential liability under the Comprehensive Environmental, Response, Compensation and Liability Act (CERCLA)[1] when participating in transactions secured by property meeting the definition of a CERCLA facility. First, since most lending transactions confer legal title to the lender, the lender faces potential liability as a CERCLA owner. In some circumstances, such as during a loan workout, the lender becomes so involved with the operations of its borrower that the lender may also face potential liability as a CERCLA operator.

CERCLA, as enacted in 1980, specifically exempted certain holders of a security interest in property from liability as owners or operators. The secured creditor exemption exempts from liability the person who:

> without participating in the management of a vessel or facility, holds indicia of ownership primarily to protect his security interest in the vessel or facility.[2]

Under the exemption, those parties holding only a legal interest in the property as part of a loan or financing mechanism should not face liability as CERCLA owners or operators.

Unfortunately, in cases interpreting CERCLA's secured creditor exemption, the courts reached confusing and often contradictory results. In particular, courts grappled with the meaning of the phrase "participating in management" as applied to lenders that took some affirmative steps in relation to the contaminated property. Some of the questions raised by the courts included whether the secured party lost the exemption upon foreclosure and at what level the lender could engage in loan management activities and still not be found to participate in management.

In *U.S. v. Fleet Factors*,[3] the 11th Circuit Court of Appeals reached the most alarming interpretation of CERCLA's secured creditor exemption. The court held that the lender's "capacity to influence alone, without direct involvement in hazardous waste decision making" was sufficient for it to lose the secured party exemption. Thus, what began as a provision designed to protect lenders involved with contaminated property became, after this ruling, a potential liability-expanding provision reaching beyond the boundaries of traditional CERCLA liability.

In 1992, the U.S. Environmental Protection Agency (EPA) responded to lender concerns about the *Fleet Factors* decision, and the CERCLA liability of lenders in general, by promulgating the Lender Liability Rule as part of the National Contingency Plan.[4] The rule set forth EPA's interpretation of the secured party exemption and identified specific actions that, if taken by a lender, triggered loss of the liability exemption.

Under the Lender Liability Rule, a lender could be liable under CERCLA only if it (1) took direct action to control hazardous waste management decisions; (2) took over operational control of the facility prior to foreclosure; or (3) failed to dispose of the property after foreclosure. More importantly, EPA clarified those activities, previously unsettled by the case law, that a lender could take in connection with the property or the borrower and still take advantage of the liability exemption.

In 1994, the Lender Liability Rule was successfully challenged by various interest groups in *Kelley v. EPA*.[5] The court held that EPA did not have the rule-making authority to interpret CERCLA's liability provisions and therefore struck down the rule. After *Kelley*, lenders once again found themselves facing uncertainty with respect to transactions involving contaminated property.

EPA did successfully promulgate a rule providing lender liability in a ruling based on the Resource Conservation and Recovery Act (RCRA).[6] After *Kelley*, EPA needed a specific grant of statutory authority in order to promulgate rules in this area. Fortunately, RCRA provided that authority and EPA enacted a version of the Lender Liability Rule applicable to the underground storage tank program under RCRA.[7]

EPA made one more attempt to address lender liability concerns in 1995, when it issued the Lender Liability Rule as enforcement guidance.[8] Although somewhat helpful, guidance does not bind the government, nor does it provide any protection for lenders from private party claims. At most, the guidance confirmed EPA's post-*Kelley* practice of following the Lender Liability Rule.

Congress ultimately resolved the lender liability situation by changing the language of CERCLA itself. In the final days of the government's fiscal year ending in September 1996, Congress enacted the Asset Conservation, Lender Liability and Deposit Insurance Act of 1996[9] ("the Act"). The Act amends CERCLA to clarify the scope of the secured party exemption by specifically rejecting the *Fleet Factors* approach to lender liability and adopting EPA's approach as outlined in the Lender Liability Rule.[10]

The Act amends CERCLA's definition of *owner or operator* to include specific terms and activities that were previously subject to varying judicial interpretation. CERCLA now defines *participation in management* to mean actual participation, not merely the capacity to influence standard from *Fleet Factors.* A lender now knows that it will lose the protection of the secured creditor exemption if (1) the lender exercises decision-making control over the borrower's environmental matters such that the lender controls hazardous waste practices or (2) the lender exercises day-to-day operational control over the facility comparable to that of a manager.[11]

In addition to specifying what a lender cannot do, CERCLA now includes a long list of specific activities lenders are authorized to take without losing the security exemption. The term *participate in management* does not include the following actions: holding, abandoning, or releasing a security interest; including a condition relating to environmental compliance in a loan instrument; monitoring or enforcing the terms and conditions of a loan instrument; monitoring or inspecting the facility; requiring the borrower to address a release of hazardous substances; providing financial advice to the borrower to cure a default or prevent diminution in the value of the collateral; restructuring or renegotiating the terms of the loan; exercising available remedies for breach of a loan condition; and conducting a response action under the direction of state or local officials.[12]

Finally, the Act addresses the situation of lenders foreclosing on contaminated property securing a loan transaction. CERCLA now excludes from the definition of *owner or operator* a lender that forecloses and takes title to or possession of collateral property so long as the lender seeks to dispose of the property in a commercially reasonable manner.[13]

In sum, the Act provides a better demarcation of the activities a lender can take with respect to a loan secured by contaminated property and still take advantage of the CERCLA liability exemption. Although the Act provides more clarity, property with environmental contamination will continue to present legitimate concerns for a lender. For example, the environmental condition of the property may negatively impact the borrower's financial condition or the marketability of the property. For the most part, however, lenders can now approach such transactions with less fear of direct CERCLA liability and focus instead on the specific economics of the transaction.

Federal Brownfields Initiatives

Over the past several years, the federal government, in particular the U.S. Environmental Protection Agency (EPA) has steadily increased the amount of resources available for brownfields redevelopment. In fiscal year 1998, for example, Congress appropriated $86.4 million to EPA for brownfields activity.

Since concern over liability under the Comprehensive Environmental Response, Compensation and Liability Act (CERCLA or Superfund) presents one of the primary barriers to brownfields redevelopment, EPA initiatives arose in the context of its administration of the federal Superfund program (for a discussion of recent statutory changes to Superfund regarding lender liability, see earlier in this section). These brownfields initiatives can be divided into the following categories: institutional changes to EPA's administration of the Superfund program; issuance of new EPA policies and guidance; support for state voluntary cleanup programs; assistance to local government; and tax incentives. Each category will be discussed in further detail.

Changes to the Superfund Program

As part of the federal Superfund program, EPA maintains a site tracking database known as the Comprehensive Environmental Response, Compensation and Liability Information System (CERCLIS). EPA automatically places all sites with a reported potential for Superfund action in CERCLIS. Once in the CERCLIS system, a property will often carry a stigma from its association with environmental contamination, despite the fact that response action under Superfund may not even be necessary. In response to these concerns, in 1995 EPA announced the removal of approximately 30,000 sites from CERCLIS that required no further federal action under Superfund.[14]

EPA response action at a CERCLIS site can come in two forms—a *removal action* or a *remedial action*. A removal action, initiated under EPA's authority to respond to an "imminent and substantial endangerment," addresses the immediate threats posed by the site in a short time period. During a remedial action, which can take place following a removal action or on its own, EPA conducts an extensive investigation of site conditions and evaluates several alternatives for cleanup. With either type of response action, EPA's determination of the most appropriate cleanup plan for the site is known as the *remedy selection process.*

EPA also began to incorporate consideration of brownfield redevelopment into its administration of the Superfund program. In 1995, EPA issued a policy requiring consideration of land use issues during the remedy selection process.[15] When EPA conducts a Superfund cleanup, it selects the "remedy," or the most appropriate plan to address the contamination present at the site. EPA received criticism for selecting remedies with unrealistic cleanup goals. For example, EPA would select a remedy based on residential cleanup goals (the most stringent and protective), even if there were little likelihood of residential development occurring at the property. As a result, Superfund remedies were often unnecessarily expensive and lengthy.

Under the new policy, EPA must consult with local officials to determine the reasonably anticipated future use of the property. Based on local plans for the property and the surrounding area, EPA's remedy should then conform to the identified future use. When EPA selects a remedy consistent with future land use, the more realistic remedy can return the property to industrial or commercial use in a shorter time period.

EPA Policies and Guidance

Beginning in 1995, EPA issued a series of policy statements and guidance documents designed to clarify certain aspects of the Superfund program. EPA's policies describe the parameters of enforcement under the Superfund program, with the expectation that increased certainty about the potential for Superfund enforcement action will encourage brownfields redevelopment. The guidance documents discussed in the following text describe EPA's policies with respect to owners of property above contaminated groundwater, municipalities acquiring contaminated property, and parties interested in the Superfund status of particular sites.

The Contaminated Aquifer Policy

EPA issued the *Final Policy Toward Owners of Property Containing Contaminated Aquifers*[16] ("the Contaminated Aquifer Policy") to clarify the potential Superfund liability of owners of property above contaminated groundwater. As discussed in Sec. 5.2, CERCLA liability accrues with property ownership regardless of fault. CERCLA's liability structure forces a particularly harsh result for owners of property located above contaminated groundwater when the contamination migrates from a source off the property.

In the Contaminated Aquifer Policy, EPA states it will refrain from taking enforcement action against a property owner for contamination of groundwater beneath the property under the following circumstances. First, the hazardous substances beneath the property must be the sole result of subsurface migration in an aquifer from a source outside the property boundaries. In addition, the Contaminated Aquifer Policy does not apply to an owner that caused the release of the hazardous substances. Finally, the property owner cannot be in a contractual relationship with the party that caused the release.

The Contaminated Aquifer Policy indicates that it could apply to purchasers of property above contaminated groundwater. However, EPA requires an innocent landowner analysis of the purchaser to determine whether the purchaser knew, or had reason to know, about the presence of hazardous substances at the property. As discussed in Sec. 1.2, it is extremely difficult to establish the innocent landowner defense to Superfund liability. Therefore, the Contaminated Aquifer Policy will have limited applicability to purchasers of property above contaminated groundwater (see Sec. 1.2 for a discussion of prospective purchaser agreements with EPA).

Since the policy arises from EPA's exercise of its enforcement discretion, it does not bind EPA, nor does it provide protection against private party claims under CERCLA. In the event a property owner faces private party claims, the Aquifer Policy provides that EPA will consider entering into a de minimis settlement[17] with the property owner, which can include protection from EPA against private claims.

The Acquisition Policy

In 1997, EPA issued the *Policy on Interpreting CERCLA Provisions Addressing Lenders and Involuntary Acquisitions by Government Entities* ("the Acquisition Policy").[18] Like the liability exemption for lenders (see Sec. 5.2), CERCLA also contains a liability exemption for government entities that involuntarily acquire property subject to Superfund activity. Section 101(20)(D) of CERCLA[19] provides the following exemption from the definition of *owner or operator:*

> [A] unit of state or local government which acquired ownership or control involuntarily through bankruptcy, tax delinquency, abandonment or other circumstances in which the government involuntarily acquires title by virtue of its function as a sovereign.

As discussed in Sec. 5.2, CERCLA contains a third-party defense to liability. So long as the third party is not in a contractual relationship with the defendant, that defendant can assert the defense. Government entities receive another liability exemption at Sec. 101(35)(A) of CERCLA,[20] which excludes from the definition of *contractual relationship* certain instruments where:

> The defendant is a government entity which acquired the facility by escheat, or through any other involuntary transfer or acquisition, or through the exercise of eminent domain authority by purchase or condemnation.

Despite the existence of these exemptions, government entities still faced uncertainty when considering whether to proceed with acquisition of contaminated property. Almost any acquisition by a government entity requires a voluntary or affirmative act. For example, the government entity may need to initiate a tax sale, a foreclosure proceeding, or another affirmative legal proceeding to acquire a property with delinquent taxes.

EPA attempted to address these concerns when it issued its Lender Liability Rule,[21] which included a test for evaluating when government action remained "involuntary" for purposes of the CERCLA liability exemption. The Asset Conservation, Lender Liability and Deposit Insurance Act[22] explicitly affirmed EPA's approach to involuntary acquisition contained in the Lender Liability Rule. EPA's method, as stated in the Acquisition Policy, provides the following test:

> [A]ny acquisition or transfer in which the government's interest in, and ultimate ownership of, a specific asset exists only because the conduct of a non-governmental party—as in the case of abandonment or escheat—gives rise to a statutory or common law right to property on behalf of the government.[23]

When this test is applied to the tax delinquency example, the government's right to acquisition results solely from conduct of the former property owner. Therefore, the government entity can take affirmative steps to obtain the property and still take advantage of the liability exemption.

The Comfort Letter Policy

EPA issued the *Policy on the Issuance of Comfort/Status Letters*[24] ("the Comfort Letter Policy") in response to frequent requests from property owners for some form of assurance that EPA would not pursue an enforcement action at a particular site. EPA normally does not provide specific assurances that it will refrain from enforcement due to an earlier enforcement policy prohibiting such assurances outside the context of an enforcement action.[25]

EPA concluded that the majority of concerns about Superfund enforcement could be addressed by providing information known to EPA about a particular site, with an explanation of that information's relevance to EPA. Under the Comfort Letter Policy, EPA can issue four types of letters tailored to the following situations:

1. *No previous federal Superfund interest letter*—EPA will issue this form of letter when there is no historical evidence of federal Superfund involvement at the property.

2. *No current federal Superfund interest letter*—EPA will issue this form of letter when the property has been removed from the CERCLIS inventory, when the property has been deleted from the National Priorities List, or when the property does not fall within the boundaries of a CERCLIS site.

3. *Federal interest letter*—EPA will issue this form of letter when the property is subject to federal Superfund action; EPA should inform the recipient of the current status of federal involvement and highlight applicable Superfund policy or regulations.

4. *State action letter*—EPA will issue this form of letter when EPA has deferred action at the site to the appropriate state agency.

EPA's decision to issue a comfort letter is discretionary, and the agency generally avoids involvement with purely private real estate transactions. The Comfort Letter Policy recommends issuance of a letter only when the requesting party faces a realistic probability of CERCLA liability and when the letter may facilitate redevelopment of the property in question.

Finally, the content of EPA's comfort letter varies depending upon which type of letter applies to a particular situation. Some of the information EPA may provide includes CERCLIS data, the location of the administrative record, state or federal contacts, the status of the EPA response action, and identification of applicable policies or regulations.

EPA Support of State Voluntary Cleanup Programs

State voluntary cleanup programs, an alternative to the traditional Superfund approach to remediation of contaminated property, provide the most effective means to address the majority of brownfields cleanups. Currently, there are more sites in voluntary cleanup programs within EPA Region 5 than there are Superfund sites on the National Priorities List. These programs succeed because states offer participants flexible cleanup standards and relatively efficient oversight and review procedures. More importantly, most state programs provide some form of final liability resolution at the end of the cleanup process.

EPA provides financial support to state agencies to develop and implement voluntary cleanup programs as part of EPA's annual disbursement of Superfund program assistance. EPA's most visible support, however, comes in the form of agreements with state agencies about the status of sites within voluntary cleanup programs.

Superfund Memoranda of Agreement

A memorandum of agreement (MOA) is an agreement between EPA and a state agency that sets forth the roles and responsibilities of each agency with respect to delegated federal programs such as the federal Superfund program. EPA discovered that state voluntary cleanup program participants were concerned about potential federal enforcement or second-guessing of the state cleanup decisions. Under CERCLA's broad liability provisions (see discussion in Sec. 5.2), a release issued by a state agency following a voluntary cleanup would not prevent a federal enforcement action.

In order to reassure state program participants, EPA entered into MOA addenda with state agencies to clarify the status of sites in the state voluntary cleanup programs and encourage the use of such programs.[26] EPA Region 5 entered into the first brownfields MOA addendum with the Illinois Environmental Protection Agency in April 1995. The MOA addendum contains a statement from EPA that, for sites within the state program, absent an imminent and substantial endangerment situation, EPA does not anticipate taking Superfund action. EPA Region 5 completed MOA addenda with five of the six states in the region.[27] Nationally, EPA entered into 10 MOA addenda through 1997.

The brownfields addendum creates no enforceable rights against EPA or the state, nor does it function as a release of federal liability for the state voluntary cleanup program participant. The MOA addendum does, however, serve as EPA's public affirmation that EPA does not intend to become involved with individual state voluntary cleanup program sites.

EPA did begin, however, to scrutinize the state voluntary cleanup programs themselves for certain minimum elements. In 1996, EPA issued a policy[28] regarding use of MOAs that set forth criteria for regional evaluation of state voluntary cleanup programs. The policy provides criteria that must be met before EPA can enter into a brownfield MOA addendum. The policy establishes the following criteria for state voluntary cleanup programs:

1. The state program must provide the opportunity for meaningful community involvement in the cleanup process.

2. The state program must ensure that voluntary response actions are protective of human health and the environment.

3. The state program must have adequate resources to ensure that voluntary actions are conducted in a timely and appropriate manner.

4. The state program must provide a mechanism for written approval of response action plans and documentation indicating that the cleanup is complete.

5. The state must provide adequate oversight.

6. The state should have the capability (through enforcement or other authority) to ensure completion of the response action if the participant fails or refuses to complete the voluntary action.

Some states expressed concern that EPA, by issuing this policy, was beginning to dictate the elements of a successful state program, functioning with few of the problems associated with the federal Superfund program. These state concerns were confirmed in 1997 when EPA issued a draft policy for public comment that significantly increased the amount of federal scrutiny and oversight of state voluntary cleanup programs.[29] Although EPA withdrew the guidance after receiving negative comments, EPA review and oversight of state voluntary cleanup programs will remain a controversial subject.

EPA Assistance to Local Government

EPA's brownfields initiatives recognize the primary role local government plays in the successful redevelopment of brownfields. The majority of EPA's brownfields resources are directed toward local governments in the form of direct funding and technical assistance. Several of those initiatives are discussed in the following text.

Brownfields Assessment Demonstration Pilot Program

EPA's Brownfields Assessment Demonstration Pilot Program began in 1995 as a means to increase awareness of brownfields redevelopment issues and develop local solutions for national replication. Each pilot community receives a $200,000 award over a two-year period through a cooperative agreement with EPA. The cooperative agreement describes the intended use of the pilot grant and may include reporting and documentation requirements.

The assessment pilot funds may be used for any activity preliminary to cleanup of contaminated property. Acceptable uses of the funding include brownfields surveys or site identification, site assessment and investigatory sampling, and preliminary design activity. Local governments can also use pilot funds for community outreach and public education. Since the statutory basis for EPA's issuance of the assessment pilots is the same one that authorizes EPA's investigatory authority,[30] pilot funds may not be used for actual cleanup or response action at the site.

To date, EPA has awarded 228 assessment pilots totaling $42 million. In Illinois, nine local governments or other entities have received assessment pilot grants. The Illinois pilots are described in further detail here:

Calumet City—Calumet City received a $200,000 grant in May 1998 to address a portion of a redevelopment project area established within the city. EPA funds will be applied to assessment and investigation of three former industrial properties—the Marble Street Dump, the Auto Salvage Yard, and the Burnham Rail Storage Yard.

Chicago—The city of Chicago received a pilot award in April 1997. The city plans to update the results of the 1995 Brownfields Forum. Chicago originally convened the Brownfields Forum by bringing together all interest groups associated with brownfields to develop recommendations for brownfields redevelopment. The EPA pilot funds will be used to assess the effectiveness of Forum suggestions and conduct outreach at Chicago brownfields sites.

Cook County—The County received a $200,000 pilot award from EPA in April 1997. The county plans to work with the South Suburban Enterprise Communities, a redevelopment organization made up of Harvey, Ford Heights, Phoenix, and Robbins. The county has identified the former Wyman-Gordon manufacturing facility in Harvey as site for assessment and investigation.

East St. Louis—East St. Louis received a $200,000 pilot award from EPA in April 1997. The city plans to focus on 220-acre former Alcoa Aluminum site in East St. Louis. Given the size and environmental condition of the site, the city plans to develop and test innovative remediation technologies for recovery of gypsum and other materials on site.

West Central Municipal Conference—The West Central Municipal Conference includes 36 communities in Cook County representing the inner ring suburbs around Chicago. The conference received a $200,000 pilot award from EPA in September 1995. The main focus of the pilot has been on coordinating brownfields redevelopment activities among the member communities. The conference has also developed a team of experts to evaluate identified brownfields. In addition, Canton, East Moline, Lacon, and Waukegan were also recently awarded pilot grants.

Brownfields Cleanup Revolving Loan Fund Pilots

In light of the restrictions on use of the assessment pilot funds, namely the prohibition against using those funds for cleanup, in 1997 EPA developed the Brownfields Cleanup Revolving Loan Fund Pilot program. The program allows local governments to utilize revolving loan funds for cleanup and remediation of brownfields. Under the pilot program, EPA provides initial capitalization of the fund in the amount of $350,000. Specific projects can be funded by low-interest loans from the revolving fund, which, when repaid, can be used to fund additional loans for cleanup. EPA limited the availability of this program to the first 29 entities awarded Brownfields Assessment Demonstration Pilots.

Brownfields Showcase Communities

EPA's most recent brownfields initiative involved the mobilization and coordination of other federal agencies[31] with resources that can be devoted to brownfields initiatives. The Showcase Communities program creates a partnership of federal agencies offering technical, financial, and other assistance to local governments. The selected communities will serve as models demonstrating the benefit of collaborative federal activity. In March 1998, EPA selected 16 communities for the program.[32] EPA estimates that over $28 million will be committed by the various federal participants.

Federal Tax Incentives

A property owner's expenses associated with cleanup of contaminated property can be treated in two ways for tax purposes—as depreciable capital costs or as deductible expenses. Treating such expenses as depreciable capital costs means that the owner cannot deduct the expenses from income in the year the expenditure occurs. Instead, the costs can only be depreciated over the life of the property. If the cleanup costs are treated as deductible expenses, however, the taxpayer can deduct those costs in the year they are incurred.

A 1994 Internal Revenue Service ruling further complicated the tax treatment of environmental costs. The IRS did find that certain costs for remediation of land and groundwater were deductible expenses. The finding only applied, however, to cleanup costs incurred by the same taxpayer that caused the contamination. The preferred tax treatment therefore did not apply to new owners purchasing brownfields.

1997 Remediation Tax Incentive

As part of the Taxpayer Relief Act of 1997, Congress revised the federal tax law to provide incentives for redevelopment of brownfields. The

Brownfields Tax Incentive attempts to correct the disparity between owners and purchasers by making all environmental cleanup costs fully deductible if incurred at property in certain areas. Any taxpayer may treat "qualified environmental remediation expenditures" as deductible expenses. A qualified remediation expenditure is one incurred in connection with a contaminated site within certain targeted areas.

The tax incentive targets the following areas: (1) a population census tract with a poverty rate of 20 percent or higher; (2) a population census tract with a population of less than 2,000, 75 percent zoned for commercial or industrial use, and adjacent to a poverty area; (3) any empowerment zone or enterprise community; and (4) any site within EPA's Brownfields Pilot program prior to February 1, 1997. In order to claim the incentive, taxpayers must obtain approval from the state that the site falls within a targeted area.

Conclusion

Federal brownfields initiatives encourage redevelopment of brownfields in a variety of ways. Through internal changes and issuance of guidance, EPA attempts to reduce the uncertainty surrounding Superfund enforcement. Other federal initiatives, such as the brownfields MOA addenda, support state efforts to encourage participation in state voluntary cleanup programs. Finally, the majority of federal resources are properly focused on local government efforts to address brownfields. Ultimately, brownfields redevelopment must succeed from the ground up rather than through federal mandates from the top down.

References

1. 42 U.S.C. §§ 9601, *et seq.*
2. 42 U.S.C. § 9601(20)(A)
3. 901 F.2d 1550 (11th Cir. 1990), *cert. denied*, 498 U.S. 1046. (1991)
4. Lender Liability Rule, 57 *Fed.Reg.* 18344 (April 29, 1992). The National Contingency Plan, which governs the conduct of response actions under CERCLA, is found at 40 CFR Part 300.
5. 15 F.3d 1100 (D.C. Cir. 1994), *cert. denied, sub nom. American Bankers Ass'n v. Kelley,* 115 S.Ct. 900. (1995)
6. 42 U.S.C. §§ 6901, *et seq.*
7. 60 *Fed.Reg.* 46692. (September 7, 1995)
8. 60 *Fed.Reg.* 63517. (December 11, 1995).

9. 110 Stat. 3009–462 (1996)

10. The Act also amended RCRA to validate EPA's lender liability rule for the underground storage tank program.

11. 42 U.S.C. § 9601(20)(F)

12. 42 U.S.C. § 9601(20)(F)(iv)

13. 42 U.S.C. § 9601(20)(E)(ii)

14. Technically, the sites removed from CERCLIS are "archived" for historical tracking. The sites have been removed from consideration for further response action under the Superfund program.

15. *Land Use in the CERCLA Remedy Selection Process,* OSWER Directive No. 9355.7-04. (May 25, 1995)

16. *Final Policy Toward Owners of Property Containing Contaminated Aquifers,* 60 *Fed.Reg.* 34790. (July 3, 1995)

17. Section 122(g)(1)(B) of CERCLA, 42 U.S.C. § 9622(g)(1)(B), authorizes de minimis settlements with landowners.

18. *Policy on Interpreting CERCLA Provisions Addressing Lender and Involuntary Acquisition by Government Entities.* (June 30, 1997)

19. 42 U.S.C. § 9601(20)(D)

20. 42 U.S.C. § 9601(35)(A)

21. 57 *Fed.Reg.* 18344 (April 29, 1992). For a discussion of the Lender Liability Rule, see Sec. 5.3.

22. 100 Stat. 3009-462 (1996). For a discussion of lender liability, see Sec. 5.3.

23. 57 *Fed.Reg.* at 18372 (April 29, 1992).

24. *Policy on Issuance of Comfort/Status Letters.* (November 8, 1996)

25. *Policy Against No Action Assurances.* (November 16, 1984)

26. The MOA is actually an addendum to the existing MOA between the state and federal officials responsible for administering the Superfund program.

27. EPA Region 5 entered into brownfields MOA addenda with the following states: Illinois (April 6, 1995); Minnesota (May 3, 1995); Wisconsin (October 27, 1995); Indiana (December 4, 1995); and Michigan (July 10, 1996). Region 5 and Ohio have not yet entered into an agreement regarding Ohio's voluntary cleanup program.

28. *Interim Approaches for Regional Relations with State Voluntary Cleanup Programs.* (November 14, 1996)

29. 62 *Fed.Reg.* 47495. (September 9, 1997)

30. Section 104 of CERCLA, 42 U.S.C. § 9604.

31. The federal agencies participating in the Showcase Community program include: Department of Agriculture; Department of Commerce; Department of Defense; Department of Education; Department of Energy; Department of Health and Human Services; Department of Housing and Urban Development; Department of the Interior; Department of Justice; Department of Labor; Department of Transportation; Department of the Treasury; Department of

Veterans Affairs; EPA; Federal Housing Finance Board; General Services Administration; and Small Business Administration

32. The selected communities are: Baltimore, MD; Chicago, IL; Dallas, TX; East Palo Alto, CA; Eastward Ho!, FL; Glen Cove, NY; Kansas City, MO and KS; Los Angeles, CA; Lowell, MA; Portland, OR; the State of Rhode Island; St. Paul, MN; Salt Lake City, UT; Seattle, WA; Stamford, CT; and Trenton, NJ.

Andrew Warren is an associate with Schwartz & Freeman in Chicago, Illinois. At Schwartz & Freeman, his practice includes counseling owners of contaminated property, advising participants in the Illinois Site Remediation Program, and assisting clients with general environmental compliance issues. Mr. Warren also teaches a course on hazardous waste regulation issues at IIT Chicago-Kent College of Law.

Prior to joining Schwartz & Freeman, Mr. Warren served as an assistant regional counsel for the U.S. Environmental Protection Agency, Region V, based in Chicago. While at EPA, Mr. Warren handled enforcement cases under the various federal environmental statutes. He also actively promoted federal brownfields policies. Mr. Warren's brownfields activity included negotiation of prospective purchaser agreements and service in the regional brownfields initiatives. He has also spoken at numerous conferences related to brownfields development.

Mr. Warren is a 1991 graduate of Chicago-Kent College of Law, where he participated in the Energy and Environmental Law Program. He received his undergraduate degree in geology from Lawrence University in 1986.

5.4   Lenders' Viewpoints

John M. Oharenko

Background

During the past decade, great strides and progress have occurred in providing mortgage capital for brownfields properties. Since EPA and state and local governments now offer clearer guidelines for environmental compliance, lenders follow suit by financing such assets based on various rules and regulations.

In fact, most lenders are reasonably aware of brownfields issues such as underground storage tanks, asbestos, and hazardous waste. However, the burden of responsibility for addressing environmental issues remains with the property owner/developer. The owner is required to provide necessary documents including—but not limited to—clear title, evidence of legal conformity, and engineering and environmental reports.

Naturally, lending institutions will decline to finance any properties that do not legally conform to all laws including those on zoning, land use, and environmental compliance. If the environmental reports indicate existing or potential problems, the lenders will use internal underwriting guidelines to determine funding feasibility.

Qualifications

Lending qualifications and restrictions for environmentally contaminated properties still remain a challenge. At the very least, as mentioned earlier, properties must legally and environmentally conform to all laws.

Additionally, virtually every loan approved by a lending institution will be subject to an environmental and engineering report. The Phase I and Phase II (if necessary) environmental reports prepared by reputable third-party consultants are generally sufficient in most instances. These reports are the backbone for determining the extent of environmental issues beyond legal compliance.

Existing properties with environmental issues are financeable as long as an economically feasible and reasonable remediation program justifies project funding under specified loan terms and conditions.

As for new construction, if the site and location is not environmentally curable on a cost-effective basis, financing is a nonissue.

Positive Legislation

The Asset Conservation, Lender Liability, and Deposit Insurance Protection Act of 1996 is favorable legislation for lenders seeking to finance environmentally challenging assets. More specifically, this national legislative act is designed to protect lenders from brownfields liability as long as lenders do not participate in management or influence operations of a contaminated property.

The Protection Act of 1996 also covers pre- and postforeclosure activities, including exemptions, while attempting to dispose of property, maintaining business operations, and/or protecting the property or preparing for sale as long as the lender is attempting to divest the facility at the earliest practical and reasonable time and on the best terms.

Underwriting

Brownfields properties are typically underwritten using the same guidelines as conventional properties, except for the following additional requirements:

- *Reserves*—ample cash flow reserves established as per the remediation plan. The reserves may be funded over the term of the loan.

- *Higher rates*—rates are typically 15 to 50 percent or more higher than conventional mortgage rates, and as such, require sufficient cash flow to cover additional risk(s).

- *Guarantees*—personal recourse and guarantees are required from "warm bodies" (the borrowing group) who are substantial entities able to offer reasonable protection for the lender against environmental lawsuits.

- *Higher fees*—higher underwriting fees (typically 1 percent or more) are charged because of additional risk, loan processing, and documentation review.

- *Insurance*—environmental indemnification insurance underwritten by a major insurer may be required as necessary, depending upon the scope of the environmental issue(s).

In any case, environmentally challenged projects typically demand more equity than conventional projects since the aforementioned loan restrictions usually impact project net operating income available for debt service. As a result, loan proceeds tend to be lower than for the usual 75 percent loans.

At a minimum, brownfields properties are evaluated based on conventional loan underwriting. Conventional lending guidelines for first mortgages typically require the following base standards:

- *Net operating income*—net operating income is the most important figure for understanding a project's economic performance and capitalized value. Simply speaking, net operating income is the gross income adjusted for vacancy, bad debt, operating expenses, real estate taxes, insurance, and other noncapital, recurring expenses.

- *Property types*—this encompasses most forms of income-producing realty, including, but not limited to, apartment, office, retail, industrial, lodging, health care, parking, recreational, and mixed use.

- *Loan-to-Value*—loan-to-value is generally restricted to a maximum loan amount of 75 to 80 percent based on the lesser of (1) project purchase price or (2) appraised value for conventional properties such as apartments, offices, and retail and industrial spaces. Management-intensive properties such as those used for lodging, health care, self-storage, and recreational activities are more conservatively underwritten to limits of 70 percent or less.

- *Debt coverage ratio*—debt coverage is limited to 120 percent of net operating income for conventional properties. Properties secured by credit tenants with long-term leases can achieve debt coverages that are close to breakeven. At the other extreme, management-intensive projects require debt coverages of 140 percent or more.

- *Rates*—typically priced over comparable-term government treasuries. Most properties are funded at rates reflecting pricing of 1 to 3 percent over the comparable term treasuries.

Table 5-2 illustrates a checklist of items needed for processing a loan with brownfields issues.

Funding Sources

The supply of real estate capital is at an all-time high. Numerous lending sources aggressively compete for funding all types of income properties, including brownfields projects.

Generally speaking, mortgage conduits, real estate investment trusts, and other various public/private loan syndicates are the most common sources of funds for financing existing brownfields projects. Banks and credit companies are popular funding sources for new construction projects. These select lending groups will compete based on higher yield preferences in exchange for more risk.

At the other end of the spectrum, risk-averse lenders such as life insurance companies, pension funds, and/or most foreign lending institutions are less likely to finance brownfields properties, instead favoring conventional loan opportunities.

Closing Thoughts

Beyond a doubt, financing options for brownfields properties are available. However, a careful understanding of the risks and rewards in relation to project economics and other incentives is of paramount importance. Furthermore, the ample numbers of funding sources also create opportunities for finding the optimum financing terms and conditions.

Regardless of how challenging a conventional mortgage program is for funding a brownfields project, the borrower/developer should also aggressively explore alternative funding options. For example, funding options available via federal, state, and local programs also offer a broad range of pricing and terms that are not as strictly tied to the economic performance of a project after adjustments to brownfields issues are made.

John Oharenko is senior vice president of GMAC Commercial Mortgage Corp. (GMACCM), the nation's largest commercial realty funding source. Mr. Oharenko is responsible for structuring various capital transactions including loans, joint ventures, and sales.

Table 5-2. Project Financing Checklist

Project, Location, and Economics Items

Environmental/asbestos reports, all levels including Phase I and II
Project description
Survey and legal description
Current photographs
Plans and specifications, including description of materials used
Appraisal report
Project marketing brochures
Certificate of occupancy
Property easements, restrictions, etc.
General area description/economic data
General area map
Immediate area description, including description of possible contamination-
  producing uses in the area
Immediate area map
Aerial photographs
Market and comparable property survey
Zoning information and land use maps
Income and expense pro forma
Leases and/or lease summary copies
Anchor tenant information/credit rating
Detailed rent roll statement
Delinquency/bad debt report
Operating history—3 years
Real estate tax bills—3 years
Utility bills—3 years
Vendor contracts
Ground lease documents

Borrower Background

Financial statements
Ownership background information
Information on past experience and expertise in brownfields projects
Bank and other lending references list
Project architect and general contractor information
Managing agent information
Form of ownership documents (i.e., corporation, partnership)

Financial Underwriting/Misc. Items

Financing request
Purpose of financing
Existing financing documents
Purchase contract
Original cost breakdown
Recent capital improvements list with special emphasis on remediation program
Future plans for remediation, including a timetable

Mr. Oharenko's career spans nearly 20 years in real estate finance. Prior to joining GMACCM, he worked at Cushman & Wakefield Financial Services Group and Baird & Warner. He has consistently ranked among the highest performers within these organizations.

Mr. Oharenko holds a master's degree in real estate investment analysis from the University of Wisconsin-Madison and an undergraduate degree in business administration from De Paul University.

5.5 Community and Nonprofit Institution Involvement in Brownfield Redevelopment

Donna Ducharme

Since brownfields projects are often unusually complicated from a redevelopment standpoint, they can require significant cooperation and even formal partnerships between the private sector, government, and the community to succeed. This section explores the role of the community and nonprofit institutions in the brownfields redevelopment process.

Overview of Community Roles

The community is most often thought of as the various stakeholders that are interested in the redevelopment of an area or site. These stakeholders represent diverse concerns and usually are encountered in groups—such as various resident, business, landowner, environmental, religious, health, and other organizations. These groups range from longstanding formal organizations to spontaneous, informal coalitions formed around a single issue or cause. The community is really many individuals and groups that may or may not have identified their common interests and may or may not have figured out how to work together. For example, different community groups may have competing visions for the future of the community, they may compete for funding to provide similar services, or they may be allied with different community leaders or politicians. Or, they may be a vocal minority and not represent mainstream thinking in the community at all.

Some common types of community groups, and the interests they represent, are listed here:

Property owners—concerned with any factor that will affect property values, safety, health, schools, traffic, noise, jobs, and so on.

Resident activists—concerned with improving living conditions in the neighborhood from their point of view. They aggressively pursue improvements and oppose what they perceive to be detrimental development.

Community development professionals—concerned about facilitating or initiating real estate, economic, and human development activities in a specific geographic area.

Industrial businesses—concerned with a local supply of labor, security, truck access, public transportation, parking, land use compatibility, and property values.

Retail businesses—concerned with maintaining a residential community of customers and with security, parking, disposable income levels, competition, local labor, and traffic congestion.

The community can play many important roles throughout the brownfields redevelopment process. These range from planning to implementation and from advising others to participating actively in the process themselves. It is important for brownfields redevelopers to sketch an accurate picture of who is in the community, who represents whom, how capable different groups are, and to what extent collaborative forces outweigh competitive forces, resulting in clear synergistic roles among the groups. Figuring this out is critically important to many projects. In fact, this aspect of the development process (like many other aspects) can make or break a project.

Community readiness, capacity, and support can have critical impacts on whether brownfields redevelopment projects move forward or not. Communities that are not ready for redevelopment projects may at best be unable to help facilitate a project, and at worst may oppose an otherwise good project out of fear or uncertainty. Communities that *are* ready have organized a working consensus among the stakeholders, have clarity about their vision for the future, and have created the institutional vehicles needed to implement their plans. This enables them to provide important additional assistance and—in the more sophisticated communities—to act as partners throughout the process, from planning to implementation.

Communities can support development, in both the planning and redevelopment stages, by taking the following measures.

Planning Stage

- Organizing a working consensus:

    Being specific about community concerns so they can be addressed

    Organizing input and feedback

    Organizing political advocacy and support

- Setting the ground rules:

  Identifying and facilitating community benefit

  Forging consensus about land use and redevelopment guidelines

- Obtaining and analyzing key data:

  Identifying redevelopment sites

  Identifying potential users/markets

- Doing the up-front work:

  Creating redevelopment strategies for the surrounding area

  Facilitating access to government incentives

  Assembling developable sites

- Creating a marketable community:

  Catalyzing public and private improvements to the surrounding area

  Creating strategies to address area-wide problems

Redevelopment (Implementation) Stage

- Assisting or undertaking land acquisition:

  Facilitating access to specific sites

  Providing alternative site ownership options

- Providing or facilitating access to capital:

  Investing equity capital

  Providing access to lower-cost financing

  Facilitating access to government incentives

- Accessing or providing technical expertise:

  Identifying end users

  Providing access to pro bono technical and legal assistance

  Serving as project manager for specific components of the redevelopment plan

  Delivering political support

- Realizing community benefit:

  Training residents to connect the local labor force with jobs created

  Providing access to minority partners and contractors

  Functioning as a vehicle to earmark a portion of profits for other community redevelopment activities

Community readiness takes time and resources to develop. Nonetheless, it is critical to establishing an environment that is conducive to appropriate development rather than antagonistic to all development. Developers in communities that have worked through these issues do not have to carry the burden of belated attempts to address them on the backs of their individual projects. Instead, in ready communities there are clear guidelines and tools in place that facilitate the redevelopment process.

Community and Nonprofit Institution Involvement in the Planning Stage of Brownfields Redevelopment

Community readiness is a continuum, with some communities having created consensus around more issues and having taken more steps to implement their vision than others. The work that is done—regarding acceptable land use, site access and planning, environmental concerns, identifying and addressing barriers to redevelopment, understanding the market, assessing the potential benefits of redevelopment to the community, and so on—provides a platform to be built upon. Unfortunately, in some cases the platform must be built from scratch. In the absence of previous community planning and other preparatory activities, community participation and input processes must be used as the starting point to identify community interests. Then, with assessment of how well a project responds to these interests, areas of mutual benefit and potential conflict will emerge to help determine the appropriateness of the project.

The following sections describe how community planning, when done up front and prior to consideration of a specific development project, provides assistance that goes far beyond simple project acceptance.

Organizing a Working Consensus

In the planning for redevelopment, a working consensus develops about general strategies and goals as well as about specific sites/projects. This shared vision is needed (and will often be tested) by government officials who gather community input as part of their decision making about entitlements, such as zoning changes and financial incentives needed for a project to proceed. In addition, clear strategies and goals help everyone identify ways to strengthen projects so as to better address local concerns and connect project benefits to the surrounding community.

In some cases this working consensus can be developed with little effort because earlier or ongoing community planning activities have identified common interests, obtained substantial buy-in from various community interests as well as city and elected officials, and established clear redevelopment guidelines and decision-making channels. In other cases, consensus must still be organized and is hard won. Elected officials and community organizations can be valuable allies in developing a unified voice when one has not already been achieved. If consensus does not exist, even small opposition groups can cause lengthy and costly project delays.

Setting the Ground Rules

Those communities that have developed community plans and created a broadly held common agenda have also, in the process, clarified the ground rules and established the context for evaluating redevelopment projects in the area and at specific sites. This eliminates much of the uncertainty facing developers by identifying acceptable uses and parameters for site planning.

Community planning can also illuminate the community's interests and the outcomes the community hopes to achieve from development. For example, is a community looking to increase affordable housing, to meet specific unmet retail needs, to create new living wage jobs, to increase the tax base, to foster local entrepreneurship, to attract green industry, or to eliminate environmental hazards? Planning will help to clarify and determine how to balance these various goals. Some goals will be balanced by identifying different sites that are appropriate for different uses, such as residential or industrial applications. Others, such as choosing between an environmentally friendly industry that provides fewer jobs than an employment-intensive industrial user with greater environmental impact, must be balanced on a single site.

The public's willingness to contribute financial or other assistance to a development project is often determined by its understanding of how such contributions will help achieve community goals.

Obtaining and Analyzing Key Data

In addition to providing a forum for identifying and synthesizing the interests of various stakeholders, community planning can also provide a structure for collecting key data that can assist in redevelopment planning and implementation. For example, site control and assembly are often the most time consuming (and costly) barriers to brownfields redevelopment. Many brownfields sites are burdened with property tax, environmental cleanup, demolition, or other liens. Tax scavengers often buy rights to properties (or strategic portions of properties) that last for years (during redemption periods), but have no corresponding responsibilities for the properties. These scavengers must be bought out or brought in for development to proceed. Effective redevelopment planning identifies available/underutilized sites and collects information about their ownership status—who are the lien holders? Are the sites abandoned, tax delinquent, government owned, or privately owned? This information is needed to craft strategies to gain site control and to identify sites that, if assembled with other sites, would create exciting new development options.

Information collected on the expansion needs of nearby businesses, on their suppliers or customers who might relocate, and on market needs and potential, can help to identify potential end users and lead to viable development opportunities.

Doing the Up-Front Work

Community planning can identify the up-front activities needed in an area so that they can be initiated in a timely manner to facilitate development. For example, at the urging of community interests, a tax increment financing (TIF) district, enterprise zone, redevelopment area, or other geographically based development aid can be put in place by government officials up front so that a developer does not have to wait months for such a district to be established once a project site has been identified.

Similarly, the lengthy process of acquiring abandoned and tax-delinquent property for assembly and reuse could be initiated so that sites are available for redevelopment within months rather than years. Likewise, in many jurisdictions, capital budgeting is done many years in advance of implementation. Infrastructure improvements, such as viaduct or street improvements needed to make a site viable for redevelopment, must often be requested years in advance of their actual construction. By identifying and advocating for development districts/incentives, site assembly, infrastructure improvements, and other predevelopment activities such as demolition or environmental testing, communities can make a significant impact on the redevelopment potential of a site.

Creating a Marketable Community

Community planning can lay the groundwork for creating a marketable community, which in turn makes each redevelopment project more viable. Community plans can create the momentum and vision needed to enable redevelopers to see their projects as part of a larger whole. Community plans can also address issues such as safety problems, infrastructure needs, beautification, and amenities needed in order to attract and sustain development in a community. These issues can rarely be addressed effectively on a property-by-property basis.

The Model Industrial Corridor Example

The city of Chicago's Model Industrial Corridor initiative is an example of a program that was designed to increase community readiness through planning and then taking, up front, the steps needed to implement the plans. Through this initiative, local industrial groups working in 12 industrial corridors identified by the city (see Fig. 5-1) were funded to run community planning processes. Each process brought together the key stakeholders in the corridors and the neighborhoods that surround them to forge a consensus about how to make these corridors competitive industrial areas in the future. The groups were directed to complete plans that addressed the city's objectives for its industrial corridors (see Table 5-3).

Table 5-3. Model Industrial Corridor Objectives

Safety

- Get employees, customers, and suppliers to and from these areas safely
- Change corridor and individual property layouts establish control over them

Functionality and Accessibility

- Improve transportation access and circulation within and to the corridor to meet modern industry needs
- Make property improvements to meet the needs of modern industry

Marketability and Competitiveness

- Give each corridor an identity as a unique urban industrial park
- Put government incentives, such as TIFs, enterprise zones, and empowerment zones in place to facilitate redevelopment
- Assemble parcels to create marketable redevelopment sites

Attractiveness

- Make public improvements to eliminate eyesores such as graffiti, illegal dump sites, and abandoned buildings
- Improve public and private building facades and landscaping

Manageability and Sustainability

- Forge consensus between government, business, and the community about the corridor plan, the implementation strategy, and priority actions
- Identify anticipated community benefits resulting from the plan and assign roles to institutions capable of achieving them
- Confirm and stabilize corridor boundaries and land use
- Establish ongoing problem-solving and review mechanisms for each corridor and plan

The community planners were directed to identify strategies and spending priorities to achieve these objectives in their corridors. Twelve of Chicago's 22 industrial corridors now have community plans and implementation strategies to accomplish the Model Industrial Corridor program objectives in their corridor. These plans are in various stages of implementation. Each has established an identity. Infrastructure improvements have been budgeted and some have been made. Development sites have been assembled. Streets and alleys have been vacated to increase security. Public and private properties have been landscaped. Many boundary and land use questions have been resolved. Seven of the corridors have established tax increment financing districts to help fund implementation of their plans. Today these local industrial development organizations and the city are much more capable of facilitating redevelopment in these corridors than they were when the program began.

Community and Nonprofit Institution Involvement in the Redevelopment Stage of Brownfields Redevelopment

As noted earlier, some communities are ready and able to assist not only in the planning stage, but in the redevelopment stage as well. Communities that are ready to facilitate project implementation are sophisticated communities that have done up-front planning and have laid the groundwork for strategic redevelopment. Institutions in these communities have not only forged a working consensus among key stakeholders about what should be done, but have also begun to build the capacity to help carry their plans out.

Communities can facilitate brownfields redevelopment by taking direct action and/or through indirect action—actions that influence the actions of others. They can take these actions themselves through community institutions or through partnerships with nonprofit intermediary organizations that exist to facilitate brownfields or urban redevelopment. Organizations such as the Local Initiative Support Corporation (LISC) and the Chicago Community Loan Fund provide low-cost loans to urban development projects that have community involvement/support.

In some places new nonprofit intermediary organizations are being formed to facilitate aspects of brownfields redevelopment. Their involvement ranges from accepting donated property (or using other mechanisms to gain ownership) to providing predevelopment funds for site investigation and analysis to offering technical expertise. Phoenix Land Recycling Corporation (Pennsylvania), Consumers Renaissance Development Corporation (Michigan), Trust for Public Lands (California), Northern Indiana Center for Land Reuse (NICLR) (Indiana), and ChicagoLand Redevelopment Institute (REDI) (Illinois) are all examples of these new intermediary organizations. Each has its own mix of services depending on its market and funding sources. For example, Consumers Renaissance provides assistance to public and private sector developers in structuring deals and trains municipalities around Michigan to facilitate brownfields redevelopment in their communities. ChicagoLand REDI and NICLR are willing to come into ownership of properties. Both have resources to provide up-front assistance in assessing the redevelopment potential of sites. ChicagoLand REDI also has funds to invest as an equity partner in redevelopment projects. Phoenix has up-front money to assess redevelopment potential and uses options to gain site control without taking ownership.

Community organizations, alone or in partnership with intermediaries, can play many roles that add value to projects in the redevelopment stage.

Acquiring the Land

Communities can assist with land acquisition directly by establishing site control or indirectly by facilitating market access to sites. Facilitation can include advocating with public officials or private owners of land to stop mothballing properties and move forward with redevelopment. Mothballed sites are often contained in large inventories and are ignored as low priorities because of their relatively small size or limited marketability/ profitability compared with other sites. In other cases, these sites are intentionally being held off the market by corporations concerned about extensive cleanup costs and future liability for the environmental problems on their properties. Redevelopment plans, proposals for cooperative solutions that limit risk, or, on the other hand, strong advocacy and government pressure can sometimes free such sites for redevelopment. For example, the threat of condemnation has been used by the city of Chicago to push some reluctant landowners to put mothballed sites in key redevelopment areas on the market.

Finally, acquisition can be facilitated by communities through advocacy regarding land use and development. For example, on Goose Island in Chicago a local organization, the New City YMCA Local Economic and Employment Development (LEED) Council, successfully advocated that the area remain industrial in the face of residential development pressure. The LEED Council accomplished this by organizing a working consensus of support first among manufacturers, unions, churches, and community groups and then ultimately among city staff and elected officials for establishing a new industry-protective zoning classification for the area called the Planned Manufacturing District (PMD). Having obtained PMD designation, the organization wanted property owners who were mothballing sites (and waiting in case the PMD legislation would be revoked) to move forward with industrial development. To accomplish this goal, the LEED Council identified companies in the surrounding area that needed expansion space and were willing to relocate to Goose Island. Enough interest was identified to more than completely fill the site. Armed with information about real users, investment, and job creation, the city was persuaded to pressure the owners to move forward. Today, the industrial park is well under way. Over $122 million in private investment has created 1,464,000 square feet of new industrial space and 800 new jobs.

Communities can also acquire land directly. Nonprofit organizations can provide much potential assistance in this regard. First, they can accept full or partial donations of property. In some cases this provides the owner with the best return, and it has the added advantage of reducing project costs because the federal government, rather than the redevelopment team, pays the land costs (in the form of an income tax deduction). Second, in many locations, nonprofit organizations are afforded the same opportunities as local governments to come into ownership of parcels with property tax liens from county governments at no or low cost. Establishing ownership through a tax deed has the advantage of wiping out the remaining back property taxes and penalties, and has the added advantage, since these tax liens are superior liens, of eliminating most other liens (except state and federal) on the parcel.

Accessing Capital

Some community organizations, often through government grants or nonprofit intermediaries, have the ability to provide direct predevelopment funding to projects or to find other sources of patient capital. These predevelopment funds can shift early risky activities such as environmental assessments, market analyses, engineering studies, project design, and so on, which establish project feasibility to the community (which in turn would own the information and be able to use it in the future as other options are weighed if the initial project doesn't go forward). Such funds also provide much-needed equity in projects where equity investors are difficult to find, at returns that can be supported by the project.

The Greater North Pulaski Development Corporation (GNPDC) is an example of a community organization that invests its own funds in projects. GNPDC received a grant from the U.S. Department of Health and Human Services Office of Community Services to invest as a limited partner in the redevelopment of the former PlaySkool plant in its community. This investment produced a $290,000 return of the principle, which GNPDC then used to create a community reinvestment fund to support other redevelopment activities.

Community organizations can also have access to financing through intermediaries, socially responsible funds, and foundations. For example, the Chicago Community Loan Fund has $4.3 million in funds from various sources to lend predevelopment funds to projects with nonprofit partners. The LEED Council received a forgivable loan of $50,000 from the Local Initiative Support Corporation (LISC) to undertake the up-front work on its Goose Island redevelopment strategy. Repayment of this money is required only if the development goes forward. The LEED Council was successful in getting the landowner to repay the $50,000 to LISC when the owner finally agreed to move the industrial park development forward.

Accessing Technical Expertise

Technical expertise is available in many different forms from community and nonprofit organizations. Staff members of the more sophisticated organizations have considerable expertise of their own. In addition, they have access to resources through their wide-ranging relationships.

For example, in Chicago, many local industrial organizations have detailed knowledge of their service areas and the environment in which they operate. The most successful of these are aware of what properties are on the market, who owns what land, which companies need expansion space, how to work with city and elected officials, who is who in the community and can make what happen, and where controversy lurks and where it doesn't. They also have strong relationships with companies in their corridors and have access to the key personnel and, in some cases, suppliers/contractors of these companies. As with other nonprofits, their board members are often selected for their expertise and willingness to provide it—personally or through their companies—at no charge to the organization.

In addition to their own internally developed resources, community and nonprofit organizations have access to assistance through other not-for-profit groups. Some are issue-specific organizations, such as the Community Economic Development Law Center in Chicago, which provides pro bono assistance through its own network of Chicago law firms. Other organizations provide assistance in structuring deals, packaging loans, obtaining media coverage, and so on. New brownfields intermediary organizations can partner with community organizations, developers, and government as contractors, partners, or investors as needed to provide a wide range of technical assistance—redevelopment strategies; site, developer, and end user identification; environmental assessment management; community and government relations; packaging government incentives; project management; and so on.

Realizing Community Benefits

Community and other nonprofit organizations can play important roles in realigning the benefits of development to meet community goals. They can do this indirectly by brokering benefits or directly by providing the institutional vehicles to implement the benefits. Some examples are provided in the following text.

A common benefit of brownfields redevelopment is that new jobs are created. Community stakeholders will be concerned about whether the types of jobs created are likely to benefit them (for example, engineering jobs are unlikely to benefit residents with less than a high school education). They will also be concerned about the number of jobs created. A manufacturing plant might create 1 job per 450 square feet; a distribution facility might create only 1 job per 1300 square feet. Even if the number and type of jobs are acceptable, community stakeholders will be concerned about whether the jobs will actually benefit them. Job training and placement programs that target and prepare community residents for both construction and permanent jobs can make a significant impact on the benefits that actually accrue to the surrounding community.

Training and placement can be provided directly by qualified community institutions or brokered from other organizations such as community colleges. At the very least, community organizations should be engaged in recruiting the trainees. Paying for recruitment, training, and placement can be expensive. In Chicago, tax increment financing revenues are being used to repay businesses that front fund these activities. Some community partners already have funding (or access to it) to provide similar services and can use it for these purposes, especially if combined with resources generated by the development project itself.

Another common benefit of development is profits—direct profits from projects themselves and indirect profits from business (construction, landscaping, insurance, maintenance, security, material suppliers, etc.) generated through the development. The indirect profits can be realigned to benefit the surrounding community by purchasing goods and services there during construction and ongoing operations. Community organizations/nonprofits can assist by identifying potential local suppliers for the project and, in some cases, can even create these services in the form of business/training programs. For example, the city of Chicago has contracted with the Chicago Christian Industrial League to maintain all of the landscape planters on public streets. The League has used this contract to establish a training business for unemployed and homeless people.

Direct profits can be realigned as well through the structure of the development deal. For example, ChicagoLand REDI, one of the new brownfields intermediaries, is part of a proposed limited liability corporation (along with a developer and a bank) that is responding to a Request for Proposal (RFP) from the city of Chicago to redevelop a brownfield site in a low-income community. If this project is selected, ChicagoLand REDI will invest 45 percent of the equity (with money received from two local foundations) in the project, receiving 45 percent of the profits in return. One-third of ChicagoLand REDI's share of the profits will go to a community employment network to help prepare residents for jobs. Another third will go into a fund at ChicagoLand REDI earmarked to support additional redevelopment in that community. The final third will go to support the general operations of ChicagoLand REDI.

Community organizations and nonprofit organizations can also be helpful in identifying and delivering/earmarking other potential benefits such as open space, affordable housing development, tax revenues, or amenities such as child care centers that would increase the benefits of redevelopment for a given community.

Summary

Community and nonprofit organization involvement can and does take many forms in brownfields redevelopment. It can range from positive or negative reactions to proposed developments to highly sophisticated, value-added assistance with planning and implementation of brownfields projects. Early, thoughtful community and nonprofit organization involvement can lead to actions that:

- Reduce risk by organizing a working consensus and setting clear, workable guidelines

- Help create developable sites and communities

- Bring information and relationships that increase project feasibility

- Contribute value-added financial and technical resources

- Clarify and maximize community benefits from development

While this is not a new idea, much more needs to be done to educate developers, businesses, government, and community stakeholders about the contributions community and nonprofit organizations can make to facilitate redevelopment. And, much more needs to be done to build community and nonprofit capacity to understand and bring valuable assistance to the redevelopment process. Nonetheless, the examples discussed here provide a glimpse into the potential benefits that can result when community and nonprofit organizations help by providing direction early in the process and then develop the capacity to forge solid partnerships with government and brownfields developers and to facilitate redevelopment.

Donna Ducharme is the director of community economic development programs for the Delta Institute and serves as president of Chicago-Land REDI and the Northern Indiana Center for Land Reuse. Prior to establishing the Delta Institute, Ms. Ducharme was deputy commissioner of Chicago's Department of Planning and Development and the founder of a community development organization. Most recently, Ms. Ducharme worked as a private consultant, providing community and economic development expertise to a variety of clients.

While with the Department of Planning and Development, Ms. Ducharme worked with the Department of Environment on the creation of the city of Chicago's $50 million Brownfield Redevelopment Initiative to assemble, clean up, and redevelop contaminated and abandoned industrial sites. She also played key roles in establishing the Model Industrial Corridors Program and the Mayor's Business Express Program.

Ms. Ducharme also founded the Local Economic and Employment Development (LEED) Council in 1982 and developed it into an organization with over 100 business and 20 community-based partner organizations devoted to community development. The organization was vital in creating the first protected industrial zoning districts in Chicago as well as programs to connect local residents and vocational high schools with area manufacturers through literacy and job training programs.

Ms. Ducharme is the past president of the Chicago Association of Neighborhood Development Organizations (CANDO) and past first vice president of the Chicago Workshop for Economic Development, and was selected as one of the "40 Under 40" young business leaders in the Chicago area by Crain's Chicago Business.

Ms. Ducharme earned a B.A. in urban studies from Carlton College in Northfield, Minnesota. She has also earned a master's degree in city planning from the Massachusetts Institute of Technology.

5.6 Introduction to Technical Sections and Getting Started

Harold J. Rafson

In one sense the environmental issues are a continuing process with certain key landmarks. At the early stages, a Phase I assessment, which relies on historical information, will be performed. A decision has to be made about whether or not to proceed with the project. If needed, a Phase II investigation requires that holes be drilled, samples be taken, and analyses of the soil and groundwater be made in order to form a much more precise determination of contamination levels. Again a decision must be made as to whether to proceed.

Next, a feasibility study is performed that selects an approach to remediation and estimates cost and time to completion. If the project proceeds, the property is purchased, and the decision to remediate is made, then a specific plan is worked out. The remediation project requires project management as well as further sampling and analyses to confirm the effectiveness of the remediation.

In this presentation, which is divided between investigation, due diligence, and remediation, we have included details of the relevant technologies, analysis, sampling, and conformance requirements. All of these studies and decisions form the background for the environmental engineer as he or she pursues the project. These descriptions are separated into the most relevant sequential action period.

It should also be recognized that by *technology* we mean a lot more than just engineering or construction management. The subjects involved are not only those of environmental engineering but of geology, laboratory chemistry, statistics, occupational health and safety, and technical aspects of the law, plus additional specialties such as biochemistry, botany, and others that will be apparent from the discussions.

The technological issues will be discussed as they answer several questions.

1. Is there a problem? What is the problem?
2. What are the realistic options for eliminating the problem?
3. How can the remediation requirements be satisfied?
4. How can the project be managed to meet economic, time, regulatory, and community requirements?

These questions come up at numerous stages of a project: question 1 in the investigation stage, questions 2 and 3 in the due diligence stage, and question 4 in the remediation stage. Each question is itself a grouping of subjects.

It should also be noted that in this book we deal with many issues that could emerge. In most cases one project will not have all these problems. For example, a site where the ground is contaminated but the groundwater is not is much less complicated to remediate than a site where both are contaminated.

Is There Contamination, and What Is the Problem?

The developer should not deal with this question, or other technology issues, without qualified technical assistance. He or she should hire a consulting environmental engineer, or run the risk of either rejecting what could be a profitable project or accepting a project that could turn into a nightmare.

The American Society for Testing and Materials (ASTM) has written a guideline in ASTM E 1929-1998, "Assessment of Certification Program for Environmental Professionals; Accreditation Criteria." This presents the issues before the certifying body. Included in this is an evaluation of experience and technical knowledge. Therefore, you can have some confidence that an engineer who has been certified by a recognized group will have adequate capabilities as an environmental professional. This is not meant to suggest that only individuals who have been certified have the capabilities to perform the work professionally.

How do you to go about selecting such a specialist? EPA has a listing of environmental engineers, and there are listings of consultants by many trade associations and professional societies. Check out their qualifications and references. There is always a choice between a large consulting company and a small one. A large consulting company may have many resources available to support a project engineer, but the engineer assigned to your project may have limited experience. When reviewing the engineering qualifications of the firm, also interview the engineer who will be assigned to your project.

A small consulting firm generally centers on an individual or several individuals with extensive experience, but is their experience the expertise required for your project? Check references on other projects and ascertain that the senior person will have time for your project when you need it, since a small firm often has difficulty in handling larger or priority jobs.

Whomever you choose, you should enter the relationship knowing that this will be for the long term; even a single project may last for years.

It is equally important to select an excellent testing laboratory; sometimes the two functions are combined within one company. A very good laboratory will have people adept at the testing. It is very important to have a skilled laboratory that does continuous business. Results from a lab that does many gas chromatography analyses a day will be much more reliable than those from a lab that does one series of tests a month. Check out the laboratory, because analyses are significant cost items and incorrect results are a terrible waste of money, time, and effort. Sometimes the environmental engineer will have a working relationship with a particular lab. This choice of labs will reflect on the quality and attention to detail of the engineer.

Another team member who may be required particularly for complex projects is a statistician.

The first step in answering the question of whether there is a problem is to perform a Phase I assessment.

Phase I Assessment

A Phase I assessment is an inquiry into prior ownership and use based on available historical information and present observations and interviews. Phase I does not get into sampling and testing—that is Phase II.

The Phase I assessment process has been standardized by the American Society for Testing and Materials (ASTM) in ASTM E-1528 and 1527. Standard E-1528 is the transaction screen process, which can be done by the owner or a consultant. Standard E-1527 is the environmental site assessment process, in which a Phase I is to be performed by an environmental professional.

Standard E-1528 asks about:

- Use of the property and adjoining properties
- Hazardous substances, tanks or drums, or PCBs on the property or adjoining property
- Drainage from the property or wells on the property
- Environmental liens or lawsuits, notifications of contamination, or prior assessments
- Site visit observations on the current or previous use of the property and adjoining property
- A review of governmental and historical records for pertinent information

E-1527 provides the information from the owner or other knowledgeable party as to the potential environmental issues at the site. This is only one part of the overall investigation that makes up the Phase I assessment. This Phase I assessment is done by a professional, and investigates sites in a wider circle around the property and deeper into the past. It explores:

- The physical setting in more detail (by city records and site reconnaissance)
- Historical records (minimum 75 years)
- Aerial photographs

- Fire insurance maps (minimum 75 years)
- Property tax files
- Land title records (minimum 75 years)
- U.S. Geological Survey (USGS) topographic maps
- Building and zoning records
- Site reconnaissance—while limiting itself to what can be visually or physically observed, it is more detailed than in Standard E-1528
- Interviews with owners, occupants, neighbors, and governmental officials

At the conclusion of Phase I, you should be able to identify whether there is likely to be contamination. We will assume that there is a likelihood of contamination due to past use. At this point, the developer will decide either to drop the project or to investigate further. Continuing the project leads to a Phase II investigation. There is a very big difference between Phase I and Phase II. Phase I never involves drawing a sample or doing a test, but relies on existing information and nonintrusive site reconaisance. Phase II involves penetrating the soil and building materials and testing for hazardous and toxic chemicals, and provides definite answers about the actual site condition. In addition, the developer may also hire a geologist to determine soil loading characteristics, and/or structural engineers to review the structural integrity of the building.

For biographical information on Harold Rafson, see Sec. 1.1.

6

The Private Developer— Due Diligence

The project continues, and requires more in-depth analysis.

In Sec. 6.1, Robert Rafson, as the developer, continues to check out the preliminary cost estimates and the profitability of the project, and to assemble a project team.

In Sec. 6.2, John Russell discusses the function of due diligence in the investigation stage leading toward a decision concerning a project. Russell gives a detailed analysis and checklist going through the analysis of the financial validity of the project. This chapter serves as a good checklist for the inexperienced developer.

In Sec. 6.3, Noah Shlaes leads a discussion of all aspects of the appraisal process from the appraiser's point of view. The appraiser and developer have to work together closely from the start, or the appraisal may be misdirected.

In Secs. 6.4 through 6.8, several technical experts present the different aspects of the environmental, engineering, and technical efforts. Because the developer must now know costs, the environmental engineer must do his or her investigations thoroughly, efficiently, and professionally. This involves a Phase II study, the proper design of sampling, analysis of samples, and interpretation of results. The complexities of the site geology and hydrology must be considered. These technical subjects are directed toward helping the developer understand the issues involved in dealing with the environmental engineer, laboratory, statistician, and geologist. Readers who wish to pursue technology issues in detail are directed to many textbooks on specific technical subjects.

In Sec. 6.7, Christopher French discusses hydrology, which is important because groundwater contamination is the underlying basis for environmental cleanup regulations. The groundwater pathway is the most efficient method of delivering contaminants in soil to human beings.

In Sec. 6.9, Robert Rafson, the developer, has to make a decision now that all the facts have been gathered.

6.1 Due Diligence for Developers

Robert Rafson

We have a project that, based on our initial review, is a workable concept. A due diligence investigation is the last chance to determine whether a project will likely be profitable or not. At this point, we need to confirm our initial project estimates, and that there are no other restrictions to redevelopment. We also need to assemble the team members for the project.

Confirming the Project Estimates

Project estimates often include many assumptions gleaned from previous experience on completed projects. These assumptions can come from site review or subcontractor bids. In either case, the costs for each part of the project need to be confirmed. The estimate should be reviewed for detail complexity, cost, and potential delays. The result of this review will prioritize the due diligence investigation.

Even though the final building plans may not be drawn up, the conceptual drawing must be complete enough for the contractors to bid the individual work to be completed. Once due diligence is complete and the contracts for purchase are completed, then final contract drawings will be completed. This set of final drawings will be filed for permits almost immediately upon ownership in an effort to save time.

For the larger construction items, such as roofing, concrete, structural elements, electrical work, and plumbing, actual bids for the anticipated work need to be collected. Often general contractors will collect at least three bids on each major construction item after the final drawings are in for permitting. At this earlier point, we are looking for confirmation of the project estimates. One contractor is often all that is required to determine whether the estimate is close.

It is important that these final walkthroughs and reviews be as thorough as possible. You may find additional defects in the building or building plans. Project costs may deviate at this point, but the accuracy of these final estimates must come out close to the final project cost.

Simply, if the original estimate is within 20 percent of project costs, the due diligence should reduce that potential error to approximately 5 percent. This provides much more assurance that the project will not run over in cost or time.

Environmental issues are now addressed in more detail. Up to this point, we have the Phase I report, which gives us the historical review of the property and hopefully some assurances of the property's condition from the owner. This is not enough to avoid unknown remediation costs.

A Phase II environmental report provides an underground investigation of the site and can also include testing for lead-based paints, asbestos, and other contaminants that are stored in the building. These studies are expensive and can take the entire length of your due diligence period. Be sure your environmental engineers are aware of your due diligence time constraints when they are hired. They must carve out the time in their schedules to do this drilling and sampling. The lab must complete the analytical testing and the engineer must write up the report within the time available. Having the testing plan and time scheduled at the time of signing the letters of intent or contracts is recommended.

If there is significant contamination, the project may be dropped or there may be additional testing required to better characterize the extent of the contamination. Industrial revitalization requires in-depth investigation as there is the expectation that contamination exists and there may be a big problem that could jeopardize the project.

Hopefully, the Phase II investigation will uncover the existing condition adequately, provide recommendations for the cleanup, and aid in estimating the time and costs. These estimates are often conservative and can be improved by reviewing cleanup options and timing.

Depending on the cleanup options suggested, these may significantly impact the overall project. An example is when there is a leaking tank under the building to be redeveloped and the contamination levels are high enough to require cleanup. Since the building must remain in place, a longer-term pump and treat or vacuum extraction strategy may be necessary, affecting both construction financing and, more importantly, lease or sale options. This property may be required to be leased until the it is cleaned sufficiently that an NFR letter can be obtained. This may not be acceptable to the intended tenant or subsequent owner. It is important to consider this option because the entry into the state site remediation program may also extend your project redevelopment process.

On a positive note, some types of environmental problems have specific local and state funding programs. Most of these will require that the cleanup be completed prior to refund of a portion of the expended funds. Some programs, like the LUST fund, may or may not ever fund the costs that are expended; presently in Illinois that fund is paying claims on projects completed six years ago. Other incentives pay back a portion of the remediation costs in reduced real estate taxes over time.

Therefore the Phase II investigation can significantly alter the project both in the actual construction cost and schedule and in the overall project use and ownership. There are no certainties on these issues until the Phase II assessment is analyzed and compared to the development objective. The cleanup may be just another construction cost item, or it could vastly alter the entire redevelopment plan.

Finally, the project timeline and critical path must be worked out at this point. This will be critical to the subsequent part of the due diligence process—financing and building the project team.

Building the Project Team

The project team consists of both the group within the development company and outside experts as key contributors in the process. Within the development entity there needs to be personnel with experience to fill these roles:

Environmental management

Project/construction management

Capital management

Marketing management

The additional experts needed to support the project team include:

Lender

Legal counsel (contractual and environmental)

Environmental engineer

Real estate broker

In some cases the significant assistance of the city or state will be needed for the redevelopment to go forward; city or state personnel assigned to the project may be considered a part of the team.

The most important function of the team is getting funding and managing those funds. The lender and financial manager must have a good understanding of the project and also expect that they can get their portion of the project done. A significant part of the due diligence process consists in securing the lender's assurance of funding the project. This often is provided by a bank commitment letter.

There are many issues that must be addressed before the lender will fund the redevelopment of a project, especially one that includes any remediation activities. There must be a significant amount of comfort between the lending institution and the developer in order for the lender to enter into a mortgage and fund a redevelopment on distressed collateral. A lender may require additional collateral, higher equity involvement, and higher value-to-loan ratios. Each of these added costs must be allowed for.

The developer may be required to have a high percentage of equity initially in the redevelopment process. The developer may have to purchase and remediate the site prior to the bank's commitment, and this may be too high a risk for the developer to enter into for that project. There may be a middle ground between standard debt-to-equity ratios and full front funding by the developer. This negotiation is one of the most important factors in a project going forward. The developer has to determine whether he or she has sufficient equity to take on this project alone. Finally, the economics of the returns required for equity investment are significantly different than those for bank lending. Equity may require more than 20 percent return on investment and significant portions of the project. It is obvious that small changes in equity dramatically change the economics of the project.

For this reason the financial manager on the project is a key person on the redevelopment team. He or she not only brings to the table financial expertise, but can help enlist the support of some lending institutions. These relationships will be key to the funding of the project.

Even in the best of relationships with the lending institutions, brownfields redevelopment may still have funding problems. There is a reasonable fear by the banks, equity partners, and other lending partners in a project that they will be stuck with a property that will have extensive contamination, that the funding provided will be consumed by the remediation costs, and that the property will still have no net value. Even if the project goes well, there still is the potential that the completed redevelopment will be more difficult to market and thus the developer will fail to carry the loan. It is important to think about the sale of the property prior to beginning the process of redevelopment, but marketing and sale of the project are also important elements of the critical path. Since the sale of the property provides the income to offset the expenses of purchase and redevelopment of the property, thus creating profit, the sooner the sale can happen the more profit will be made. Carrying the cost of the project is very expensive and continues until sufficient income is generated or the sale of the property is closed.

The marketing manager and outside brokers must therefore be on the team early. If possible, the marketing manager should sign on the brokers shortly after property acquisition to allow the longest possible marketing time. This will hopefully reduce the time from completion of the project to sale or lease.

From the perspective of estimating the project, the marketing and brokerage fees can be estimated at this time. Brokerage fees are based on a percentage of the sale or lease value created. Marketing expenses are usually fixed and are based on the complexity of the project and the time necessary to assemble the information to explain the project to potential customers. Some of this information gathering will have been done to explain the project to the lenders and equity partners. The information provided to the public is different in tone and perspective. The inside marketing manager will often do a more complete job than the broker of presenting the project environmental information, and the time and money required to pull together that information should be included in the project estimate.

The project manager is the lead person in the preparation of the estimate of the project costs. That person will have the responsibility of taking the estimates and delivering the project within those estimated costs.

The project manager will have the understanding of the overall project and project objectives. He or she must direct the environmental engineer to the best path for the project and incorporate the engineer's recommendations into the project estimating and scheduling.

The environmental engineer is obviously critical on contaminated site projects. His or her estimates and determination of the best cleanup method can greatly affect the success of the project. The Phase II assessment will provide the technical information needed to determine the extent of the problem and should indicate the best path to mitigating the environmental risk.

Often longer-term cleanups are cheaper. If the project can support a long-term cleanup—a building or facility built for lease, for instance—then this might be the best option. If the project is driven by the ultimate sale of the property, a faster cleanup may be the best option. This is why the project manager will have to work with and direct the efforts of the environmental engineer.

On some sites, there will have to be a feasibility study. These studies are often done when the cleanup costs are extremely high; they help determine the project redevelopment path. A feasibility study should review all available options on the cleanup, and the project manager then will have to weigh the options in relation to the project costs and scheduling.

Since the results of environmental engineering studies are among the most uncertain portions of the project, extra care in this portion of the planning should be taken. Even though modern testing techniques have reduced the uncertainty in ascertaining the site conditions, there are still many cleanup options and much potential for improvement of profits and scheduling.

Lastly, all project teams need legal support. In brownfields projects there are environmental legal concerns in addition to the normal contractual and corporate advice. Lawyers think differently than project managers. A project manager will plan for how the project will go, develop the path, and try his or her best to make sure there are no obstructions to that path. Conversely, a lawyer will start by looking for the worst possible thing that could happen and find a way to protect against that happening, then look for the second worst eventuality, and so on. This way of thinking has protected many projects from destroying the developer when the projects went bad, but it is a mind-set that should not drive the development planning. This perspective will hopefully help keep the project away from some of the precipices on the development path.

The environmental lawyer has an unusual additional function. Unlike deterioration in a building that comes from neglect, environmental contamination cleanup is a potentially recoverable expense. Many times the corporation or individuals that caused the contamination either do not exist anymore or are insolvent, making recovery impossible. Other times, the negotiations of cost recovery will be integral to the purchase and redevelopment of the property. Legal advice and language are very important. Ultimately, it is the developer's decision to determine whether the risks of taking on these projects are worthwhile, but the environmental lawyer will help put these risks in perspective.

In Sec. 5.2, an environmental attorney discusses legal risk when taking on this kind of project. In this chapter, an environmental engineer discusses the process of doing a Phase II investigation; Chap. 7 deals with how to develop and implement the remedial plan. It is critical that the developer and the team have a good grasp of the details of the environmental contamination, cleanup, and related risks.

For biographical information on Robert Rafson, see Sec. 5.1.

6.2 Due Diligence

John Russell

Prior to buying a commercial real estate property, the purchaser will usually investigate the property in several respects. Typically, these investigations include an analysis of the physical conditions, economic conditions, and legal conditions. The purchaser hopes to avoid unpleasant surprises prior to ownership that could radically alter the presumed function or economics of the acquisition. This process is called *due diligence.*

The due diligence process for a commercial real estate purchase transaction and a brownfield purchase transaction have many of the same components. However, they differ in two important ways: the timing of when the due diligence is conducted, and the depth of the analysis.

To illustrate the differences in process, two sample transactions are outlined here. In the first scenario, we assume that the purchaser is considering the acquisition of an industrial building as an investment, and detrimental environmental conditions are not obvious or anticipated. In the second scenario, the purchaser has reason to believe that detrimental environmental conditions exist, but still has an interest in acquiring the brownfield property as an investment.

First Scenario: "Typical" Industrial Property Purchase Transaction

Our purchaser will find it relatively easy to review a variety of acquisition alternatives available for purchase by working through established channels. For example, these properties are commonly listed for sale by real estate brokers. The purchaser may see a sign posted on the property advertising its availability. The seller may be an investor or corporation that is solvent and marketing the property through relationships.

The purchaser typically has previously determined the investment characteristics sought, and will tailor the property search according to those guidelines. For example, investors who require an initial cash return of 11 percent on equity invested typically will not diligently pursue transactions that provide an 8.5 percent yield on equity. This is a preliminary form of due diligence, and is usually not limited to anticipated economics. Typical investors will also have strong preferences regarding physical characteristics, such as building depth and clear height, and will eliminate those properties that do not meet their criteria.

The impact of this screening process is an important consideration, because it creates one of the key differences in due diligence between brownfields and other types of property acquisitions. The bulk of the likely purchasers for this property type will have many of the same criteria, which means that those assets that are attractive to the market do not remain available for long. Financing and private capital are typically obtainable for these assets. The purchaser may be competing not only against other investors, but also against corporate purchasers who may pay a premium for a key facility. A purchaser of higher-caliber assets must quickly evaluate the anticipated economics, the location, the condition of the property, the functional suitability, and the eventual exit strategy. For higher-caliber assets, this process happens over a few days or weeks at most. Those who do not move quickly to present an offer to purchase will not be included for consideration by the seller.

As a result of this competitive environment, typically a purchaser will make a quick review, make some educated guesses where the information is lacking, and submit a letter of intent, offer, or contract to purchase the property, subject to:

1. Confirmation of physical conditions
2. Title, survey, zoning, and code compliance
3. Financing
4. Environmental reports
5. Other limiting conditions or concerns of the purchaser

The tension for the purchaser is to offer the best possible price with the fewest limiting contingencies that could prevent or delay a closing on the sale. For example, a seller may prefer an offer that has a slightly lower sale price, but that does not require a financing contingency, on the theory that the purchaser who has arranged financing already has one less barrier to closing the purchase.

Once the seller and purchaser reach agreement on the contract, detailed due diligence begins for the purchaser. The contract usually stipulates that 30 to 60 days are provided for the purchaser to confirm or challenge through investigation any of the characteristics of the property. The purchaser has typically deposited funds (earnest money), but these funds are not typically at risk during this due diligence period. In addition, the purchaser usually begins to expend money on reports and assessments, which may include studies of:

1. Existing physical conditions (structural, mechanical systems, roof, floor)
2. Appraisal and environmental reports for financing
3. Legal review of title, survey, zoning, and loan documents

At the conclusion of the due diligence period, the purchaser has to make a decision whether to proceed to closing on the acquisition, attempt to renegotiate based on previously undisclosed information that is now known, or terminate the contract. Notable benefits to the purchaser in conducting due diligence on these higher-caliber assets include the following:

- The operating history, rent roll, site and building plans, maintenance records, and building system specifications and their capabilities are usually available for review.

- The higher-caliber assets tend to be well maintained in general and are usually constructed with modern techniques and good grades of materials. They tend to feature more modern roofing, sprinkler, heating, ventilation, and air conditioning (HVAC), electrical, and plumbing systems. There are exceptions, but when contrasting a 10-year-old facility with 50-year-old brownfield facility, enhancements desired by the purchaser to existing conditions may be elective as opposed to mandatory.

- The seller usually can provide information to a purchaser regarding the property and its history.

- If construction or permanent financing has recently been placed on the property, some of the due diligence items, such as the condition of survey, title, zoning, and environmental liabilities, may have recently been evaluated by the lender to ensure that the property is sound loan collateral.

When contrasted with a typical brownfield acquisition, due diligence is a matter of a review of readily obtainable data, and falls easily within a 30- to 60-day time frame.

Second Scenario: Brownfield Property Purchase Transaction

Our purchaser will find it relatively difficult to review a variety of acquisition alternatives for brownfields properties available for purchase by working through the established channels. For example, these properties are infrequently listed for sale by real estate brokers due to the great difficulty in finding purchasers and to liability concerns. The purchaser may see a sign posted on the property advertising its availability, but the contact information may be hopelessly outdated. The seller may be an investor or corporation that is solvent, but equally as often is a defunct, bankrupt, or absentee owner who cannot be located.

More typically, the brownfield property may be identified by local community interests. When in a vacant, decaying condition, it is often a source of neighborhood trouble, and contributes to a perception of blight in the area. The investor usually will have a greatly elevated expectation regarding investment return on the equity deployed. Financing, both from lenders and private capital sources, is difficult to obtain. The property, due to functional obsolescence and decay, may require a very creative approach to reuse, and often the proposed use will be of an entirely different nature than the original one. For example, in Chicago, older multistory industrial buildings are often best adapted to reuse as residential structures, assuming this does not conflict with zoning and the concerns of neighboring industrial uses.

There is not usually a competitive pressure to quickly generate an offer, as the bulk of the investment community is focusing on the type of transaction outlined earlier in the first example. This is good news in one sense, since the purchaser typically has more time to evaluate the opportunity more fully prior to submitting an offer. The difficulty is, however, that the purchaser must invest substantial time, energy, and resources to understanding the existing conditions before even determining whether the project is commercially reasonable.

In a nutshell, the comprehensive due diligence of the purchaser of a brownfield property is done before even submitting an offer, or attempting to gain title.

It is rare when a brownfield property consists of a higher-caliber asset with an environmental liability that limits value. More commonly, the property is of an older vintage. Current environmental housekeeping practices were not followed. The environmental liability created a difficulty in refinancing or frightened off replacement capital. The facility is commonly abandoned, decaying, and stripped of any building material of value by vandals and scrappers. Multiple liens from lenders, taxing authorities, and secured creditors are common.

Under these conditions, the purchaser should conduct a detailed analysis first before making an offer to purchase or assuming title. This analysis includes an evaluation of:

1. Future use once rehabilitated, considering functional suitability.
2. The best path to gain title and an exit strategy.
3. Environmental conditions: is the existing data sufficient?
4. Investment expectations and capital sources.
5. Property condition: what building systems are functioning?
6. Title report and survey: what liens, easements, encumbrances, deed restrictions, unpaid taxes, and secured notes exist?
7. Governmental considerations (zoning, permits, approvals).
8. Neighbor issues, public impact, and public relations.
9. Owner status: is the owner ready to sell, or still in denial?
10. Does the purchaser have access to all the disciplines necessary to make the project work?

Only after considering these and other issues should the purchaser consider entering a legal contract for the purchase of the property. Any one of these considerations, if not considered up front, can derail a redevelopment due to unexpected cost increases and delays in redevelopment time.

If a selling entity does exist, then a purchaser can submit an offer or contract to purchase, subject to confirmation of the remaining variables or any assumptions made regarding the issues just outlined. Common remaining contingencies might include a commitment for financing, if it is available, or further environmental testing, if required.

Once an agreement is reached on the contract, the purchaser can begin confirmation of the remaining due diligence items. Due to the difficult nature of these projects, a purchaser can typically can get a longer period than 30 to 60 days for due diligence on the remaining items. For example, if traditional bank financing is proposed, the lender may make the commitment subject to Phase II testing, which alone can easily require 60 days.

At this point in the due diligence process, the transaction begins to more closely resemble a traditional commercial purchase transaction. Earnest money may be posted, but is typically not at risk during this period. The purchaser continues to spend money on assessments and reports. At the conclusion of the due diligence period, the purchaser is again faced with a decision:

1. Proceed to closing on the acquisition

2. Renegotiate based on undisclosed information now known

3. Terminate the contract

In the brownfield transaction, the due diligence is more difficult due to its scope and the need to understand complex difficulties that might normally be taken for granted prior to having control of the property. Before purchasers have any assurance they can gain marketable title, they will have expended considerable energy, effort, and expense. Issues that might be taken for granted regarding legal and physical conditions of the property cannot be left for cursory review after a contract is signed.

Included later in this section is a sample due diligence case study utilizing a checklist and analysis template designed to assist in the financial analysis of a potential brownfield acquisition. This checklist does not include all of the legal and construction issues, but does provide an initial indication as to whether the renovation will make sense on an economic basis. The checklist assumes that an industrial property is under review and could be easily adapted to other property types by including issues of importance to those product types.

The checklist begins with a summary of existing conditions, and leads to an evaluation of anticipated renovation expenses. Current property value is then compared with the estimated value created by the proposed redevelopment. The analysis estimates total capital requirements and projected returns for the capital. This type of analysis should not be the sole tool in determining whether to proceed, but is an important component.

Due Diligence Case Study

In this case study, we will assume that a three-story industrial building consisting of 120,000 square feet in an urban setting is beginning to show evidence of neglect. The building is currently leased to a variety of small commercial and industrial tenants, but badly needed roof and elevator repairs have not been completed.

The reason for the neglect is twofold: the owners are located out of state, and the property has environmental contamination that was discovered when the owners attempted to refinance. The lender requested a Phase I report, and the environmental engineering firm found strong evidence of previously undisclosed underground storage tanks located beneath the building, as well as small amounts of friable asbestos. The current lender will not make the requested loan for roof and elevator repairs, and is unwilling to refinance the loan, which is now past its maturity date. As a result, the borrowers are in default on the mortgage, the leaking roof and inoperable elevator are driving out tenants whose leases are expiring, and the owners do not have the cash to stabilize the property, much less to pay for a Phase II assessment.

We assume that a brownfield developer is interested in maintaining the property as a commercial and industrial facility. Based on market knowledge, the developer believes that significantly higher net rents are possible if improvements are made to the roof and the individual tenant spaces. Currently, the tenants are paying an average of $4.50 per square foot. Once improved, the spaces should command $6.50 per square foot. The seller is requesting a sale price of $2 million. The developer has access to a traditional lending source that will lend 80 percent of project costs if the contamination is minimal.

The brownfields developers begin the process by estimating environmental expenses. In addition, they estimate the required interior and exterior improvement costs and soft costs. They have a representation from the seller that there are no unpaid liens, aside from delinquent real estate taxes for last year. They will compare the total estimated capital costs per square foot to a benchmark of value created by comparing these costs to similar sales in the neighborhood expressed in sale price per square foot. In this example, a total capital cost of $32 per square foot is a relatively safe benchmark when compared to other sale prices.

Once they have estimated the required capital expenditures, the developers will then analyze annual income and expenses to create an estimate of value based on the net operating income (NOI). NOI can be defined in simple form as rental income minus property expenses. The NOI will be divided by a market capitalization rate to determine an estimate of resulting real estate value once renovations are complete. This value is important because if the NOI generated does not create a capitalized value greater than the total capital cost, then no real estate value is being created by their effort.

The checklist includes assumed figures from this case study to assist in understanding the impact of the various issues this property faces and the value that results.

Case Study: Due Diligence Form

Environmental Expense Budget

Item	Description	Units	Cost/Unit	Total
Engineering				
Reports: Phase I		1	2,000	2,000
Phase II		1	15,000	15,000
Phase III				
Other				
Testing: (A)				
(B)				
(C)				
Lab work: (A)				
(B)				
(C)				
Legal: (A)				3,000
(B)				
NFR letter plan				10,000
Remediation: (A) Asbestos removal budget				5,000
(B) Underground tank removal budget				20,000
(C)				
Consultants: (A)				
(B)				
Contingency				5,000
Total				$60,000

Renovation Budget (2)

Building Interior Repair and Rehab

Item	Description	Units	Cost/Unit	Total
Interior Hard Costs				
Cleaning/debris				2,500
Demolition		20	500	10,000
Demising walls				
Floors/concrete		3,000	1.5/SF	4,500
Masonry				
Structural				
Windows	Assume 50' = 10 windows	10	4,500	45,000
Doors		20	1,000	20,000
Plumbing				10,000
Bathrooms				10,000
Sprinklers				
HVAC				
Electrical				
Lighting		20	2,000	40,000
Painting	Corridors			10,000
Misc. repairs				5,000
Office				
Paint				
Carpet				
Ceilings				
Elevator				20,000
Other				
Contingency (20%)				35,400
Total				$212,400

Renovation Budget (3)

Building Exterior Repair and Rehab

Item	Description	Units	Cost/Unit	Total
Exterior Hard Costs				
Roof				$20,000
Gutters				
Metal repair				
Masonry				
Lighting	Common area lighting ($2,000 plus)	150	150	$24,500
Paving				
Dock work				
Stairs				
Ramps				
Structural				
Fencing				
Landscaping				
Debris				
Cleaning				
Painting				$5,000
Repairs				
Signage				$15,000
Demolition				
Other				
Other				
Other				
Contingency				$10,000
Total				$74,500

Renovation Budget (4)

Project Soft Costs

Item	Description	Units	Cost/Unit	Total
Legal: Acquisition				$5,000
Leasing/Re-sale				25,000
Financing				5,000
Brokerage fees (Leasing)				$40,500
Engineering	Sprinkler/structural/roof reports			15,000
Architects	As builts/construction drawings			20,000
Marketing	Advertising/fliers			5,000
Security				
Carry: Taxes: Redevelopment estimated vacancy budget			$= 15\% \times 86,930 =$	13,040
Insurance: Redevelopment estimated vacancy budget			$= 15\% \times 22,471 =$	3,371
Maintenance: Redevelopment estimated vacancy budget			$= 15\% \times 38,252 =$	5,738
Utilities: Redevelopment estimated vacancy budget			$= 15\% \times 122,449 =$	18,367
Other				
Financing: Fees			1%+	28,000
Interest	1 year at 9.5% × $2,600,000			247,000
Appraisal				2,500
Equity Interest				
Permits				20,000
General conditions (5%)				55,000
GC fees (5%)				38,000
Other				
Contingency (15%)				102,377
Total				$784,893

Redevelopment Expense Summary

Item	Description	Units	Cost/Unit	Total
Redevelopment Expenses				
Environmental				60,000
Interior hard costs				212,400
Exterior hard costs				74,500
Soft costs				784,893
Total				$1,131,793
Liens				
Unpaid taxes				125,000
Unpaid gas				
Unpaid electric				
Unpaid water				
First mortgage				
Second mortgage				
Third mortgage				
Other notes				
Municipal violations	Sprinkler/demising in hard costs			
Mechanic's liens				
Federal tax liens				
Demolition/debris liens				
Other				
Total				125,000
Transaction Expenses				
Title/survey				5,000
Other				
Total				5,000
Total redevelopment expenses				$ 1,261,793

Income and Expense Pro Forma

Assumption: Redeveloped @ $6.50/SF Gross

Tenant	SF	Net Rent	Tax	Insurance	CAM	Expires	Options
1st floor	40,000	260,000					
2nd floor	40,000	260,000					
3nd floor	40,000	260,000					
Total		**780,000**					

Net Operating Income

Operating Income

Gross rent 780,000

Other income _____

Tenant reimbursements 67,714

Vacancy (5%) (42,386)

    Total operating income 805,328

Operating Expenses

Structural reserves
   ($.15/SF) (18,000)

Management (4%) (32,213)

Real estate tax (125,000)

Insurance (22,471)

Common area
   maintenance (161,083)

Leasing reserves

  Tenant improvements* (23,800)

  Commissions** (37,800)

  Total operating
   expenses 420,367

Net Operating Income $384,961

*Tenant improvements = (office @ $5/SF × 12,000 SF) + (warehouse @ $.50/SF × 118,000 SF) ÷ 5 years

**Commissions = ($5.25/SF net × 120,000 SF) × (20% × 150%) ÷ 5 years

Capital Requirements

Uses

Redevelopment expenses ......(1,261,793)	
Payment to seller ....(2,000,000)	
Other ....(None)	
Total uses	$3,261,793

Sources: Predevelopment

Component		Participant	Amount	Rate	Annual Cost	Percent of Total	$/SF
Debt:	(A)	Bank	2,600,000	9.5%	247,000	80%	21.67
	(B)						
	(C)						
Equity:	(A)	Developer	661,793			20%	5.51
	(B)						
	(C)						
		Total sources	3,261,793				27.18

Sources: Postdevelopment

Component		Participant	Amount	Rate	Annual Cost	Percent of Total	$/SF
Debt:	(A)	Same as above					
	(B)						
	(C)						
Equity:	(A)	Same as above					
	(B)						
	(C)						
		Total sources:	$_____		$_____		$____SF

Redevelopment Value Creation

Resale Pro Forma

Resale Price	$ 3,850,000	($ 32.08/SF based on a 10% capitalization rate)
Broker fee (3%)	(115,500)	
Roof/maintenance credit	(None)	
Capital required	(3,261,793)	
Net sale proceeds	$472,707	
Equity invested	$661,793	
Years to sale	2	
Equity return/year	35.7%	

Lease Pro Forma

(A) Cash Flow Basis

NOI	384,961	
Annual debt service	(288,540)	$2,600,000 loan, 20-year amortization, 9.5% interest
Cash flow =	96,421	
Equity invested	661,793	
	Annual return =	14.6%

*Loan amount = lesser of three underwriting tests:

Coverage: (NOI) 384,961/(coverage) 1.2/(constant) .1110 = 2,890,097

Percent of value: (NOI) 384,961/(cap rate) 10%/(LTV) 80% = 3,079,688

Acquisition cost: 3,261,793 × 80% = $2,609,434

Assumed loan = $2,600,000-

In this scenario, there is a positive return for the capital deployed, and the development costs are within a reasonable level when compared to similar property sales. Unfortunately, this is not always the case. But this example is only intended to provide a summary of the due diligence process and the key economic considerations.

If we were to change a few key variables in this example, the transaction would not make economic sense for a private developer in view of the considerable amount of effort required.

One such variable is the availability of a traditional financing source. In this case study, the developer benefits greatly from the availability of a traditional financing source to fund the majority of the redevelopment costs. If the contamination were more extensive, or if the property were completely vacant, it would be much more difficult to obtain financing from a traditional source such as a local bank. The developer would have to provide all the funds required, and unless the developer could refinance the funds out quickly, the return for the effort would be minimal.

In this case study, if traditional financing is not available, estimated project costs required to be funded by the developer total $3,261,793. The net income generated is estimated to be $384,961, which yields a return of 11.8 percent. This is a respectable return for more traditional real estate investments, but may not provide an adequate return for the effort and risk of a privately funded brownfield redevelopment.

Alternatively, a developer with approximately $3.2 million of available equity could theoretically leverage that equity into a much larger, more conservative investment property valued in the range of $12 million to $16 million without the risks of contamination, and with a comparable yield in the 11 percent range.

Another variable that could radically change the desirability of the project is timing. In this example, if it took an additional 12 months to complete the redevelopment due to regulatory issues, such as delays in obtaining an NFR letter from EPA, the additional year of loan interest and taxes would significantly increase the capitalized costs (by approximately $400,000) and reduce the projected return from 14.6 percent to 10.5 percent.

These potential risks are highlighted to demonstrate the importance of conducting comprehensive due diligence up front on these projects. For example, if a developer assumes that traditional financing will be available, or that a regulatory approval will be achieved more rapidly than is actually possible, the economics may not ultimately make sense relative to the risk. If the developer does not discover these potential difficulties until under a contractual arrangement to close on the purchase, there will be a risk of loss of earnest money, legal fees, and potential enforcement of the contract.

John B. Russell is a licensed commercial real estate broker in the state of Illinois, and has spent the last 15 years providing a range of commercial real estate services in the Chicago area at CB Commercial, Grubb & Ellis, The Philipsborn Company, and The Trammell Crow Company. His commercial real estate expertise is focused on industrial property financing, leasing, and investment sales.

John is also a founding partner of Greenfield Partners, Ltd. Greenfield was formed in 1996 to acquire commercial real estate with environmental contamination or liabilities. Greenfield Partners handles all aspects of the brownfield redevelopment process, including remediation or other mitigation of environmental liabilities, property redevelopment, and arranging a suitable reuse for the property. Greenfield has successfully redeveloped several brownfields projects in the Chicago area, including two Superfund sites.

John has a B.S. in finance from the University of Illinois, Champaign-Urbana.

6.3   Appraisal of Brownfields Properties

Noah Shlaes

Why Brownfields Present Special Issues

True brownfields transactions are not yet common. For the most part, they differ from traditional real estate transactions in that they involve EPA, the seller, the ultimate user, and local municipalities, all before the transaction takes place. Often the property is without value (to the original owner) before the transaction begins.

This is the type of transaction envisioned in this chapter. Of course, conventional deals involving contaminated property have gone on for years, and are routinely financed, bought, and sold, often without a significant discount for stigma. Appraisals for these properties are typical, but those for brownfields are not, and the issues discussed in this chapter remain.

What Is a Brownfield?

Though discussed at length earlier in the book, it is worth revisiting this question. A brownfield is a property with real or perceived contamination that affects value and causes a barrier to redevelopment. From an appraisal point of view, looking forward toward a determination of property value after remediation, a brownfield can be defined as an environmentally contaminated property that, with government approval, is put back into service *without a complete cleanup or removal of contaminants*. This is as opposed to a greenfield site, which was devoid of contamination to begin with, or a typical contaminated property, in which the contaminants are completely remediated before the property is reused.

This distinction is important in valuation, because it limits the possible uses of the property. A former metal plating plant that has been partially cleaned up may be suitable for heavy manufacturing, but not for agriculture, day care, or other exposure-intense applications. In any event, brownfields properties are often older urban industrial properties reused for processes that require safety equipment and monitoring.

Two Audiences

This chapter is geared toward two audiences: first, the brownfields practitioner, who hopes that the appraisal (and the rest of the transaction) goes smoothly and fits into the larger plan, and second, the appraiser looking to understand the special challenges of brownfields.

The Appraiser's Role

According to the Appraisal Institute, the appraiser can perform any or all of several functions when dealing with contaminated (not specifically brownfields) properties, including:

1. Observe obvious environmental hazards on the property or in close proximity

2. Recommend a Phase I ESA or suggest that an environmental professional's services are required

3. Provide an estimate of property value, disclosing to the client the appraiser's lack of knowledge of or experience with environmental hazards

4. If within the scope of the specific appraisal assignment, value the property in light of its disclosed or obvious hazards; if a full environmental investigation is conducted, the appraiser should be able to use the time and cost information from the investigation in the estimation of property value.[1]

Typically, appraisers are unwilling to take on the environmental issues facing property without seeking opinion from, and deferring to, an environmental expert. Other than observing and noting the obvious (leaking barrels, crumbling insulation, fill and vent pipes) and handing the issue off to an expert, the appraiser can and should stay clear of the specifics of environmental contamination.

In many cases, the appraiser will simply parrot the cost estimates given by the environmental expert. However, this can cause a problem if the estimates are not appropriate for the assignment. For properties where total cleanup (to the no further action letter state) is the highest and best use, this may be reasonable, but for a brownfield—where the most expensive cleanup may not be practical—this can seriously understate value.

What Is an Appraisal?

An appraisal, as defined by the Appraisal Standards Board, is:

> The act or process of estimating value; an estimate of value[2]

In practice, the definition of an appraisal is an unbiased opinion of value or the act of rendering the same. Though the definition does not mention bias, the requirements of the Appraisal Institute, the Appraisal Foundation, and most other governing bodies require that an appraiser certify the absence of bias in rendering an opinion. This differs from an *appraisal report*, defined as:

> The written or oral communication of an appraisal; the documentation transmitted to the client upon completion of an appraisal assignment.[3]

So, the appraisal is the opinion, whereas the report is the way the opinion is communicated to the reader.

Identified Real Estate

By *identified real estate* an appraiser means that not only the property (identified by legal description) but the property *rights* must be identified.

Now, let's look at the types of opinions and how they can be communicated (see Table 6-1). This will help you, when hiring an appraiser, to make sure that you are getting the product you expect.

Complete or Limited

This distinction refers to the appraisal, not the report; that is, it refers to how the appraiser arrived at the value conclusion.

In a *complete* appraisal, all appropriate methods and approaches are used, without restriction. The *limited* appraisal, by contrast, either leaves out part

Table 6-1. Summary of Appraisal and Report Types

	Reports		
	Self-contained	Summary	Restricted
Appraisals			
Complete	Strongest		
Limited			Weakest

of the analysis or makes an assumption about the property that is somehow different from its current state. For example, a limited appraisal may:

- Disregard existing environmental contamination
- Assume property taxes to be abated
- Fail to apply the cost approach (at the client's request)[4]
- Assume that future improvements (such as curb cuts and highway ramps) are already in place

Users of appraisals assume that appraisals cover the property as they see it before them. Therefore, the term *limited* is a red flag, an indication that the appraiser is addressing some other question that is either simpler or different from the property as it exists.

Limited appraisals are common in environmental cases, and often call out the value of property as though it were clean (a hypothetical assumption), or as though special financing or zoning is in place, when in fact it is not.

However, a limited appraisal is not a deficient appraisal. It simply answers a different question from what is asked in a complete appraisal.

Types of Reports

Self-contained reports have, in theory, every piece of analysis or information used in reaching the opinion of value. A reader should be able to tell simply by looking in the report how each step of the valuation was accomplished. This is the most expensive kind of report, and the most authoritative, taking the longest to produce. However, for a complicated transaction, a self-contained report may be the only way to determine that the appraiser fully understood the project.

Financial Institutions Reform, Recovery and Enforcement Act (FIRREA) reports are a subset of the self-contained class, and typically are required by lending institutions that are federally insured. Beyond what is always present in a self-contained report, a FIRREA report generally meets more stringent requirements for documenting verification of sales.

Summary reports do not contain the entire analysis, but lay out the basic assumptions and facts of the assignment. These are the "state of the market" for many transactions, and are less expensive than self-contained reports.

The problem with summary reports—the lack of disclosure of *how* or *why* the appraiser did what he or she did, and the specialized information on which he or she relied—shows itself when there is a disagreement. For an ordinary property, this is seldom an issue, but for a property facing environmental, legal, or financing concerns, or based on a specific development plan, this can mean trouble.

The appraisal user (a lender, borrower, or developer) may have to explain how and why the document addresses the specifics of the situation: "Where did these costs come from?" "Why is this building better than the competition?" "How long will these remediation costs go on?" Often, this information will be absent from a summary report.

Restricted reports, which used to be referred to as *two-page letters,* are the minimum standard for communicating a written opinion. The restriction comes from the fact that these reports are so brief as to be usable only by those who are completely familiar with the property and the appraisal assignment. The minimum contents of a restricted appraisal include identification of the following key factors:

- Real estate
- Property rights
- Definition of value
- Date of value
- Value conclusion
- Special assumptions
- Required certifications

Note that most of these factors are identifying in nature as opposed to persuasive. So, although the underlying complete or limited appraisal may be thorough and well supported, the restricted report may not get this across. Still, restricted reports serve a purpose; they are economical, because the appraiser spends no effort writing a narrative report, and they may be sufficient to document that an investor was duly diligent. But they do not answer many questions on their own, and are not intended to.

Preparation for Later

File Memoranda

Any time an appraiser makes an appraisal, he or she is required to prepare a file memorandum, which documents the opinion, the circumstances, and the basis for the opinion. In the case of a self-contained appraisal report, the report *is* the memorandum. A summary report should be backed up by a file containing facts and analyses that were not deemed necessary in the initial document.

The restricted report should have the largest file, since the report essentially contains only the conclusion. Basically, a restricted appraisal is a commitment by an appraiser that he or she can, on demand (at additional cost), prepare a self-contained report, and a certification as to what the conclusion of that report would be. Simply stated, it is an unbiased opinion of the value, as of a specific date, of identified real estate including a legal description, and the rights therein.

Unfortunately, few appraisers view their responsibility this way. When called upon to do justice to the restricted report by issuing the self-contained document, many appraisers are unable or unwilling to comply. In any case, the total cost of post facto preparation of a report is often much more than the price of having the report prepared at the time of the original opinion.

Communication Blocks

One of the principal barriers to a successful brownfield transaction is the communications gap between the appraiser and other parties to the transaction, especially the seller. Like other professionals, appraisers have their own nomenclature and unique, esoteric practices, which may be as unintelligible to the seller as a foreign culture is to a first-time traveler.

Making the process work requires understanding these differences—even if you never come to understand the language—and approaching the situation openly and carefully. Just as you wouldn't go into a lumber yard and just ask for some wood, you must know what to ask and what to tell an appraiser.

Understand the Appraiser

Left on their own, appraisers answer a single question: what is the market value of a property, on a given date, under normal circumstances? For most brownfields properties, this is the wrong question. The fact that under normal circumstances the property may well be worthless is what makes the property desirable to a brownfield developer in the first place. So, how do you ask the right question? Typically, it is a matter of changing one point of the assignment, as follows:

Date: is it prior to cleanup, or after, or at stabilized lease-up? Circumstances: property as is, or instead with contamination under control?

Market: is the potential buyer a hypothetical, off-the-street individual, or a particular investor with a specific use in mind?

This last, value to the market at large versus to a single investor, is the driving factor behind many transactions, not just brownfields transactions. It matters less what the rest of the world thinks of your property's value if a particular investor or group of investors values it differently. And, if your knowledge of rental markets, environmental issues, and user needs creates value for you, that can be enough to advance the deal and your prospects for success.

Highest and Best Use— What Is It?

Appraisers value property under a concept known as *highest and best use*. Simply put, this means that they make assumptions as to how a prudent, informed owner would use the property to achieve the best possible result.

For the appraiser to arrive at the number being sought—or one that can be lived with—he or she has to understand what the potential user plans to do with the property. For ordinary property with obvious or consistent uses, this can be very simple: "The highest and best use of the property is continued use as a multitenant office building."

But if you plan to alter a property's use or develop a new property, you have to inform the appraiser of your plans. In some circumstances, divulging a limited amount of information may be sufficient; in others, you may have to tell all. That means letting the appraiser know things that are critical, but that he or she might not otherwise ask about, such as "I have a tenant already," "Yes, there's a contamination on the property, but we have a plan in place to contain it," or "The city has already approved of this use." If you carefully consider this beforehand, you might even be able to suggest a specific use, for example, "This used to be a paint plant, but here's why I think it has a future as a movie theater complex. . . ." Of course, this requires an adequate knowledge of the market, demographics, economic climate, the risk appetites of potential lenders and investors, and other factors.

The Importance of Sales and Rent Data

Again, to share your view of the property's value, the appraiser needs to understand and share your vision for its use. That means seeing beyond the things that made the property a good deal for you, to how another buyer could or would use it.

The price you paid for the property may have nothing to do with its value. But do not expect appraisers to figure that out on their own—tell them! Appraisers are required by law to disclose the sale history of the property, and if you hide all or part of that history, appraisers will likely dig it up and interpret it, perhaps in a way that is damaging to your case.

Who Is the Client?

If a bank is lending on the property, it usually will hire the appraiser. That means that the appraiser cannot tell you his or her conclusions (or much of anything else) without permission from the lender. In this case, try to get the lender's approved appraiser list and assist in the lender's selection. Because you are almost certainly paying for the appraisal, you do have some leverage in the selection.

If it is your deal, be assertive. Nose around, and call the appraisal firm to ask about its experience. When an appraiser is hired, make all relevant information available to him or her, and clearly communicate your vision, your specific ideas, and the reasoning behind them.

Draft Reports—Get an Early Look

If, on the other hand, you hire the appraiser directly, arrange to have the report issued in draft, subject to your review, before the final version is produced. It will not cost much extra and could avoid problems arising from a choice of wording. Also, ask to see a sample appraisal done for brownfields purposes, which will typically contain the theoretical basis that will be applied to your property.

Do Not Just Talk. Listen!

You cannot tell if your idea has been understood if you do not listen. The questions and restatements that come from the appraiser should clue you in that you need to provide more information or more effectively communicate your vision. Never forget that the appraiser is a professional, often with a broad range of experience in property types. He or she will be looking for respect for his or her expertise and experience, and may offer insights that save—or even make—money for you.

The Environmental Engineer's Role

Traditionally, appraisers expect to piggyback on the environmental engineer's work. Some appraisers will incorporate cost estimates and overall descriptions of contamination directly from the executive summary. Few will question the necessity of cleanup, or indeed will understand the entirety of the report, as they are not in that business.

An environmental engineer, in the absence of additional information, is likely to inspect the property, identify all the contaminants visible or detectable within the scope of the study, measure the levels of each, and estimate the cost to completely remove them. This disregards a lot of what makes a brownfield deal work. It ignores, for example, these key questions:

Are these levels of contaminants atypical?

Is it necessary to completely remove them?

What is the cost of containing or encapsulating the contamination, or of other, lesser measures?

What remediation is typical for this situation?

By nature, engineers prefer to measure and report rather than recommend. The typical engineer's report contains a short executive summary listing levels and contaminants, backed up by test results, soil and laboratory reports, and photographs, which reinforces the notion that you should obtain a first draft of the report. This is not so that you can alter the report's conclusions, but rather so that you can communicate them to reflect any development plans. Since most users only read the executive summary, it should contain clearly stated, relevant conclusions and recommendations, especially because the appraiser may piggyback on these conclusions.

Robert Rafson, an environmental engineer with Greenfield Partners, a brownfields redevelopment company in Chicago, explains that "Environmental stigma is not well addressed by either the appraisal or site investigation processes. But the stigma clearly affects financing, contracts, and even continuing liability. Because of that, the appraised value may not reflect actual value—for example, if you cannot get financing, the property may effectively be worthless."

The Best Surprise Is No Surprise

The key to success in brownfields transactions is to divulge information, ask relevant and probing questions, and avoid surprises. Communicate the issues to the appraiser beforehand, and put them in context, along with your vision for the property. Examine samples of presentation, and finally, get a draft of the plan.

Using Appraisals

A discussion of a few basic rules of appraisals is in order, and will be helpful. Although anyone can hire an appraiser, it makes sense to do some basic fact-finding before doing so.

For example, if you do not have access to the property, the appraiser likely will not, either. In that situation, he or she might be able to develop information on the property using public records and other sources, but has no special powers and is not legally empowered to trespass on the property for the purpose of performing the appraisal.

In addition, it is important to consider what the appraisal will be used for. Many users, such as federally regulated lending institutions, cannot rely on an appraisal commissioned by anyone but themselves. In that event, your having already commissioned—and paid for—an appraisal may still require another one to be performed, at additional cost.

Obviously, an appraisal is commissioned to find out what the property is worth, but that figure can be very helpful (or unhelpful) in securing mortgage financing, as a reporting number for accounting purposes and/or paying estate tax, and in other situations in which it is useful to have an accurate value regardless of an intent to buy or sell. Other uses include establishing a base level for TIF subsidies and estimating damages in a lawsuit. Fighting a property tax bill is also a popular cause for hiring an appraiser.

Do You Need an Appraisal?

Appraisals may be focused on a question you do not care about: the value to a hypothetical buyer on a specific date. More likely, your buyer is not hypothetical but real, with a genuine use in mind for the property. Also, the proposed transaction date may not be certain—or imminent—because some transactions, brownfields or not, can take months or years to close. Moreover, you might be looking for other information such as local land costs for clean sites, market rents, or absorption, and these may not come from an appraisal. In any event, you may be able to get better data another way.

Sometimes, the last thing you want is an estimate of value. Corporations may have book values based on other measures (depreciated cost, for example) that would not be defensible in the face of an appraisal, or litigation concerns may dictate that it is better not to know. Keep this in mind.

When Do You NOT Want an Appraisal?

The case of Slippylube, a fictional maker of home "grease-it-yourself" kits, offers guidance and insight about when not to perform an appraisal.

Slippylube has its old plant on the books for $1.2 million, based on a $2.5 million acquisition in 1980 and significant depreciation plus capital infusion. The property has three leaking underground storage tanks (LUSTs) and asbestos that requires removal. Cleanup cost, including soil removal and treatment, is about $2.6 million.

However, Ike Slippy, CEO and owner of the company, is working with you on an alternative. This brownfield transaction involves Slippy's selling you the property for $1.4 million, after completing a $1.35 million partial cleanup and containment. Slippy retains liability and title, and puts up another clean property to secure your mortgage. You in turn lease the (sufficiently) clean property to It's Blue!, Inc., which makes spray paint in a rather limited range of hues. It's Blue! loves the building, and can live with the level of toxicity in the soil, provided it is documented. The village economic development office is sponsoring half of the cost of the cleanup, because It's Blue! will bring 45 wage-earning, taxpaying employees to the village.

If the appraisal shows the value (net of cleanup costs) as a negative number, Slippy may have to write off the property. In this case, the appraisal necessitates Ike's taking the hit all at once, and in doing so, he loses his freedom to use creative accounting. Actually, Ike wants to know what the property is worth to It's Blue!, but only for negotiating purposes. Absent an appraisal, he can look like a hero for his sale of the property, and treat the cleanup cost as a cost of doing business.

Hypothetical Appraisals

Items do not have to exist to be appraised. Hypothetical situations abound, including fee simple appraisals of leased property (for tax purposes), market value as if still standing appraisals (for fire insurance or lawsuits), and with/without contamination analyses for environmental work.

An appraisal as if clean for a property with known contamination is usually a base case for comparison, or a starting point for valuation considering that contamination. But in some cases, when estimating damage or whether it is worthwhile to clean a site, the as if clean estimate stands on its own.

Other Types of Value

In addition to market value, there are other types of value that may be relevant.

- *Business value* reflects those components in addition to real estate that can exist in an operating business—for example, machinery, operating accounts, patents, and goodwill.

- *Investment value* reflects value to a particular party (such as a single investor) as opposed to the market at large.

- *Financing value* is often a component of investment value, and reflects the effect of below-market or above-market financing. This is not usually real estate value, but may be relevant to the property or investor.

Appraiser Fear

Appraisers are afraid of contaminated property—and rightly so. Many have been sued for failing to identify or understand environmental issues, or have been the scapegoats for lenders who did not make adequate investigations of their own. As a result, appraisers shy away from environmental issues.

Where does this fear come from? The litigious nature of our society means that liability often falls to those who can afford it. Though an appraiser may declare his or her lack of expertise in an area, he or she can unknowingly assume liability, which can present ruinous costs eventually if not immediately. However, there are two ways around this: indemnification and fee. Most appraisers require that the client indemnify them (protect them from liability) when taking on these assignments, but, naturally, they invariably charge more.

In such a case, using a range of value might be the answer. Uncertainties in the market and property may dictate that the value can be expressed as a range, within a certain percentage.

If not, try asking a different question—the property may be so unusual that it defies valuation. But simpler questions, which may serve your needs and those of the lender, may work as well. For example: "Is the property worth more than $5 million postcleanup?"

Where to Find an Appraiser

Apart from the telephone directory, there are many places to find an appraiser:

Approved lists—lenders maintain lists of appraisers that are acceptable to them. This may end up being your most important limiting factor.

Professional designations—the Member of Appraisal Institute (MAI) designation has long been regarded as the premier appraisal designation for commercial property. The institute has a directory, an online presence (http://www.appraisalinstitute.org), and several education and certification programs.

While not strictly an appraisal organization, the Counselors of Real Estate (CRE) is a tightly knit organization of real estate advisors, many of whom are also appraisers. This is a great place to start for complex property issues and non-appraisal advice. CRE can be reached at http://www.cre.org.

Identifying a Qualified Appraiser

The competency provision of the appraiser's Uniform Standards of Professional Appraisal Practice (USPAP) dictates that the appraiser must declare his or her competency and reveal any special steps or procedures needed to make him or her competent for the assignment at hand. For example, if it is the appraiser's first assignment involving environmental contamination, he or she must state the sources he or she turned to for methodology and data. These are all questions worth asking at the time of engagement.

Appraisal Regulation

Appraisers are covered by several sets of regulations. *The Financial Institutions Reform, Recovery and Enforcement Act* (FIRREA) was passed in 1990 as part of the savings and loan bailout. Among other things, it imposed strict requirements on appraisals and appraisers.

The Uniform Standards of Professional Appraisal Practice is a broadly endorsed set of standards and procedures promulgated by the Appraisal Foundation, and serves as the basis for most appraisal regulation.

State certification is required in all 50 states, and current certification in the state in question is a must. Few appraisers are certified in all states, though some have impressive collections of certificates. Reciprocity and changes in appraisal regulation may change this, but if there are any questions, ask.

The Engagement Letter

Appraisers neither want to be biased nor to appear so. For this reason, you need to tell them in writing which specific assumptions should be factored into the appraisal. This is accomplished by the engagement letter—an agreement between appraiser and customer that precisely dictates the assignment and any special circumstances or factors to be considered in the opinion. If you are unsure how to word the engagement letter to serve your purposes, or precisely what you wish to have appraised, ask the appraiser for help.

Lenders who hire appraisers often have standard engagement letters of their own, which reflect their particular concerns about liability, standard certifications and language, and representations the appraiser needs to make. However, the small section describing the assignment may need some editing for an unusual transaction.

If the assignment calls for an as if clean valuation, the appraiser should lay out any special assumptions in the engagement letter. This establishes the parties' understanding of the assignment before the work even begins.

Indemnification and the other issues shown here also belong in this letter.

Theoretical Issues in Appraisal

A 1994 article by Albert R. Wilson lays out a method for estimating impaired value.[5] Wilson describes the value of an impaired property as follows:

$$I = U - C_{NCP} - C_R - C_F - M_U$$

where I = impaired value
           U = unimpaired value
           C_{NCP} = cost to implement the National Contingency Plan (NCP)-defined remediation plan
           C_R = cost of restrictions on use and/or environmental liability prevention
           C_F = impaired financing cost
           M_U = intangible market factors

This approach is best suited to assignments calling for a fee simple value of the property. Speculative developments, litigation, and establishing damages are all areas where this might be useful. A discussion of these components is in order.

Unimpaired Value

Unimpaired Value reflects the value of the property in the absence of environmental contamination issues. This "before" situation is somewhat hypothetical, and is best supported with information on similar properties that are in the same or similar markets, have the same highest and best use, and/or are of similar utility.

This is a familiar problem to most appraisers. For a fungible property, such as a single-family residence or small industrial building in a neighborhood surrounded by residences, the appraiser can simply examine sales or rentals of nearly identical, clean properties and prepare the analysis. For a brownfield transaction, however, determining unimpaired value may be difficult.

Cost to Implement Remediation

The most reliable source of information on the cost of remediation is from an engineering study of the particular property; however, issues can arise in this arena as well. It is important to ask—and answer—key questions about the program, including:

- Whether it is an approved program
- Range of costs
- Costs over time (a cash flow consideration)
- Timing (when the cleanup can take place, and how it will affect plans/ uses for the property)
- Resulting level of cleanup
- Suitability for proposed use

Understand the Environmental Report

The appraiser will often rely on the environmental report for degree of contamination, cost and method of cleanup, and a host of other issues. But it is not sufficient to simply rely on the report without understanding it. The appraiser must understand the costs laid out, and the contamination and remediation described.

Just as in an appraisal report, the scope, qualifications, and methods of environmental work must be adequate. In appraisal, an MAI designation or state certification is not a guarantee of appropriate skills for the assignment at hand, nor is an engineering license. Ultimately, if the appraiser knowingly relies on environmental engineering work that is later determined to have been flawed, he or she may be liable for the results and subject to censure or dismissal. Under the competency provision of USPAP, if the appraiser is unable to tell the difference, the same may apply.

The appraiser is seldom asked to determine whether the cleanup is adequate for a given purpose. In this case, the environmental report should contain representations about this use. Another common mistake is to rely on cleanup programs that are overkill for a certain use.

Here is a real-life example: As part of a due diligence assignment, the author recently reviewed a property report that stated that levels in excess of EPA guidelines existed on the property and described the nature of the contaminants. By itself, the report presented a scary picture. However, it left out that the contaminants were limited to a tiny, 1/2-acre piece of a 900-acre parcel; that the contamination was typical of waste found at gas stations; and that this was probably no issue in any event, because the property would be developed as a truck stop. The report also failed to mention that the site had no neighbors. By making this information apparent, we were able to turn a terrifying property file into an attractive bid package.

Cost of Restrictions on Use

Contamination can limit a property's use permanently, preventing it from becoming the site of a school, day care center, or retail store regardless of cleanup. If this changes the highest and best use of the property, then the cost can be significant.

Cost of Environmental Liability Prevention

The remediation program you put in place might involve a continued commitment to protect the environment. Pumping programs, periodic inspections and recertifications, and special air and waste handling systems may all add substantial costs when compared to an ordinary transaction. These are major considerations that can impact the user financially far into the future.

Impaired Financing Cost

Simply put, financing for a brownfield deal costs more than for other properties. Lenders generally do not understand brownfields, cannot sell them on the secondary market, and do not receive the same security as under a regular loan.

The security issue is crucial, given that in a clean transaction the lender's ultimate security is the right of foreclosure. For a contaminated or formerly contaminated property, however, the lender may not want that right, because accepting it will remove the firewall protecting the lender from liability. In essence, lenders are protected only as long as they do not take ownership of the property; after foreclosure, the protection disappears. At the same time, without the right or desire to foreclose, the lender undertakes more risk, which can mean higher interest rates, higher equity levels, and a shorter loan term. This figures into the appraisal in one of two ways; as a lump sum subtraction for financing cost, or in the rate of return.

Lump Sum Financing Adjustment

This method has advantages in that it segregates the effect into a single number. It makes the appraisal easier to understand in that it does not cloud the remainder of the valuation. Further, it permits the appraiser to examine the financing issue in light of alternative investments, instead of blending it in with property yields.

Capitalization and Discount Rate Selection

Others choose to change property yields to reflect environmental risk. This can be appropriate, provided it is supported by data on this differential. This is a new area of research, and is only now becoming feasible.

Intangible Market Factors

Stigma

Stigma is derived from perception, not from actual threat, but reflects the concerns of rental and purchase markets about the uncertain effects of contamination. These concerns can include uncertainties or fears about:

- The direct health risks and dangers posed by toxins that may be present
- The adequacy of government standards for acceptable levels of a given toxin
- The success or extent of cleanup
- Unfamiliarity with remediation procedures
- Other fears stemming from rumors

In some cases, stigma can appear in the absence of documented environmental threat. For example, consumer concern about electromagnetic radiation from power transmission lines has already affected some markets, despite the absence of a scientifically acknowledged risk. Unfortunately, there is no clear measure of the effect of stigma. But a brownfield practitioner can use comparable transactions in other markets to give the appraiser added insight to estimate the value and discount that are appropriate.

Another Approach: "Take the Bull by the Horns"

The concept of impairment was developed mainly to estimate damages due to contamination. Therefore, if the appraiser is attempting to determine the value of a completed (postcleanup) property, this process can be greatly simplified.

First, consider highest and best use. Because it is not your intention to clean the property completely, the highest and best use—as if clean—is moot. If the property, after remediation, is only suitable for heavy industrial uses, there is no need to start from all-purpose property and take discounts. You can compare the property directly to heavy industrial property in terms of rent and sale prices, which primarily affects the sales comparison approach.

Using the income approach, again you can avoid analyzing hypothetical alternatives and move directly to the planned development. Your income approach will model the actual lease, the income, and the expenses of this project, not those of a property in a perfect, clean world. This is a much simpler task, and requires only an understanding of the deal at hand and the special yield requirements of the market for this type of property.

As to the cost approach, it is again applied to the same highest and best use. If your use is heavy industrial, then this is what you will cost out. If there is ongoing obsolescence arising from monitoring and control costs, then adjust for it; otherwise, this is a conventional deal.

Environmental Impact on Value—Is There Any?

There may be little or no effect on value from environmental contamination, depending on what is determined to be the highest and best use of the property.

For example, consider a downtown parking lot in a Midwestern city, contaminated by fill taken from an adjacent coal gasification plant.

Though the plant had been closed for nearly a century, the fill used to level this site remained, and contained carcinogens at high levels. The state EPA had issued a no further action letter stating that if the site were left in its current use (surface parking) or another use that did not disturb the surface, no further action was needed.

Because this was a downtown site, it appeared that the high cost of cleanup had a huge effect on value. But further investigation revealed that five sewer lines, the largest of which was more than 10 feet in diameter, intersected beneath the site. All were old, fragile, and close to the surface. A construction engineer estimated the costs of relocating the lines or building over them, both of which were prohibitive. Ultimately, the highest and best use of the site, regardless of contamination, was not to build on it. It became a parking lot, and the contamination had no impact on that use.

Where to Get Comparable Rentals and Sales

For a precleanup valuation, the appraiser needs comparable sales and, if possible, rentals. Rentals are unlikely, since most are initiated after cleanup is complete. But sales do exist. Remember that for difficult or unusual property, the reach for sales may be longer and deeper. In this instance, comparable properties may be several states away, and may be older than those the appraiser uses for clean property. When faced with such a property, use data searches and online tools to gather this information. Understand the whole story—the nature of contamination/cleanup—and remember that the unit of comparison may not be the price per square foot or building, but instead the cap rate or other measures. Of course, for a postcleanup valuation, the selection of sales is of clean property, or ideally, of cleaned property.

This Is a Specialty Field

Unlike conventional appraisal, this is not a geographically defined area of specialization. Appraisers who have taken the time to master the regulatory, financial, engineering, and technical aspects can work in a territory that is regional or national.

Income Approach

Tenant credit is central—the willingness and ability of a particular tenant to stick by the lease is more important because brownfields tend to be leased or converted to special uses specific to a party. There may be large differences between contract rent to this user (who sees special business value in the property) and to the market at large, which may see the property as a white elephant.

Consider key issues—is the lease long enough? Is it executed? (If it awaits execution at the closing of a deal, consider a draft report or special condition.) How good is the tenant's credit? A credit report from one or more credit reporting agencies may be in order, as well as a review of security deposits and guarantees. Is the tenant installing equipment in a permanent manner? All of these can affect the risk of the deal.

Cleanup costs require special handling. They may occur over time, and may overlap the operation of the property. Again, review the environmental report and consider the timing and possible variation of these costs. Are you assuming appropriate growth? The engineer may not have.

Financing

Remember to figure in the cost of financing—this might take the effect of a premium as opposed to other property. Equity requirements for brownfields deals can be high, affecting required yields. Surveys are beginning to emerge that support these selections, and the band of investment method remains appropriate.

Stigma

There may remain, even after cleanup, a perception that the property is tainted or less desirable. This comes from the cost of changes in financing, differences in rent, and time to market property. However, measures of stigma are unclear, and data are sparse.

Conclusion

In just a few years, brownfields transactions have become more prevalent and better understood by all parties to the transaction. In many urban areas, potential new tenants and a resurgence of industrial and commercial users have made markets take another look at properties that were considered useless until recently.

Therefore, the appraisal of these properties will only get easier, as comparable sales and rental information is more commonly available and lenders move toward clear-cut definitions of requirements for loans on brownfields. As this continues, appraisals of brownfields will become cheaper, faster, and better.

References and Notes

1. Colangelo, Robert V. and Miller, Ronald D., *Environmental Site Assessments and Their Impact on Property Value: The Appraiser's Role.* Chicago: Appraisal Institute. (1995)

2. Appraisal Foundation, *Uniform Standards of Professional Appraisal Practice* (1999 Edition). Washington, DC: Appraisal Foundation. (1999)

3. *The Dictionary of Real Estate Appraisal.* Chicago: Appraisal Institute. (1993)

4. If the appraiser justifies a decision that the cost approach (or other approaches) is not appropriate for the assignment, this may not necessarily classify the appraisal as limited. If, instead, the appraiser omits the approach solely at the request of the client, this limits the appraisal.

5. Wilson, Albert R., "The Environmental Opinion: Basis for an Impaired Value Opinion." *The Appraisal Journal.* (July 1994)

Noah Shlaes is a manager of real estate advisory services for Arthur Andersen and has 15 years of background in all property types, including downtown and suburban office buildings from 30,000 to more than 2 million square feet, hotels, apartment buildings, regional and local shopping centers, downtown vertical malls, industrial buildings, and vacant land.

Mr. Shlaes has provided guidance in negotiations of ground and office leases, including the disposal of more than 50 leases in a bankrupt chain. He recently prepared and taught a course in basic real estate for the Bulgarian Privatization Agency. He has managed due diligence teams in the disposition of portfolios of major life insurance companies and the Resolution Trust Corporation. Other due diligence assignments have included analyzing the real estate of a major bankruptcy, preparing marketing materials for several property types, underwriting loans of all sizes, and reviewing portfolios of investment real estate for life insurance companies and others.

His extensive background and work in real estate–related computer applications, including a pioneering microcomputer-based cash flow forecasting package for multitenant properties, has made Mr. Shlaes an industry expert. He has developed in-house systems for the valuation and loan pricing systems for a major real estate lender.

Mr. Shlaes has testified as an expert before the Illinois Property Tax Appeal Board and the Federal Bankruptcy Court. He has also developed precedent-setting methodology for the appraisal of federally subsidized low- and moderate-income housing, based on an after-tax analysis of the costs and benefits of ownership.

Mr. Shlaes holds a bachelor of administration in economics degree from the University of Michigan. He lectures on real estate valuation at DePaul University's Kellstadt Graduate School of Business. The

Counselors of Real Estate have awarded Mr. Shlaes the CRE designation. He is also a member of Lambda Alpha.

6.4   Technical Investigation

Harold J. Rafson

Introduction

This section discusses Phase II. The environmental site assessment work is discussed in detail in the following sections, which discuss planning of site assessment sampling and scope of work (Sec. 6.5), test methods (Sec. 6.6), and a geological section (Sec. 6.7) dealing with subsurface conditions. Finally, there is a discussion of the interpretation of results, decision making, and the next steps to be taken.

If the Phase I site assessment indicates the likelihood of contamination, it is necessary to investigate further and to perform a Phase II site assessment.

Phase II

The American Society for Testing and Materials (ASTM) in Standard E 1903-97 has provided an excellent guide for Phase II site assessments, and the various aspects to be considered. However, it cannot be precise, because each site and environmental condition is different. Therefore, the environmental professional is required to complete the assessment, but this standard is useful to the developer in that it highlights the various points to be covered.

The scope of work of Phase II as detailed in E 1903-97 includes the following:

- Review of existing information
- Potential distribution of contaminants
- Sampling
- Health and safety
- Chemical testing
- Quality assurance/quality control techniques
- Field screening and field analytical techniques
- Environmental media sampling
- Sample handling

- Evaluation of data
- Interpretation of results
- Report preparation

For biographical information on Harold Rafson, see Sec. 1.1.

6.5   Environmental Due Diligence

Keith R. Fetzner, Lawrence Fieber, and Frank H. Jarke

Introduction

Environmental issues are inherently associated with brownfields properties. These issues not only affect a property physically; they also affect its value, the cost and logistics of redevelopment, and the liability of current or future owners and lenders that are involved. Therefore, it is necessary to identify and understand the environmental issues as much as possible when proceeding in the evaluation and development of brownfields properties. This requires environmental investigations that are designed to identify potential or known environmental conditions followed by investigation(s) as necessary to assess the impact of the conditions on the property.

Environmental assessments of properties are typically performed in the following steps:

- *Phase I Environmental Site Assessment*—an evaluation of the current and historical ownership and use of a property as well as surrounding properties for known or potential environmental issues. This investigation step typically does not involve invasive sampling or analysis.

- *Phase II Environmental Site Assessment*—typically involves physical testing of air, soil, or water at or around the property potentially affected by the environmental concerns identified during Phase I. It could also include additional research of records or other information sources that are beyond the scope of a Phase I assessment. The scope of this step will vary for each situation depending on the conditions identified and the goals of the parties involved. It may also require multiple testing steps to complete the necessary investigation goals.

- *Phase III Environmental Site Assessment*—may involve remediation of identified environmental impact or an assessment and reduction of the risk of exposure to environmentally impacted material at a property.

Typically, a prospective purchaser or owner of property who wants to perform these assessments will hire an experienced and qualified environmental engineering or consulting firm. The consulting firms should have the expertise, experience, qualifications, and insurance applicable to the tasks you need them to perform and preferably through all phases. Proposals should be obtained that are clear and specific as to scope of service and that have reasonable and applicable terms and conditions. The consulting firms commonly provide the personnel and equipment necessary but will subcontract the drilling and analytical services. Some engineering firms may provide the drilling services because they have existing geotechnical services. However, environmental and geotechnical drilling methods are different and it is critical that the firm understand the investigation needs and that applicable environmental investigating techniques be used. The analytical services are mostly subcontracted because the equipment is specialized and expensive and requires trained and qualified personnel, making the cost of operations higher than it is worth for a consulting firm. Subcontracted laboratories should be qualified, experienced, and licensed where applicable. This will be discussed later in this section.

Often the Phase I assessment is bid as a lump sum, unless the project is larger or more complex, in which case it may be bid as time and expense for a defined fee that is not to be exceeded. Typical Phase I costs are $1500 to $2500 or higher depending on the property size, location, and condition; the number of structures; and the scope of the assessment requested. The Phase I assessment will typically take two to four weeks to complete. The cost is primarily for personnel, with some expenses for information database fees, government information fees, and expendables such as film and vehicle mileage.

Phase II investigations are typically performed on a time and expense basis due to the higher price and the uncertainty in the scope of work. When it comes to Phase II investigations, a scope of work will be tailored to the objectives of the client unless there is a regulated environmental issue on the property that may require a scope of work geared to the governing law, regulation, or regulatory authority. The amount of money spent will be proportional to the certainty of the investigation's conclusions. In other words, the more you spend, the more solid your conclusions will be. This is why it is best to perform a Phase II investigation in steps, so that each step is performed because it is necessary only to meet the objectives. Each step will typically take three to four weeks, depending on laboratory analytical turnaround time (one to two weeks, depending on the laboratory) and the scope of the work. Costs can range from $5,000 to $25,000 and up, depending on the environmental issues being investigated and the objectives of the investigation.

One-third to half of the Phase II cost will be for drilling, equipment (i.e., monitoring wells, screening tools, and sampling equipment rental), and analytical costs. It is critical to have a proposal that clearly states the investigation objective, how it will be completed, and to whom and how the information will be reported. Phase II investigations will commonly generate such waste as drilling cuttings, water from wells, and unused samples that will be owned by the generator (the client) and not the consultant. These wastes are typically stored appropriately (i.e., drums, covered stockpiles) but may require disposal at an additional cost.

Phase III, or remediation, when necessary, is tailored to the type of environmental issue and the objective of the remediation and is based on the Phase II information. Remediation costs can vary greatly, from a few thousand dollars to millions. This step can be as simple as paving an impacted area to prevent exposure or as complex as groundwater and soil remediation systems that will operate for decades. Typically, the need for expensive remediation will outweigh the benefits or options of developing a brownfield property. The trick is to perform Phase II investigations economically and in a timely manner to determine whether remediation is necessary and what the cost may be. Be prepared to walk away from the property if the economic redevelopment cost threshold is exceeded. The remediation methods vary widely, are typically designed by the engineering or consulting firm, and are subcontracted to contractors experienced with the technology. This step will commonly involve a regulatory agency to ensure that remediation is properly completed and ultimately will obtain regulatory approval after completion. The following sections provide more explanation of the typical Phase I and Phase II investigation steps.

Phase I Environmental Site Assessments

The primary use of a Phase I assessment is to allow parties to a real estate transaction to cost-effectively meet their due diligence or all appropriate inquiry obligations. The due diligence and all appropriate inquiry objectives in a Phase I assessment developed from the innocent purchaser defense introduced in 1986 in the Comprehensive Environmental Response, Compensation, and Recovery Act (CERCLA, also known as Superfund).

The premise of the Phase I Environmental Site Assessment is that reviewing certain reasonably available information concerning the past or current use and activities on the property may indicate the presence of potential recognized environmental conditions (RECs). If a recognized environmental condition is identified, then due diligence may require that further inquiry be made to adequately quantify the impact on the property. The Phase I assessment will also generate information that is useful when evaluating redevelopment options and the scope of Phase II or Phase III operations that may also be needed.

In the initial days of due diligence, the definition of an adequate Phase I assessment depended on the requirements of the person who requested it and what the consultant thought was necessary for the client. Over the years, larger institutional lenders and various trade groups developed fairly comprehensive guidelines for Phase I. In late 1989, a number of real estate groups, lenders, owners, and attorneys initiated a project to standardize the due diligence process. In 1990, the American Society for Testing and Materials (ASTM) formed a subcommittee on environmental assessments in commercial real estate transactions. After several years of development, ASTM published its first environmental due diligence standards in 1993.

Today, many Phase I assessments are performed in accordance with ASTM Standard E1527-97, "Standard Practice for Environmental Site Assessments: Phase I Environmental Site Assessment Process." Certain states also have their own due diligence–type laws that should also be considered when performing this investigation step.

A Phase I Environmental Site Assessment typically consists of the following components:

- A records review, typically including examination of title records and environmental business records, if available, to determine the former ownership and use of the property. This also includes review of a topographic map, historical aerial photographs, and fire insurance maps, if available, to investigate past property conditions. Reasonably available building, assessors', and fire department records, and/or department of environment records for permits, citations, and reports connected to the property, may also be examined. Specific government lists regarding environmental activities for the property and local area properties may be reviewed.

- A property inspection by a trained professional to investigate the current use of the property and identify the presence of hazardous substances, wastes, underground storage tanks, or other areas of environmental concern. This will also include an inspection of adjoining properties from reasonably available public viewpoints to identify the current use of these properties. The inspection may also include surveys of building materials at the property suspected of containing asbestos, lead, or other materials that may be potential environmental issues.

- Interviews with current and possibly past property owners and operators.

- Preparation of a report that specifically lists recognized environmental condition(s), if any, identified during the Phase I Environmental Site Assessment. Recognized environmental conditions identified during the Phase I Environmental Site Assessment are used to perform the initial steps in evaluating the redevelopment potential of a brownfield property. Additional investigation of the recognized environmental conditions conducted in a Phase II assessment is discussed next.

Phase II Environmental Site Assessments

A Phase II Environmental Site Assessment is performed if information discovered in the Phase I assessment suggests that further investigation is needed to evaluate RECs. When conducting the Phase II investigation it is important to understand the following:

- Compounds of concern (i.e., petroleums, solvents, pesticides, or metals)
- The nature or source of the RECs
- The physical and geological setting of the property
- Pathways of exposure to compounds of concern
- The existing or potential regulatory framework
- Redevelopment plans that relate to or will be affected by RECs

Identifying the compounds of concern is the first step in designing a Phase II investigation. More often than not an REC will include materials such as gasoline, paint, or solvent mixtures, which contain numerous different chemicals and can vary depending on the manufacturer. The key or primary chemicals associated with a material are commonly referred to as *indicator contaminants* or *indicator compounds*. The list of indicator compounds is commonly preset by a regulatory agency. However, these lists are not always comprehensive enough for all situations, which is why you should have as much knowledge as possible about potential compounds of concern. Generally, there are the following basic compounds of concern and the desired analysis:

- *Volatile organic compounds* (VOCs)—petroleum (i.e., gasoline, diesel fuels, lubricating oils), solvents, or materials that contain solvents (i.e., paints)
- *Semivolatile organic compounds* (SVOCs)—when investigating petroleum concerns, this will generally involve middle to heavy distillates of petroleum such as diesel fuel, heating oil, and lubricating oil, and not

the lighter distillates of petroleum such as gasoline. This will include just about any material that is not volatile and can include pesticides and polychlorinated biphenyls (PCBs).

- *Pesticides and PCBs*—the pesticides are self-explanatory and will include a large list of compounds. PCBs are commonly associated with electrical transformer cooling oils in older electrical transformers, possibly hydraulic oils, and many other oils that commonly reached high temperatures. PCBs in oil mixtures may suggest the need for analysis for VOCs and SVOCs.

- *Metals*—metals are also self-explanatory. However, metal analyses are requested in three general groups that include Resource Conservation and Recovery Act (RCRA), Total Priority Pollutant (TPP), and Target Analyte List (TAL) metals. The groups differ in the number and type of metal compounds included. There are 8 RCRA metals, TPP metals include 13 metals, and TAL metals include 23 metals. Be aware that not all the RCRA metals are included in the TPP metals. The TAL metals include all the RCRA and TPP metals. There are two typical ways to evaluate metals: total metals and leachable metals. Total metal analyses determine the total content of metals in soil. Leachable metals analyses identify the amounts of metals that will leach out of a soil sample. Leachable metal concentrations are not typically as high as the total metal concentration because different soils will exhibit different abilities to retain metals within their matrices.

These are more common analytical tests. Laboratories can typically provide a wide range of analytical services that can be designed for your particular investigation.

Knowing the nature of the REC is the next step in designing the scope of a Phase II investigation. Typical sources of RECs include, but are not limited to, surface spills, underground storage tanks (USTs), underground piping, aboveground and underground disposal areas, and fill material of unknown origin. Generally, surface spill source investigation will begin with shallower sampling over the area, whereas deeper sources such as USTs or underground disposal areas will involve deeper sampling.

It is also important to know the characteristics of the compounds that are being investigated. A good example is the comparison of gasoline and chlorinated solvents such as tetrachloroethene, a common dry cleaning and degreasing fluid. Gasoline is less dense than water and is commonly detected at or near the surface of groundwater, whereas tetrachloroethene is denser than water and will migrate down through groundwater until impeded by less permeable material such as clay. This difference in density will influence where contamination is present and how it migrates.

Additionally, tetrachloroethene will break down into other chlorinated solvent compounds that also need to be tested. There are numerous other characteristics such as solubility, vapor pressure, and degradation rates that can influence not only where to investigate for the contamination but also how to sample the media and what to test for.

It is important to understand the potential regulatory framework that might be associated with Phase II investigations. A common example is performing investigation of USTs. The Phase II investigation may identify a release that will need to be reported and addressed by the owner/operator of the USTs. Additionally, more and more states are developing and operating programs tailored to the redevelopment of brownfields properties and will provide both the guidance necessary in the investigation and remediation and regulatory approval after completion. The regulatory agencies will have regulations or guidelines governing the investigation, remediation, and reporting methodologies that will need to be followed to gain their approval. Furthermore, the regulatory agencies will provide information on the acceptable levels for contaminants. Over the last few years these cleanup objectives have become flexible and are based on risk of exposure for the use and configuration of a property.

The scope of Phase II investigations and the interpretation of the results should incorporate the anticipated redevelopment plans if they are known. A good example is that an investigation may identify contaminant concentrations that are below those required for certain cleanup objectives and will not require remediation. However, if redevelopment plans require the excavation of the area, the excavated material may be considered a generated waste. The classification, handling, and disposal requirements will depend on the concentration and type of contaminants in the soil and will typically add costs to redevelopment.

Basic Steps to Consider in Designing the Phase II Site Investigation

Consider the Existing Information

Before you begin to spend any money performing work beyond a Phase I Environmental Site Assessment, review the information you have and consider other information that may be available at little or no cost. Examples of such information include geologic maps and publications, topographic maps (both current and historic), aerial photographs (both current and historic), reports of prior investigations, business records, interviews with knowledgeable persons, public records and listings, and site inspections. It may be hard to believe, but one of the most common mistakes environmental consultants make is to proceed with sampling and other expensive work before considering what information may be available cheap or free.

Geologic maps and publications are usually available from state or county geologic surveys or departments. County-specific geologic information may be the most detailed available. Staff geologists at the geologic survey may be able to provide specific information concerning your property area and are usually very willing to assist in identifying information, maps, and publications relevant to your investigation. Topographic maps may provide information useful in predicting specific information concerning your property area, and may enable you to predict the direction of shallow groundwater flow and the locations of former landfills, clay pits, or quarries. Historic topographic maps may provide information concerning former improvements on properties and changes in topographic expression created by property developments.

Aerial photographs help clarify current and historic property improvements that may be relevant to the site investigation. Features that are visible on aerial photographs include buildings, vegetation, variations in soil moisture, roads, surface drainage features, pavement and other surficial improvements, and disturbed or piled soil. Aerial photographs from various years may be used to construct sequences of development including former surface drainage and utility locations that may be preferential migratory pathways for environmental contamination. Aerial photographs may also identify former outside storage areas and building locations that represent areas of environmental concern.

Reports of prior investigations may be the only existing site-specific information you will find. Usually, the most useful prior reports are environmental, geotechnical, and groundwater investigation reports. Environmental reports may describe the former use of the property and previously identified areas of environmental concern. Phase II Environmental Site Assessment reports commonly include geologic and groundwater information, as well as specific information concerning the presence of chemical constituents in soil and water. In circumstances where industrial facilities operate or previously operated on a property, environmental compliance investigations may reveal chemical usage, storage, and disposal practices. Geotechnical reports may have been prepared before property development or before numerous preceding developments on the property. Geotechnical reports are valuable because they commonly contain information concerning soil and groundwater conditions that may be useful for identifying potential migratory pathways. In circumstances where on-site groundwater wells exist, logs of subsurface conditions may be available at the property or in department of public health or state or county water survey department files.

Other prior investigation reports may be available for municipal, county, and state-sanctioned infrastructure projects involving sewer, road, and public building improvements. Depending upon the nature and magnitude of such infrastructure projects, the reports may be limited to geotechnical information, but in other circumstances they may also include environmental impact statements.

Business records may provide information concerning chemical use, waste disposal, and construction plans. The chemical use and waste disposal information may be useful in limiting the scope of the site investigation to specific chemicals and wastes, eliminating the need for investigations of numerous chemicals that are unlikely to be present. Construction plans may provide explicit detail concerning on-site utilities, foundations, and prior improvements. All of these physical improvements are relevant when identifying potential migration pathways.

Do not underestimate the value of interviewing current and prior owners and occupants of the property as well as government agency representatives, neighbors, and community members. Former owners and occupants of the property may provide the most accurate information concerning the current and former property use. Government agency representatives and neighbors or community members may provide valuable and accurate information from their perspectives that differs from that of the owners and occupants. Remember, owners and occupants have an obvious conflict of interest in disclosing all information to you. Along the same lines, information in public records such as building, planning, zoning, fire, assessor's, treasurer's, and recorder's offices may provide data useful in the site investigation planning process.

Finally, one of the most important steps necessary in designing a site investigation is the visual site inspection. The purpose of the visual site inspection is to identify conditions that may influence site investigation methods or sampling locations, or that may change interpretation of other or existing information already reviewed. Usually the visual site inspection is performed by the person designing the site investigation; however, in circumstances where another professional has seen the property, the site inspection may become redundant, especially if the other professional is involved in designing the site investigation.

Gather and Review Information that Is Available Cheap or Free

Once you have collected the information described in the preceding text, you should review it all together and identify information that corroborates other information. It is important that the same person review the information or that a single person review notes concerning all information reviewed in order to build an overall understanding of the physical characteristics of the site. This review may result in a complete change in the understanding of the property or objectives of the investigation. It is at this point that the areas of environmental concern become apparent.

Considerations When Setting the Objectives of the Investigation

Differing interests can make setting objectives for environmental investigations difficult. The competing interests common to most environmental investigations on brownfields properties include:

- Business interests
- Technical interests
- Environmental regulatory interests

In most brownfields properties, the business interests determine the investigation objectives. The business objectives may differ if you are buying a property versus selling a property. Budget constraints, timing constraints, and future use plans also affect objectives. Objectives may also vary depending on whether the future use involves developing the property or redeveloping it, and on whether the property will be used for residential or commercial purposes. Both development and redevelopment may be influenced by current property uses. In all cases, businesses wish to make money and limit future liability.

Technical interests are those of the environmental consultants retained by the businesses wishing to purchase brownfields properties. Technical professionals try to comply with regulations by gathering enough of the right information. When gathering their information they need to be ethical, use appropriate methods, make money, limit future liability, keep the client happy, and keep their jobs. Balancing these differing interests is an ethical challenge, to say the least. Brownfields purchasers should be aware that environmental consultants will probably understand environmental regulators better than they understand brownfields purchasers. This technical-business gap is particularly critical when business owners are unwilling or unable to understand enough of the technical issues to appropriately lead the environmental consultant in negotiations with the technical representatives of the environmental regulatory agency governing the brownfield site.

It is true that most brownfields negotiations are conducted between the environmental consultant and environmental regulator. The most successful negotiators of risk-based solutions to brownfields properties are those who manage or conduct the regulatory negotiations themselves, rather than leaving it to technical professionals who are untrained and relatively inexperienced in successfully negotiating their views.

The final party with its own set of differing interests is the environmental regulator. Environmental regulators are guided by four primary interests:

- Complying with regulations
- Protecting the environment
- Protecting humans
- Practicing good science

Do not make the mistake of assuming that all regulators have a current and thorough command of the regulations they are charged with enforcing. Rather, you should assume that regulators may not be current or well acquainted with their own regulations. This lack of knowledge, though unintentional, is dangerous to your ultimate goal, which is closing the deal. Therefore, you may be well advised to begin your evaluation of project objectives by first reviewing the governing regulations and summarizing them in a quick-reference format in order to check the statements of your regulatory contact against what you know to be true based on recent research. This regulatory review process should not end after the project planning stage; rather, it should continue throughout the project.

Protecting the environment and protecting humans are honorable and worthwhile activities. They should be the call letters of your negotiations with the environmental regulator because they will hit home and engender support from the regulators. Environmental regulators who are particularly strong in their protection of humans and the environment are doing their jobs well. They should be appreciated and appropriately reminded that businesses employ humans and that fallow brownfields properties do more damage to the population overall than do redeveloped properties that are not as clean as they were the day God created the earth.

Now to the reason environmental consultants communicate well with environmental regulators: they are both scientists. Use this background to your advantage by allowing the scientists to communicate freely in order to build a relationship. Make it your job to lead the technical professionals in the desired direction.

Consider these differing interests when you begin to set the objectives for the site investigation. As the business owner, you must set the objectives to accomplish your goals. If you don't do your job and set the objectives, do not be surprised if the investigation costs thousands of dollars, takes months or years to complete, and is totally unsatisfactory to you. Do not set your objectives so high that they are unachievable. Be demanding, but be reasonable. Many unsatisfied business owners are unsatisfied because they failed to set objectives or because they set objectives that were unreasonable from the outset. Keep your head, think things through, seek input from the regulator and the environmental professional, and then set the objectives for the investigation. Thereafter, continually revisit the stated objective of the investigation and check it against what you are currently doing. If the investigation objective and the current actions do not line up, either the objectives need to be refocused or the environmental consultant or regulator has lost sight of the objectives you set.

One closing thought about setting objectives: meet early with the environmental regulator and consultant. Open a dialog with them, clearly state your needs, and continually communicate with them as the project progresses. Briefly stated: communicate with the your team members early and often.

Designing the Site Investigation

Designing the site investigation consists of the following major steps:

- Identifying the potential sources of environmental contamination
- Identifying potential migration pathways
- Identifying the contaminants of concern
- Identifying the regulatory driver(s)
- Identifying the physical evaluations you need to perform

Many of the foregoing steps overlap in various ways. However, each step is distinct and requires specific consideration. Let's consider the steps one by one. What are your potential sources? Underground storage tanks may represent both aboveground and underground potential source areas. For example, surface spills may influence surface soil and leave subsurface soil unaffected. Subsurface underground storage tank spills may contaminate only soil in the subsurface, leaving surficial soil relatively clean.

Aboveground storage tanks nearly always represent surficial sources. Landfill areas may pose source potential to surface water, surface soil, subsurface soil, and groundwater.

There is much overlap between identifying the potential sources of environmental contamination and identifying the contaminants of concern. These tasks are inseparable because the contaminants of concern are directly related to potential sources of environmental contamination. However, the thought process regarding contaminants of concern is more focused on the nature of the contaminants of concern and their physical and chemical characteristics. Ask many questions before you decide on the appropriate project methods. Some questions you might ask about the contaminants of concern:

- Do they sink or float in groundwater?
- Do they dissolve in, remain distinct from, or become miscible with water?
- Do they consist of organic compounds, inorganic compounds, or both?
- Do they degrade into other compounds?
- Do they classify as naturally occurring substances or synthetic organic chemicals?

These and many other questions should be considered when identifying the contaminants of concern and the appropriate methods for investigating them.

The next question to consider is migratory pathways. Migratory pathways can be man-made or natural. Some of the most common man-made pathways are utilities, foundations, surface drainage ditches, and filled areas. Natural migratory pathways include surface water, groundwater, and features in soil and rock. Natural migratory pathways are simpler to identify in many circumstances because they are thoroughly described in available publications such as books and maps describing regional geology and groundwater. Man-made migratory pathways are not always self-evident and may require extensive specific research and far-reaching experience on the part of the investigator. Carefully consider the locations of potential sources of contaminants and the answers to the foregoing lists of questions regarding the nature of the constituents of concern when determining what migratory pathways may govern the migration and fate of contaminants of concern.

Next, consider the regulations and guidelines governing your investigation and how they may alter the steps to achieving your objectives. Risk-based corrective action (RBCA) investigations can be as simple as collecting chemical and physical information for evaluation in various formulas. However, in other circumstances, the RBCA investigations may take a backseat to other regulations and laws such as the Resource

Conservation and Recovery Act, Toxic Substances Control Act, and the Comprehensive Environmental Response, Compensation and Liability Act. These laws and related regulations may carry specific sampling and analytic methods and reporting obligations that make RBCA more arduous and difficult—maybe impossible.

In the many circumstances where RBCA evaluations are possible, consider the kinds of information required to perform the necessary calculations to establish remediation objectives and to show that attenuation capacity and soil saturation limits have not been exceeded. Doing the appropriate sample gathering and analyses will make the calculation process simpler and less expensive. For example, performing an inexpensive test for total petroleum hydrocarbons or for fraction organic carbon may mean the difference between finishing the job or returning to the site for another round of sampling. These simple but foreseeable tasks should be considered every time you plan to spend money sampling. These steps save time and money. They make the difference between regulatory approval and disapproval. They decide whether you satisfy your project objectives.

The final step in designing the investigation is to identify the chemical and physical evaluations you must perform. Before you can effectively consider this step, you must have completed the four prior steps. In this phase, you are simply identifying the specific laboratory analyses and measurements you must perform. This is technical stuff. Examples of some physical and chemical analyses and methods are as follows:

- Fraction organic carbon
- Soil porosity
- Soil bulk density
- Aquifer hydraulic conductivity
- Volatile and semivolatile organic compounds
- Polychlorinated biphenyls (PCBs)
- Organo-chlorine pesticides and PCBs
- Priority pollutant metals
- Hazardous waste threshold analyses

In addition to the foregoing laboratory analyses methods, physical evaluations may require field tests or adjustments to determine:

- Aquifer hydraulic gradient
- Source width or plume width and length
- Lateral and vertical extent of contamination in soil and water

Developing the Sampling Plan

After you have reviewed the existing information, gathered and reviewed information that is available cheap or free, considered the differing interests that affect the project objectives, and designed the investigation, you should develop a sampling plan. When developing your sampling plan, keep in mind that there are two categories of investigation: (1) the detection investigation and (2) the delineation investigation.

The detection investigation may involve very few samples analyzed for a variety of constituents of concern near potential source areas. The purpose of the detection investigation is self-evident: to detect the presence of constituents of concern. The delineation investigation is an iterative sampling process for an increasingly narrow range of constituents aimed at delineating the extent of impacts in the vertical and lateral direction. Delineation sampling may involve grid sampling over a large area, collecting multiple vertical samples for laboratory analyses, or using geophysical or other means to test for the presence of constituents of concern over large areas. You should follow three simple rules of thumb:

- Collect no sample before it is time.
- More samples may not be better.
- Sample wisely (less is more).

If you follow these simple steps, you will be more satisfied with the outcome of your brownfields investigations and will accomplish your objectives faster using less money.

Finding a Reputable Analytical Lab

Analytical laboratories can be as variable as any consultant or other type of service that may be needed to complete a Phase II investigation. However, there are ways to qualify laboratories that can aid in choosing a laboratory that can meet all of your needs.

First, you need to make a list of potential candidates to meet with to determine whether they can satisfy the needs of your project. Usually, it is best to find a laboratory that is near the property being investigated. Many laboratories will even provide a pick-up service that can be utilized as needed to ensure that the samples arrive at the laboratory in a timely fashion. Contact each laboratory on the list and set up a meeting and tour of the facility. The questions that need to be answered are:

- Does the laboratory have the capacity to complete the work?

- Does the laboratory have all the equipment necessary to complete all the tests that are planned, or will it be subcontracting some of the tests?

- What are the detection limits and holding times for the various analytes proposed for the project?

- What will be the turnaround time for the samples?

- How much extra does it cost for more rapid turnaround time?

- Does the lab have the capability to fax or e-mail results?

- For the project planned, is the volume large enough to warrant a discount on pricing?

- What is the appearance of the laboratory?

- What types of certifications does the laboratory possess?

- Does the laboratory provide sampling services as well as analytical analysis?

The laboratory should be able to help in deciding what test methods will be used for the analytes targeted in the investigation. There may be less expensive methods that will allow a great many samples to be run less expensively than other methods. Also, the laboratory should be able to provide the necessary bottles or sample containers required for each type of test method to be performed.

A laboratory should be chosen very early in the planning stages so that its input can be utilized in designing the sampling and analysis plan. As you work with the staff at the laboratory, you should be able to get some indication of the level of expertise of the individuals you are working with, which will also be an indication of the level of performance you should expect from the laboratory.

All the questions just listed should be clarified prior to contracting with the laboratory. This will prevent any surprises from occurring at an inopportune time during the project. All of this information is usually contained in the quality manual of the laboratory. The sampling and analytical expenses are a major cost in any brownfield investigation, and it is imperative that a great deal of attention be paid to selecting the laboratory. The data will only be as good as the performance of the sampler and the laboratory, and many times a serious oversight may not become apparent until the data is being reviewed, which may be weeks after the site has been sampled.

Sampling and Sampling Plan

The most important aspect of the data-gathering phase of the project is the sampling and the sampling plan. If the samples are taken and preserved incorrectly, they are useless, and where you sample is probably more important for most projects than how many samples are taken.

Samples at brownfields properties usually are soil samples. In some cases there may be standing liquids that might be sampled, but for the most part you are trying to identify past impacts to the soil. The sampling locations should be well thought out in advance based on the information available from the Phase I investigation. Areas around former building foundations, buried or aboveground tanks, and sites of other activities where chemicals could have potentially been spilled may be sampled. The samples may consist of surface samples or samples removed from cores of soil taken several feet into the property. Whatever locations are selected, they should represent the potentially worst contaminated areas of the property.

Those who are going to take the samples must understand what analytes are going to be analyzed for. Volatile organics must be sampled in a way that preserves the compounds and prevents evaporation during the sampling process. Metals, pesticides, and semivolatile analytes are not as volatile, and it is less of a problem preserving their actual concentrations while sampling.

Sampling techniques and analytical methods are discussed in Sec. 6.6.

Conclusion

The amount of environmental due diligence required in the redevelopment of a brownfield site is enormous. There are no shortcuts, and there are myriad details that must be considered and researched. While this chapter may seem confusing to the novice, these are the steps that are inherent in every qualified environmental investigation. Find an environmental consultant you are comfortable with who is following the details outlined here in your investigation, and you are in for a much more rewarding experience.

Keith R. Fetzner earned a B.S. in geology at Northern Illinois University in 1989 and a master of science at the University of Texas in El Paso in 1992. He is a licensed professional geologist in several states, is a member of various environmental organizations, and has completed additional education in other environmental subjects focusing on risk assessments and computer modeling. He is currently employed by Mostardi-Platt Associates, Inc., as manager of environmental assessments.

His environmental consulting experience since 1992 includes several hundred environmental investigations and remediation projects involving a variety of contaminants in soil, water, and air. These investigations involve the installation of soil borings, monitoring wells, and soil gas probes; collection and analysis of soil, groundwater, and air samples; and the completion of pilot study to full-scale remediation projects including groundwater pump-and-treat, bioventing, and soil vapor extraction, stabilization, and enhanced natural attenuation.

Lawrence Fieber holds an M.S. in geology from Southern Illinois University in Carbondale, Illinois, and a B.S. in geology from the University of Illinois in Champaigne-Urbana, Illinois. He has more than 15 years of professional consulting experience, 13 years specifically focused in the environmental consulting area.

Mr. Fieber is the director of the environmental assessment department of Mostardi-Platt Associates, Inc., where he has practiced since 1988. During this time he has been involved with thousands of environmental consultations including environmental audits; Phase I and Phase II environmental assessments; underground storage tank investigations; soil and groundwater remedial investigation; design, monitoring, documentation, and reporting; regulatory agency negotiation; expert consulting and testimony; asbestos and lead-based paint inspection and abatement; and risk-based remediation methods.

Mr. Fieber is a Certified Professional Geologist (AIPG 9240), a licensed professional geologist (Illinois and Indiana), a certified asbestos inspector, a certified and licensed lead-based paint inspector, a certified underground storage tank decommissioner, and an OSHA-certified hazardous waste site worker.

A recognized leader in environmental regulatory matters, Mr. Fieber has written numerous professional articles and spoken before professional and private gatherings concerning environmental compliance matters, subsurface sampling methods, geophysical investigation techniques, environmental regulatory developments, and risk-based corrective action methods. He is a member and president (Illinois-Indiana Section) of the American Institute of Professional Geologists (AIPG), a member and director of the Lake Michigan States Section of the Air and Waste Management Association (LM-AWMA), and a member of the Association of Groundwater Scientists and Engineers (AGSE).

Frank H. Jarke is manager, analytical and quality assurance, safety director, and head of odor sciences at Mostardi-Platt Associates, Elmhurst, Illinois. He has wide experience in the environmental field going back to 1969. He has been associated with the Illinois Institute of Technology Research Institute in scientific and project management positions, including heading IITRI's Odor Science Center. Mr. Jarke was with Waste Management Corp. from 1981 to 1994, establishing facilities and programs for shipment and analysis of hazardous waste samples, groundwater monitoring, and quality assurance. He worked with USEPA in the development of Good Automated Laboratory Practices. Mr. Jarke has a B.A. from Southern Illinois University and an M.S. from IIT. He has 35 publications and 1 patent and is active in ASHRAE, ACS, and Sigma Xi.

6.6   Sampling Techniques and Analyical Methods

Frank H. Jarke

Sampling for Volatile Organics in Soil or Water

Volatile organics, as the name implies, are volatile and require a great deal of care to ensure that the analytes of interest do not vaporize during the sampling process. Recently, USEPA has approved methodology that assists in the preservation of volatile organics during soil sampling. This methodology is referred to as Method 5035, "Closed-System Purge-and-Trap and Extraction for Volatile Organics in Soil and Waste Samples," and is published in the USEPA Solid Waste Division's manual of methods known as SW-846. The method utilizes a hermetically sealed sample vial, the seal of which is never broken from the time the sample is taken until the sample is analyzed by the laboratory. Prior to this, samples of soil were simply taken and placed in 4-oz jars packed as full as possible. But the contact with the air was thought to result in some or all of the target volatile analytes escaping, especially in cases where the concentrations were low. This new sampling procedure prevents the loss of volatiles and ensures that the soil is analyzed with little or no disturbance and that the results reflect the actual concentrations of the analytes present in the soil at the property.

Liquid samples for volatile organics should be placed in what are known as 40-ml VOA vials. These vials are available precleaned with the preservative already added from a number of container suppliers or from the laboratory. The VOA vial should be made of brown glass with a Teflon septum cap. The bottle is filled until a meniscus of liquid extends above the rim, and the cap is carefully put on the bottle. If the sample is correctly taken, there should be no bubbles visible inside the bottle if it is turned upside down. If there are, throw away that sample and sample again.

Soil samples for volatile organic analysis must be taken to the laboratory and analyzed within 14 days of sampling. Water samples that are unpreserved must be analyzed within seven days of sampling. If the water is preserved with a small amount of hydrochloric acid, then the sample must be analyzed within 14 days of sampling. These holding times must be adhered to and are generally not negotiable. Therefore, in setting up the sampling plan, some provision must be made to get the samples to the laboratory as soon as possible after sampling. Proper sampling technique ensures that the results will not be challenged later.

Sampling for Other Analytes in Soil or Water

Other analytes of interest may include semivolatiles, metals, organo-chloride or organophosphorous pesticides, chlorinated herbicides, poly-nuclear aromatic hydrocarbons, and PCBs. These analytes do not generally require the care in sampling that the volatile organics do. The important issue in sampling for these analytes is to ensure that the sample taken is large enough, that it is preserved properly, and that it is taken to the laboratory in enough time to allow for the analytical process to begin before the holding time is exceeded. All samples should be taken in brown glass containers.

Semivolatiles, pesticides/herbicides, polynuclear aromatic hydrocar-bons, and PCBs are generally heavy molecules such as naphthalene, and other compounds found in materials such as tar or asphalt, that are widely used in or result from industrial activities. They have low volatility and low water solubility, so the analysis for their presence at brownfields properties is usually a given. For soil analysis, a 4-oz sample is usually large enough for the laboratory, but generally a 1-l sample is required for water. The samples for semivolatile analysis are not preserved. The first step in the analysis of semivolatiles is to extract the sample using methy-lene chloride. This extraction step must be performed within 14 days of sampling for soil samples and within 7 days of sampling for water sam-ples. Once the samples have been extracted, the analysis must be com-pleted within 40 days.

Metals can occur either in pure form or as various compounds. They have wide use in and result from many industrial activities. Since metals are quite stable, the requirements for sampling them are less stringent than those for organic analytes. For soil samples, a 4-oz jar of soil is suf-ficient; for water, 100 ml is usually sufficient for a metals analysis. The samples should be collected in plastic containers, and generally no preser-vation is required. The holding time for samples taken for metals analysis is six months.

Laboratory Analytical Techniques

There are several laboratory analytical techniques available for the inves-tigation of soil and water samples from brownfields properties: gas chro-matography (GC), gas chromatography–mass spectrometry (GC-MS), high-performance liquid chromatography (HPLC), and inductively coupled plasma (ICP). All of these techniques are very useful and produce results that are quite satisfactory in these investigations.

The majority of the chemicals of interest are organics and metals. GC, GC-MS, and HPLC provide the best solution for analysis of organic compounds, as will be explained. ICP is the method of choice for the analysis of metals.

Gas Chromatography

GC, an elegant technique invented near the end of World War II, is the workhorse of the analytical laboratory. It allows a tester to separate a complex mixture into its individual components and identify the compounds. This is a tremendous advantage over other techniques, because a mixture containing hundreds of compounds can be investigated and the components can be identified. No other analytical technique allows for this type of diversity.

The GC technique is quite simple. The instrument consists of an injector where the sample is introduced. This is nothing more than a miniature block heater with a hole down the middle. At one end is a device, usually a rubber septum, that allows introduction of the sample. At the other end is a connection to the column. The purpose of the injector is to ensure that as much of the sample as possible is vaporized when introduced into the interior of the block heater. The vaporized sample is known as the *mobile phase.*

The column can be made of any number of materials including copper tubing, steel tubing, glass tubing, or more recently quartz optical fibers of various internal diameters. Inside the column are two components, a stationary phase and a support. In the case of large-diameter tubing of glass, steel, or copper, the support is usually an inert material such as pumice or firebrick that has been ground to a fine mesh. Coated on the support is the stationary phase, usually a material like soap. With quartz optical fibers, also called *open tubular columns*, the stationary phase is simply coated on the walls of the fiber. The internal diameter of these columns is quite small, which allows the molecules to have better contact with the walls than do large "packed" columns.

One end of the column is attached to the injector oven; the other end is attached to the detector oven. The detector oven is a block heater containing the detector. The detector can be a thermal conductivity detector (TCD), a flame ionization detector (FID), an electron capture detector (ECD), a photoionization detector (PID), an electrolytic detector, or a mass spectrometer. Each detector has certain properties that allow for the detection of certain molecules. The selection of the detector and the column are quite important in the determination of the target compounds. For example, the TCD works best on permanent gases such as methane, ethane, and so on. The FID is the best all-around detector and is used for most general work. The ECD detects chlorinated compounds.

The operation of GC is quite simple. The sample is introduced into the injector. The temperature of the injector oven vaporizes the compounds that make up the sample into the gaseous state. A carrier gas, such as helium, that is introduced into the sidewall of the injector oven sweeps the gaseous molecules into the column. As the compounds are swept through the column, each compound is preferentially absorbed and desorbed from the stationary phase depending on its affinity for the stationary phase. That is, some of the molecules prefer to be in the gas phase and others prefer to be absorbed into the stationary phase. This results in the partitioning, or separation, of the compounds. The compounds that prefer the gas phase reach the end of the column first; the compounds that prefer to be absorbed reach the end of the column later. As the compounds emerge from the end of the column, they are detected. In the case of the TCD, the compounds alter the conductivity of a hot filament. In the case of the other detectors, the compounds are ionized in a flame or by a radioactive source and are detected by an electrometer. The signal from the detector is fed to a recorder that plots response versus time. The result is the familiar gas chromatogram that consists of a baseline tracing in which each compound is represented by a peak. The area of each peak can be used to determine the concentration of the compound in the original sample.

Each peak that emerges from a GC column can be identified by comparing the time it takes to traverse the column—the *retention time*—to a known set of retention times for various compounds. For a given combination of column stationary phase and injector and detector oven temperatures, known compounds will have specific retention times. The retention time becomes the marker that identifies that compound uniquely and allows the GC technique to qualitatively identify unknown compounds in mixtures. This is known as a *secondary analytical technique* because the identification is by inference. While the peak from the sample has the same retention time as the known compound under the same conditions, there is no way to verify this compound's identity. This is the major drawback of the GC technique. In the following text a technique is described that removes the uncertainty associated with not being able to positively identify unknown compounds.

For volatile organics from soil or water, the sample is introduced by passing an inert gas such as helium through the sample. The volatiles vaporize into this stream of inert gas and are then collected on a porous polymer trap. After sufficient time as passes to collect a representative sample, the porous polymer trap is heated very rapidly to a temperature of 200°C and the sample is swept into the injection port and onto the column for analysis.

Gas Chromatography–Mass Spectrometry

GC-MS is a hybrid of a separation technique—GC—and a specialized detector—MS. The coupling of these two techniques has greatly enhanced capabilities for the analysis of complex mixtures.

Mass spectrometry is a unique tool in that it produces a fingerprint of any molecule that is introduced for analysis. A mass spectrometer operates under high vacuum. Molecules exiting the end of a GC column are first introduced into an evacuated chamber. Since the GC column is at approximately 30 psi, this first chamber is primarily for the purpose of removing most of the molecules and reducing the pressure to about 1 torr. Molecules in this first chamber are then drawn into a second chamber, at much lower pressure, through a skimmer or small hole. Upon entering the second chamber, the molecules that are traveling as a beam enter the ion region of the mass spectrometer. Electrons from a hot filament bombard the molecules with 70 electron volts of energy, which breaks the molecule into a number of positively charged fragments. These fragments are repelled by negatively charged plates through a series of lenses that focus the charged fragments into the mass analyzer, which consists of four charged rods. Two opposing rods are negative and the other two are positive. The charge on these rods alternates very rapidly. The rods also have an increasingly larger DC current applied. As the charge fragments move into this region, they are alternately attracted to one rod or the other and quickly begin a spiral motion through the length of the rods. As the DC current is ramped up, the lighter fragments exit the rod chamber first, while the large fragments are destroyed on the rods. Later the heavier fragments emerge. The fragments are detected by an electrometer and this signal is fed to a recorder. The resulting mass spectrogram is a plot of the ion intensity versus the voltage applied to the rods. This fingerprint, which is unique for each compound, is then compared to a computerized library of more than 50,000 compounds and can be identified with very high precision.

The ability of GC to separate a complex mixture into its component parts and of MS to uniquely identify the individual components makes GC-MS the most versatile of all analytical techniques.

High-Performance Liquid Chromatography

HPLC is a technique used with some methods that will be discussed at the end of this section. HPLC is essentially the same as gas chromatography except that the carrier is a liquid instead of a gas. Samples to be tested are injected in a manner similar to that used in GC; however, the sample is injected into a stream of liquid. This liquid can be a single liquid that does not vary during the analytical process (isocratic elution operation), or one whose concentration can be altered during the analytical process to affect the elution of the compounds in the sample (gradient elution operation). The molecules are separated by the combination of the effects of the column packing and the carrier solution. This method is usually used with molecules that have been collected in solution and that are too fragile for the temperature conditions of the GC or that are too large to pass through a GC column.

Inductively Coupled Plasma

Metals other than mercury can usually be analyzed by a technique known as inductively coupled argon plasma (ICAP) or inductively coupled plasma (ICP). This analytical method allows the sample to be analyzed for a wide range of metallic elements in one analysis and consequently has gained wide acceptance in the analytical community. The method begins with the complete dissolution of the sample. The sample is placed into a beaker and a quantity of acid—usually a mixture of hydrochloric and nitric acid—is added. The beaker is then heated on a hot plate at about 180°C until the water has evaporated and the beaker is almost dry. An aliquot of water—usually about 100 ml—is added and the residue is dissolved. The sample is now ready for analysis. The sample is aspirated through an inert tube into the plasma torch of the ICP. The plasma torch produces very high temperatures in excess of 5000°C. The metals now contained in the solution are excited into an elevated state and, as they relax back to their ground state, give off wavelengths of light that can be detected. Each metal element has several characteristic wavelengths. One of these is selected, primarily because of a lack of interference from other elements, and this wavelength of light is detected for each metallic element. The amount of light detected is proportional to the amount of that element present in the sample and this value is quantified.

Mercury is detected in a similar manner, but the apparatus detects the absorption of a specific wavelength of light that mercury is known to absorb. The apparatus is known as a cold vapor atomic absorption spectrophotometer.

Field Techniques

Field techniques allow for the measurement of VOCs directly in the field. There are a wide variety of techniques, from simple devices to very complex devices. These techniques can be very useful in determining where the best location to sample may be or what section of a core soil sample is likely to yield the best result. It sometimes possible to do this by smell as well, but this is not recommended. In this section we will cover all of these techniques.

Organic Vapor Analyzers (OVAs)

OVAs are simply gas chromatographs without the column. Since they do not allow the separation of complex mixtures into their component parts, they only measure total organic vapors. Several types are available, and generally only differ in the type of detector used. They are quite simple to use. A gas of known concentration is used to standardize the unit. For example, a gas such as isobutane at 100 ppm can be used. The OVA is zeroed with ambient air and then the standard gas is introduced. The unit now reads total organic vapor concentrations as isobutane. A built-in pump in the OVA samples the unknown gas and the gas passes over an ionizer such as an FID, a radioactive source (rare), or a photoionization source such as an ultraviolet (UV) lamp. The ionized gas is then detected by the electrometer and the signal is read out on a meter.

These units are ideal for survey work in which the investigator is trying to isolate areas that are producing high emissions of organic vapor. While the readings do not give any indication of the organics that are causing the response, the readings can be very useful.

The detection limit of these devices is usually about 500 ppb. This can be quite useful in pinpointing areas of potentially high organic content. The device also has the ability to be very portable and the sampling port can be placed very close to the source.

Absorption Tubes

Absorption tubes are glass tubes containing an inert material impregnated or coated with a chemical that reacts preferentially with a specific chemical. An indicator is also added so that a color change is produced.

These devices are quite simple and produce semiquantitative results. A tube is selected that responds to the range of interest and air is drawn through the tube with a hand pump. If the amount of air drawn through is precisely what is recommended for the particular tube, then the calibrations on the side of the tube give an accurate reading of the concentration in the air sampled. These devices are used most frequently in the industrial hygiene area but also may be used to indicate areas of high concentrations of specific analytes.

Portable GCs and GC/MSs

There are a wide variety of commercial GCs and GC-MSs that have been designed very compactly expressly for use in field applications. These operate exactly as the laboratory instruments, but have much simpler electronics and fewer bells and whistles. These devices, however, require that a discrete sample be introduced. This is usually accomplished with a gas-tight syringe or with a valving system. Portable GCs and GC-MSs simply take the laboratory out to the field and allow the investigator to obtain laboratory-quality data without having to ship samples to the lab and wait several days for the results.

Test Kits

There are available several field kits for determining specific analytes in the field. These usually involve manipulation of the sample with various chemicals that ultimately produce a color change that can be related to concentration. They can be difficult to use and interpret, but once again the purpose is not to be quantitative but to isolate those areas that should be sampled for analysis by the off-site laboratory.

For biographical information on Frank Jarke, see Sec. 6.5.

6.7 Groundwater Flow and Transport

Christopher French

Hydrogeology is the broad scope of scientific inquiry concerning the interaction of groundwater with the geologic medium. For exposure to a particular contaminant to occur, the constituent of concern (COC) must travel from a source to a receptor. The medium of transport may be either air, water, or a solid. The principal mode of subsurface transport for buried sources—such as underground storage tanks (USTs), lagoons, disposal pits, and so forth—is via groundwater, through the porous or fractured solid medium. A thorough understanding of the hydrogeologic setting is essential to the successful evaluation, design, and implementation of a remedial action.

Hydrogeology

The subsurface medium may consist of unconsolidated sediments such as clay, silt, sand, or gravel; lithified sedimentary rocks such as claystone, siltstone, sandstone, or limestone; or crystalline lithologies such as volcanic rocks (basalt, andesite, rhyolite), metamorphic rocks (marble, schist, gneiss), or igneous rocks (gabbro, diorite, granite).

Unconsolidated Deposits

Deposits such as clay, silt, sand, and gravel are generally unconsolidated and nonindurated, meaning that the particles are not cemented or bound together. Such sediments may be transported and deposited by the action of water (rivers, streams, lakes), ice (glaciers), or wind, and are termed *alluvial, glacial,* or *aeolian* deposits, respectively.

Clay and silt deposits are very fine-grained, do not transmit water readily, and will yield only small quantities of water to wells. Even small percentages of clay in a sand matrix can affect the hydraulic properties of the medium. The physicochemical properties of clays can also significantly impact contaminant behavior in the subsurface. Sand and gravel deposits, on the other hand, are coarse-grained, transmit water readily, and readily yield water to wells.

In the site investigation phase, the degree of heterogeneity and complexity of the study area will have a direct bearing on the cost of the investigation. The nature of the hydrogeologic medium will also influence decisions concerning remedial action. Because unconsolidated deposits of sand and gravel are important sources of water supply, and contaminant transport may be comparatively rapid, a remedial action may require some measure of source removal or control, such as active pumping. On the other hand, a source located within impermeable clay deposits may be amenable to limited engineering and institutional controls in the form of a cap or fence to prevent direct contact with the environment.

Sedimentary Rocks

Sedimentary rocks include claystone, siltstone, sandstone, and carbonate rocks such as limestone and dolomite. The most significant factors in assessing contaminant behavior in consolidated sedimentary rock are the degree of control exerted by structure, stratigraphy, or a combination of both. Structural influences include folding, faulting, and fracturing. Stratigraphic relationships describe the geometric arrangement of strata. The imposition of structural and stratigraphic influences, acting alone or in concert, can result in unforeseen conditions, which may translate into costly project overruns.

The cost of investigating and remediating contaminated sites in sedimentary bedrock should be carefully considered. In general, costs for investigation will be greater because the hydrogeologic setting is more complex, the drilling methods are more expensive, and the range of rapid assessment technologies is more limited. Remedial alternatives may be more limited and/or costly.

Crystalline Rock

Samples of unfractured crystalline and metamorphic rock typically exhibit very low primary permeabilities in laboratory experiments. In the field, most crystalline bedrock settings are cut by a sufficient number of joints, fractures, and faults to impart a significant secondary fracture permeability. Some volcanic rocks can exhibit very significant secondary permeability.

The structural influences that control contaminant migration in fractured bedrock can be exceedingly complex. Intersecting fracture sets can create one or more directions of preferred groundwater flow. Crystalline bedrock aquifers can therefore be highly *anisotropic*.

Investigations of fractured bedrock settings are quite costly and require specialized equipment and methods. The behavior of contaminant transport can be difficult to predict or model, and the discovery of unanticipated subsurface conditions can significantly impact the evolving conceptual understanding of the hydrogeologic setting. Remedial action alternatives are limited and can be prohibitively expensive.

Basic Groundwater Concepts

Groundwater in the subsurface occurs in an upper unsaturated zone and a deeper saturated zone. Pore space in the unsaturated zone (the *vadose zone*) is not completely filled with water. Water is present in the pore space of the unsaturated zone as soil moisture. Pore space in the saturated zone is completely filled with water. The transition between the vadose zone and the saturated zone is the *capillary fringe*.

An aquifer is a saturated geologic formation that is permeable enough to transmit water readily and thereby yield usable quantities of water to wells. The body of the aquifer rests upon a confining bed, or *aquitard*, or may be layered within confining strata.

Groundwater in the aquifer may occur under unconfined or confined conditions. *Unconfined* groundwater occurs below a free water table. Movement of groundwater under unconfined conditions occurs due to gravity and is controlled by the slope of the water table. *Confined* or *artesian* groundwater occurs between upper and lower confining strata such as clay. Confined groundwater flow occurs in response to changes in pressure.

Groundwater is a part of the hydrologic cycle. Some understanding of the water balance and seasonal variations in the hydrologic cycle of the site setting is essential. One factor to consider is whether or not a site has been adequately characterized to account for seasonal variations in groundwater recharge, flow, and discharge.

Hydraulic Gradient

Groundwater flows from areas of high potential energy to areas of low potential energy. The measurement of the potential energy is called *hydraulic head.* Hydraulic head is measured by surveying the level of water in a well relative to specific surveyed elevation data. Data from a minimum of three wells are required to determine groundwater flow direction and gradient.

Hydraulic head data from wells is contoured to construct a groundwater elevation map, or potentiometric surface map (Fig. 6-1). This fundamental map represents a graphical configuration of the surface of hydraulic head and the variation in hydraulic gradient. Groundwater flow occurs perpendicular to lines of equal potential. As seen in Fig. 6-1, the flow occurs from areas of high hydraulic head to areas of low hydraulic head.

Examination of the groundwater elevation map will reveal important clues about subsurface migration of contaminants. The flow lines indicate the direction of groundwater transport, and the spacing of the lines of equal hydraulic head (equipotential lines) provides a measure of the magnitude of the hydraulic gradient. The convergence or divergence of flow lines may indicate changes in the subsurface geology or the presence of groundwater discharge or recharge. While the map is a two-dimensional representation of the groundwater flow, it is important to remember that groundwater flow occurs in three dimensions.

Velocity of Groundwater

Groundwater flow velocities occur on a scale of feet per year, whereas surface water flow, by comparison, occurs on a scale of feet per second. Groundwater flow velocities are estimated by multiplying the value of the hydraulic gradient i (discussed in the preceding text) by the value of hydraulic conductivity K, using Darcy's law, defined by $q = K \cdot i$.

The hydraulic conductivity K is a measure of the ability of the aquifer medium to transmit water. It has units of length over time (L/T) and is determined empirically by aquifer pumping tests or estimated from literature values. The gradient i is determined from examination of the groundwater elevation map (discussed earlier). The value is obtained by dividing

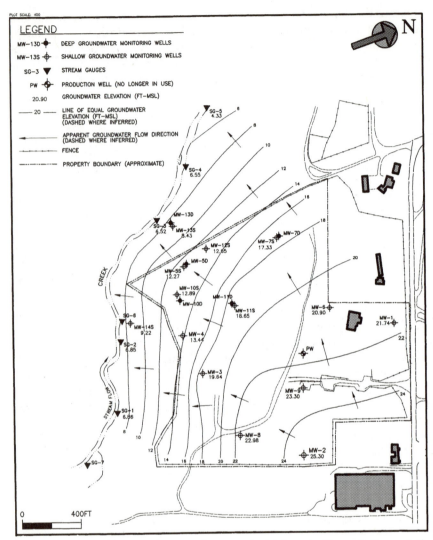

Figure 6-1. Groundwater flow map, former industrial waste disposal site, central New Jersey. (Courtesy Envirogen, Lawrenceville, NJ.)

the difference in hydraulic head between two points by the distance between the two points along the direction of groundwater flow. It is expressed in feet of rise per foot of length, and hence is a unitless measure.

A distinction is made between Darcy velocity q and pore velocity v. Darcy velocity, or specific discharge, is a measure of the volumetric flow through a cross section of the aquifer. To determine the actual pore velocity, the Darcy velocity is divided by the effective porosity of the aquifer n_e. Thus, $\upsilon = q / n_e$. The pore velocity is much faster than the Darcy velocity.

The hydraulic conductivity of an aquifer may vary by position within the aquifer and according to the direction of groundwater flow. The aquifer is *homogeneous* if the hydraulic conductivity remains constant throughout. An aquifer wherein hydraulic conductivity varies from place to place is said to be *heterogenous*. Values of hydraulic conductivity also vary as a function of flow direction because the geologic properties of the aquifer are not uniform. For example, alluvial and sedimentary aquifers are often layered so the vertical hydraulic conductivity is generally much lower than the horizontal. An aquifer with uniform hydraulic conductivity regardless of flow direction is *isotropic*. When hydraulic conductivity varies according to flow direction, the aquifer is said to be *anisotropic*. Aquifers are almost never found to be homogeneous and isotropic.

Contaminant Fate and Transport in Groundwater

A groundwater plume results from the migration of contaminants in groundwater downgradient from a source (Fig. 6-2). Three primary factors affect how dissolved contaminants will move through the aquifer. These include (1) advection of the substance with groundwater as groundwater moves through the subsurface; (2) dispersion of the chemical by mechanical and molecular means; and (3) physicochemical and biological interactions of the contaminant with the medium, such as adsorption (causing retardation), chemical transformation, or biodegradation.

Advection

A dissolved chemical migrates with the flow of groundwater. This is termed *advective solute transport*. This is the primary means by which contamination moves through the subsurface. The measure of pore velocity υ applies to advection (see the preceding text).

Advection is often used as a first approximation of the rate of contaminant migration. As discussed earlier, the value is easily estimated from the basic hydraulic parameters of hydraulic conductivity, the gradient, and the effective porosity. The principals involved are commonly understood, and the equation is easily solved.

Consider a well located downgradient from a source of contamination, such as a continuous release from a leaking underground storage tank or a slug release from a one-time spill. Theory predicts that the process of advective transport would result in the instantaneous appearance of the

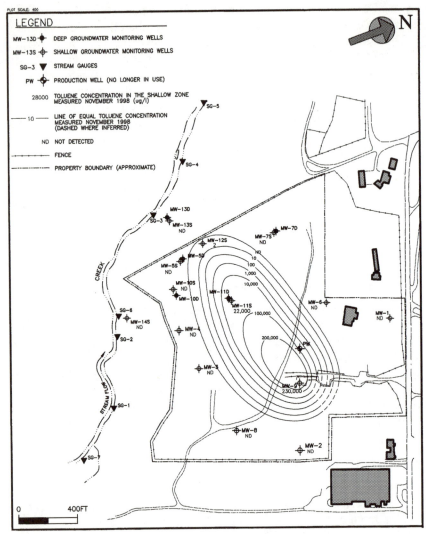

Figure 6-2. Toluene plume isoconcentration map, former industrial waste disposal site, central New Jersey. (Courtesy Envirogen, Lawrenceville, NJ.)

entire plug of dissolved contaminant at the well and the instantaneous disappearance of the plug upon exhaustion or removal of the source. This is never observed in practice. Instead of arriving as a sharp front, the plume is spread out, and concentrations within the plume vary in space and time. Other processes, such as dispersion, adsorption (causing retardation), chemical transformation, or biodegradation, must always be taken into account.

Dispersion

Investigation of contaminated sites indicates that chemicals dissolved in groundwater arrive at downgradient wells sooner than predicted by consideration of advective transport alone, and that the dissolved contaminant mass is spread out along the flow path. The spreading and mixing of the dissolved contaminant mass is due to mechanical and molecular mixing. The mechanical mixing is caused by the migration of groundwater through irregular pathways in the subsurface, and by variations in the pore velocity of groundwater. Molecular mixing is a diffusion process attributable to molecular kinetics.

Mechanical dispersion is essentially a dilution and mixing process. It causes contamination to be spread out along the plume length (longitudinal dispersion). It also causes the plume to widen (transverse dispersion). A plume with high longitudinal dispersion and low transverse dispersion will be long and narrow. A plume with high transverse dispersion will impact a broader area of the aquifer.

Molecular diffusion impedes transport as a portion of the contaminant mass diffuses from areas of high permeability into areas of low permeability. This process is known as *matrix diffusion*. Matrix diffusion has several unanticipated effects. Diffusion into small pore spaces removes the contamination from microbial processes. Hence, this portion of the contaminant mass may not be available for biological degradation, an effect that is termed *sequestration*. Matrix diffusion can be reversed, causing contaminants to migrate out of the low-permeability zone at a later time. This tends to extend the time frame required for remediation and to limit the chance of meeting stringent cleanup criteria.

The degree of dispersion is scale dependent, and quantification of dispersion for the purpose of computer modeling is very complex.

Sorption and Retardation

Chemical species migrating in groundwater are slowed in their movement by reactions with the hydrogeologic medium, such as clay particles, organic matter, and certain hydroxides. This has the net effect of slowing, or *retarding*, the advancement of the contaminant plume relative to advection. Expressed another way, retardation slows down a plume's advancement due to interactions with the subsurface materials.

Sorption occurs due to partitioning into the solid phase. Many contaminants, such as organic compounds, show an affinity for the solid phase over the aqueous (dissolved) phase. Such contaminants are said to be *hydrophobic.* The soil-water partition coefficient is the relative magnitude of the chemical concentration on solid particles and in pore water for a particular soil. Several mathematical relationships, including the *Langmuir isotherm* and the *Freundlich isotherm,* are used to describe the equilibrium partitioning between the solid and aqueous phases.

Adsorption processes can also facilitate transport of chemicals that would otherwise be bound to the solids in the subsurface. Small solid soil particles termed *colloids* move with groundwater by advective transport. Certain contaminants, such as metals, sorb on soil colloids and may be transported with groundwater flow. This transport mechanism is termed *colloidal transport.*

Adsorption is partially reversible: the reverse process is known as *desorption.* Similar to the reversal of the matrix diffusion process discussed earlier, desorption tends to extend the time frame required for remediation and to limit the chance of meeting stringent cleanup criteria.

Physical, Chemical, and Biological Degradation

Contaminant mass may be lost in the subsurface due to physical, chemical, and biological processes that transform or degrade the contaminant of concern. Persistence in the environment can be described by a parameter known as the environmental half-life of a compound. In practice, the half-life is an empirical measurement that quantifies loss of mass due to biological, photochemical, chemical, or physical degradation mechanisms.

Volatilization of benzene from groundwater into soil vapor is an example of physical loss. Chemical processes include hydrolysis, oxidation, and reduction. *Hydrolysis* is the reaction of compounds with water or the hydroxide or hydronium ions associated with water. However, organic chemicals such as halogenated aromatics, ketones, benzenes, and phenols are generally resistant to this mechanism. *Oxidation* and *reduction* can also alter and attenuate organic compounds. For most inorganic compounds, geochemical transformations, such as precipitation as a result of oxidation-reduction reactions, are the most important degradation mechanisms.

Biodegradation. Many organic compounds may be subject to biodegradation. Bacteria involved in the biological degradation of organic contaminants are for the most part *heterotrophic* organisms—that is, they require organic compounds for growth and reproduction. These organic compounds serve as sources of carbon and energy. Bacteria obtain energy by transferring electrons from oxidizable organic compounds (electron donors, or substrates) to reducible compounds (electron acceptors). This energy-producing process is known as *metabolic respiration.* These processes can result in the transformation of organic chemicals to harmless inorganic compounds such as carbon dioxide and water. This process is termed *mineralization.* Within the subsurface, biological activity is believed to be the principal cause of the mineralization of organic compounds.

Aerobic heterotrophs (a type of bacteria) use oxygen as the electron acceptor during a process called *aerobic respiration.* In anaerobic environments, microorganisms gain energy by using electron acceptors other than oxygen to drive their metabolic respiratory processes. There are numerous inorganic electron acceptors in anaerobic environments; the most common are nitrate, Fe(III), Mn(IV), sulfate, and carbon dioxide. The organisms that can use these electron acceptors are referred to as *denitrifying, manganese-reducing, iron-reducing, sulfate-reducing,* and *methanogenic* bacteria, respectively. Denitrifiers reduce nitrate to nitrogen gas. Iron reducers reduce Fe(III) to Fe(II), while manganese reducers reduce Mn(IV) to Mn(II). Sulfate reducers produce hydrogen sulfide from sulfate reduction and methanogens produce methane from the reduction of carbon dioxide.

The electron acceptor hierarchy, based on energy liberated, is as follows:

Oxygen reduction: molecular oxygen $(O_2) \rightarrow$ water (H_2O)

Nitrate reduction: nitrate $(NO_3) \rightarrow$ molecular nitrogen (N_2)

Manganese reduction: tetravalent manganese [Mn(IV)] $\rightarrow$ divalent manganese [Mn(II)]

Iron reduction: trivalent iron [Fe(III)] $\rightarrow$ divalent iron [Fe(II)]

Sulfate reduction: sulfate $(SO_4^{-2}) \rightarrow$ sulfide (H_2S)

Methanogenesis: carbon dioxide $(CO_2) \rightarrow$ methane (CH_4)

Electron acceptors tend to be used sequentially, due to the relative amounts of energy liberated by each reaction. Oxygen will be the preferred electron acceptor; however, given oxygen's low water solubility (up to 11 mg/l at temperatures typical of groundwater systems) and its rapid consumption by microorganisms, oxygen levels found in ground water are typically low, thus limiting aerobic respiration. Therefore, other electron acceptors and the bacteria they support may predominate.

The number of sites pursuing biodegradation as a component of the remedy has increased dramatically. A remedy that relies exclusively upon natural biodegradation is termed *monitored natural attenuation.* However, it must be shown that contaminant mass is being destroyed, not simply diluted or sorbed onto the aquifer matrix, in order to demonstrate the viability of natural attenuation as a remedial alternative. Three lines of evidence are commonly evaluated in assessing the viability of natural attenuation:

- There must be an observed decrease in contaminant concentration along the flow path. This is the primary line of evidence; however, it cannot document contaminant destruction on a stand-alone basis.

- Chemical/physical data at the site must correlate with decreases in parent compounds and an increase in degradation products; decreases in a parent compound should correlate with an increase in metabolic end products and daughter products. Within this second line of reasoning, sufficient data should be collected to enable a demonstration of mass reduction and calculation of biodegradation rate constants.

- Native microbiological species can degrade the constituents present in groundwater. This can be evaluated through microcosm studies.

Metabolic respiration can be inhibited by a lack of nutrients or electron acceptors, or if the contaminant concentrations in the aquifer reach levels that are toxic to the microbial community. In certain circumstances, biodegradation processes may be facilitated by the addition of commercially available formulations or by amendment or removal of highly contaminated soil (hot spots).

Nonaqueous Phase Liquids (NAPLs)

Many organic contaminants do not dissolve well in water and can be present in the aquifer as a separate phase. These nonaqueous phase liquids (NAPLs) serve as the source for many dissolved contaminant plumes and are found at many contamination sites (Fig. 6-3(a) and (b)). NAPLs less dense than water are termed LNAPLs; those more dense than water are termed DNAPLs. An example of an LNAPL is gasoline floating on groundwater. Elemental mercury and many chlorinated organic solvents, such as perchlorethylene (PCE) and trichlorethylene (TCE), are examples of DNAPLs. Many of the LNAPLs are the fuel hydrocarbons, including heating oil and the components in gasoline such as benzene, toluene, ethylbenzene, and xylene. Examples of DNAPLs include elemental mercury; the chlorinated organic solvents perchloroethylene (PCE), trichloroethylene (TCE), and trichloroethane (TCA); and many other chlorinated organic compounds.

A significant release of LNAPL will form a floating product layer on groundwater (Fig. 6-3(a)). Soluble components of the LNAPL will dissolve and migrate downgradient. Most regulatory agencies require that floating product be removed to the extent practicable. DNAPLs will sink through the water column and, if present in sufficient quantities, will collect in structural depressions of the underlying clay or bedrock surface (Fig. 6-3(b)).

The presence of a NAPL in the subsurface will drive the scope of remedial action. The NAPL will serve as a source term for a dissolved contaminant plume for many years. A fraction of an LNAPL floating product layer can be removed, but residual quantities will likely remain. There is

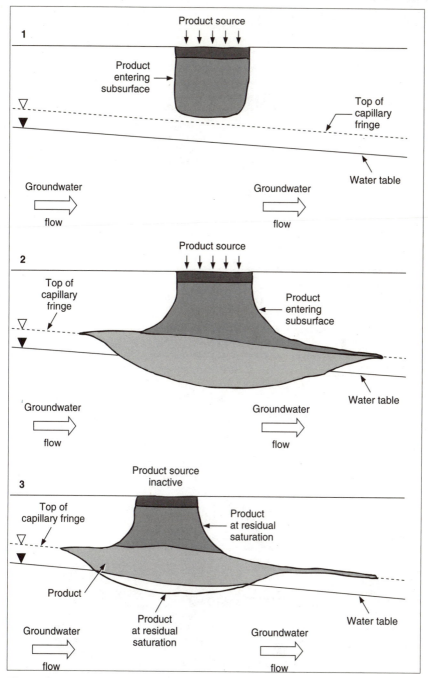

Figure 6-3(a). Movement of LNAPLs into the subsurface. (1) Distribution of LNAPLs after small volume has been spilled. (2) Depression of the capillary fringe and water table. (3) Rebounding of the water table as LNAPLs drain from overlying pore space (Palmer and Johnson, 1989b).

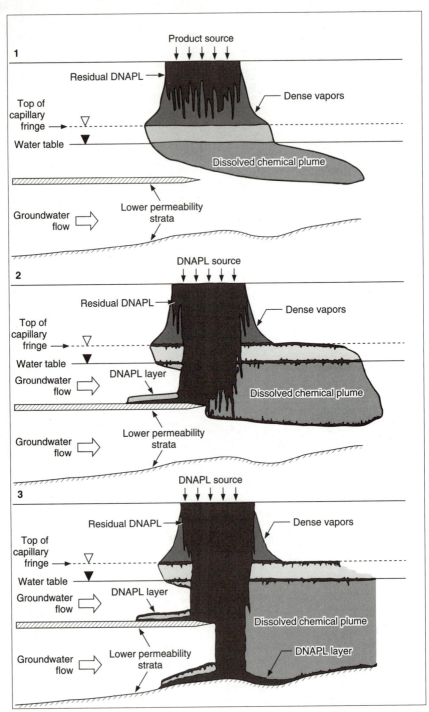

Figure 6-3(*b*). Movement of DNAPLs into the subsurface. (1) Distribution of DNAPLs after small volume has been spilled. (2) Distribution of DNAPLs after moderate volume has been spilled. (3) Distribution of DNAPLs after large volume has been spilled (Palmer and Johnson, 1989b).

no known reliable method for removal of DNAPLs. In situations where removal is impractical and plume migration is an issue, source containment options should be evaluated.

Although aquifer restoration is often a statutory requirement, the presence of DNAPLs in an aquifer makes aquifer restoration to drinking standards virtually impossible. In circumstances where a designated source of drinking water cannot be restored, the enforcement agency may require a demonstration of technical impracticability and the site may be subject to a natural resources damage claim.

Emerging remedial technologies, including surfactant flooding and steam injection, have shown some potential for addressing NAPL contamination. However, the application of experimental or unproven remedial technologies to address DNAPL contamination can result in unforeseen or adverse consequences.

Site Characterization

Site characterization is an iterative process. It is not unusual to go through several phases of site investigation for complex sites. A conceptual model for the site is developed, modified, and refined as the universe of data concerning the site grows.

Preliminary site characterization includes an evaluation of historic site activities through examination of public and private historical records. It is essential that this basic data-collecting phase be as comprehensive as possible. The historical site assessment should delineate potential areas of concern and identify potential contaminants of concern based upon the assessment of previous site uses.

One or more site investigation phases will follow. The purpose of the site investigation phase is to fully characterize the hydrogeology, chemistry, and contaminant transport processes of the site. A Phase I site investigation will involve an initial soil and groundwater characterization effort. Rapid assessment techniques such as geoprobe®, hydropunch, and cone penetrometers, and on-site methods of chemical analysis, such as portable analytical devices and immunoassay kits, should be employed in the early phases of investigation in order to gain a broad understanding of potential site issues. Soil vapor surveys and geophysical methods are also recommended in the early phases of site investigation. This approach will limit initial capital outlays and permit an early, informed decision concerning the property. Initial screening techniques should be augmented by installation of a limited number of continuously sampled soil borings in potential source areas and by completion and sampling of a minimum of three monitoring wells for evaluation of the hydrogeologic setting. Soil and groundwater samples should initially be analyzed for a full suite of organic and inorganic contaminants. Basic indicator parameters and general chemistry should also be determined.

Subsequent phases of investigation are utilized to fill data gaps, refine the conceptual site model, and develop the set of data required to assess remedial action alternatives. These assessments should include evaluation of background conditions, full physical and chemical characterization of site sources, complete delineation of the horizontal and vertical extent of groundwater contamination, complete characterization of the hydrogeologic setting and aquifer hydraulic characteristics, and risk characterization.

There is a tendency to underestimate the time, scope, and costs of site investigation and to cut corners during the early phases of investigation. For example, the initial scope of analytical parameters may be restricted to a few contaminants of concern, or the initial breadth of the investigation may be limited to installation of an insufficient number of borings. Once the decision has been made to proceed with a site investigation, it is advisable to build a comprehensive understanding of the site early in the process and refine and reduce the scope of latter phases on the basis of site-specific knowledge gleaned from the prior characterization efforts. A properly designed and executed phased investigation will greatly facilitate remedial alternatives analysis and reduce the likelihood of costly remedy failure. In some cases, a comprehensive investigation in conjunction with a detailed evaluation of human and ecological risk can be used to justify and support limited response actions.

References

Bear, J., *Hydraulics of Groundwater.* New York: McGraw-Hill. (1979)

Cohen, R.M. and Mercer, J.W., *DNAPL Site Evaluation,* C.K. Smoley ed. Boca Raton, FL: CRC. (1993)

C.W. Fetter, *Contaminant Hydrogeology.* New York: Macmillan. (1993)

Freeze, R.A. and Cherry, J.A., *Groundwater.* Englewood Cliffs, NJ: Prentice-Hall. (1979)

McCarty, P.L. and Semprini, L., "Groundwater Treatment for Chlorinated Solvents." In: *Handbook of Bioremediation.* Boca Raton, FL: Lewis, pp. 87–116. (1994)

Nyer, E.K., Kidd, D.F., Palmer, P.L., Crossman, T.L., Fam, S., Johns, F.J., II, Boettcher, G., and Suthersan, S.S., *In Situ Treatment Technology,* Geraghty & Miller, Environmental Science and Engineering Series. Boca Raton, FL: Lewis. (1996)

Palmer, C.D. and Johnson, R.L., "Physical Processes Controlling the Transport of Contaminants in the Aqueous Phase." In: *Transport and Fate of Contaminants in the Subsurface.* Cincinnati, OH: U.S. Environmental Protection Agency, Center for Environmental Research Information; Ada, OK: Robert S. Kerr Environmental Research Laboratory, EPA/625/4-89/019. (1989a)

Palmer, C.D. and Johnson, R.L., "Physical Processes Controlling the Transport of Non-aqueous Phase Liquids in the Subsurface." In: *Transport and Fate of*

Contaminants in the Subsurface. Cincinnati, OH: U.S. Environmental Protection Agency, Center for Environmental Information; Ada, OK: Robert S. Kerr Environmental Research Laboratory, EPA/625/4-89/019. (1989b)

Pankow, J.F. and Cherry, J.A., *Dense Chlorinated Solvents and Other DNAPL in Groundwater.* Portland, OR: Waterloo Press. (1996)

Heath, R.C., "Basic Ground-Water Hydrology." In: *U.S.G.S. Water-Supply Paper 2220.* Washington, DC: U.S. Geological Survey. (1983)

U.S. Environmental Protection Agency, "Dense Non-aqueous Phase Liquids—A Workshop Summary." In: *EPA/600/R92/030.* Ada, OK: Robert S. Kerr Environmental Research Laboratory. (1992)

U.S. Environmental Protection Agency, "Ground Water—Volume I: Ground Water and Contamination." In: *EPA/625/6-90/016a.* Washington, DC: Office of Research and Development. (1990)

U.S. Environmental Protection Agency, "Ground Water—Volume II: Methodology." In: *EPA/625/6-90/016b.* Washington, DC: Office of Research and Development. (1991)

U.S. Environmental Protection Agency, "Seminar Publication—Protection of Public Water Supplies from Ground-Water Contamination." In: *EPA/625/4-85/016.* Cincinnati, OH: Center for Environmental Research Information. (1985)

U.S. Environmental Protection Agency, "Site Characterization for Subsurface Remediation." In: *U.S. EPA Report EPA/625/4-91/026.* Washington, DC: Office of Research and Development. (1991)

U.S. Environmental Protection Agency, "Symposium on Natural Attentuation of Chlorinated Organics in Ground Water," In: *EPA/540/R-96/509.* Washington, DC: Office of Research and Development. (1996)

Christopher M. French, R.G., C.E.G., received a B.A. in geology in 1981 from Amherst College, and is currently completing an M.S. in environmental science, with an emphasis in Geographic Information Systems (GIS), at Rutgers University. Mr. French has 15 years of experience in consulting and corporate management of contaminated site investigation and remediation. His current responsibilities include management of site remediation and CERCLA projects with the Environmental Services Department of Cytec Industries, Inc. Mr. French is a Registered Professional Geologist and Certified Engineering Geologist in the State of California.

6.8   Interpretation of Results and Decision Making

Harold J. Rafson

The report has been prepared by the environmental professional. It has considered all the elements of good commercial and professional practices in preparing the site assessment. This includes, based on the preliminary objectives, the sampling to be done, the data quality objectives, the methods used in sampling, and analytical methods. The site has been assessed through sampling and geological evaluation. The results have been prepared, and it is necessary to draw conclusions. If the work has been planned and performed properly, decisions will be made on the nature and the extent of the problem. It should be possible to compare existing with allowable contamination quantities. With an understanding of the subsurface conditions to be addressed, the question of what must be done to remediate the site can be considered.

Let us assume at this point that the developer decides that the extent of contamination found, the site conditions, and the business aspects of the project are favorable, so that it is desirable to advance work to the next step. This is a feasibility study that will estimate the costs of remediation, or the use of alternative methods, to comply with the regulations.

Feasibility Study

In performing the feasibility study, the engineer has to consider reasonable options for work to be done, or other methods for control.

The developer has selected the subject site. A Phase I investigation has suggested the possibility of contamination. A Phase II study has identified the type and extent of contamination.

The developer has a pretty good idea of what he or she will have to pay for the property—but this is not yet fixed, because negotiations have not begun.

The developer has an estimate of what the market value will be for the cleaned-up property—but that is not fixed: final negotiations with a potential buyer are still a long way off.

The question is how much it will cost to remediate, and whether the developer will be left with an adequate profit to cover all the time, costs, and risk involved.

There are still many unknowns. What should the strategy be in developing a feasibility study? A consulting engineer working for the developer is most likely to be conservative and provide for a worst-case scenario. The consulting engineer certainly doesn't want to mislead the developer, or to be so conservative that the developer now feels that he or she has wasted money and corporate resources, by determining that remediation costs and time to complete the work will far exceed early estimates. The engineer has to balance his or her advice and the risk of errors and omissions, whether he or she gets the contract for the study, and his or her professional reputation. There is uncertainty in the work, and there is often a choice of the amount of risk the engineer foresees in the work. The engineer must be forthright and fully communicate with the developer, so the developer understands the basis of the advice the engineer provides.

It doesn't take much conservatism to throw a project from profitable to unprofitable. The developer has selected this project, and believes it to be good, for many other reasons: there may be a purchaser for the remediated property; the developer doesn't want to lose a potentially profitable project he or she has already spent time on. On the other hand, the developer certainly doesn't want to get into something that yields a loss or only a small profit, and that is a great waste of time and effort.

It is this uncertainty on the part of both players involved in this feasibility study that makes it one of the critical parts in the entire process. To be successful, a feasibility study requires a great deal of involvement on the part of the developer, because the developer has to make his or her own business decisions.

To ensure a decisive study, the engineer must involve the developer in the project all the way, and discuss with the developer all the decisions that are being made where the engineer is uncertain.

Let us take an example. There is a project in which a developer can get a site for nothing, and the developer knows that the remediated site can be sold for $2 million. This site is in a good location, and has good buildings, but it requires a great deal of remediation work. Because of the uncertainty, particularly about groundwater contamination which has been inadequately studied, the engineer's estimate for remediation of the project can be anywhere from $500,000 to $3 million. The engineer's first recommendation, of course, is that additional studies will be required, but those studies will cost $50,000 and the developer wonders whether it is worthwhile. There must be a serious discussion of the likelihood of keeping the remediation costs below $1.5 million.

The developer decides to proceed with further investigation, and it is found that the underground water contamination is limited. A pump and treat system for the groundwater is installed, and a soil cleanup by land farming is done (because of the volatility of the contaminant). The final remediation costs are about $600,000, in addition to the project's other costs of about $300,000. This concludes as a very profitable project for the developer. It should be noted that in today's climate, there are still few developers who will risk projects with the potential for high and uncertain remediation costs. There are many other opportunities for developers' time and investment that are less troublesome.

A properly done feasibility study should include the following:

1. A clear description of contamination levels and locations from the Phase II studies

2. A description of the geology and hydrology of the site

3. A review of the options for remediation, with the costs and true estimates from the most reasonable alternatives

4. A statement of the issues of uncertainty in the study

5. A discussion of the use of alternative methods, such as institutional
controls and engineered barriers

For biographical information on Harold Rafson, see Sec. 1.1.

6.9   Final Analysis for Developers

Robert Rafson

Finally, all the estimates are in and it's time to make a decision. This is the
point of no return on the project. The final cost estimates, both of soft costs
and hard costs, are gathered and the issues with the city, county, and state
are understood. Now it is time to execute the contracts and start the work.
The final analysis should include the issues listed in this chapter and consider the equity investor's perspective.

There are many issues that may not be financial issues but that have
great impact on the evaluation of the project. Following is a list of items to
be reviewed in the final analysis of the project.

Economic:

   Funding availability

   Unpaid real estate taxes

   Other liens

   Project costs (soft and hard costs)

   Project timeline (construction and financial)

   Potential end users

   Incentives

   Net returns on investment

Legal:

   Survey

   Title search

   Zoning and neighborhood issues

Regulatory/environmental:

   Superfund and other environmental liabilities

   Cleanup requirements

Environmental timeline

Stigma difficulties

If the economic analysis works, then the project will most likely work. There are several key components within this analysis, each of which has an important effect on the economic viability of the project. Funding, costs, and incentives provide the basis to determine return on invested capital. Therefore, this bottom-line analysis is a key consideration during the review of project viability.

The project timing and financing must be coordinated. The amount of money required at each step of the project must be coordinated with the project timeline. This might seem obvious, but brownfields projects have the problem of the environmental cleanup occurring prior to and during the restoration or construction phase of the project. This may cause challenges to the financing due to the greater level of risk early in the project. As stated earlier, banks and other lending institutions may be willing to lend on these projects, but they are very concerned about the liabilities and risks the project represents. Since they have no interest in potentially foreclosing on these troubled sites, they may require additional assurances to make a loan possible.

Project budgeting has been reviewed in Sec. 6.2. Projecting costs and scheduling clarifies the scope of the project. These costs, along with all the other problems imparing value, will provide a fairly good view of the project's viability. There should be a final review of these numbers and the contingency reserves needed because of the complexity and uncertainties.

Some discussions of incentives have been reviewed in Secs. 4.1 and 4.3, and there are new incentives being developed, but there are also other considerations associated with incentives. For instance, if the city provides construction financing or a block grant, there may be minority- or women-owned business contractor requirements, and this may require a reevaluation of prior cost estimates or include heavy reporting requirements that may decrease the desirability of the incentive. A very careful look at the fine print of these incentives is required prior to depending on the incentive for the project financing.

When all the issues have been taken into account and the project cost estimates are all in, the bottom line of return on investment can be reviewed. The return on investment (ROI) takes into account the timing, financing structure, and project costs to arrive at a net return amount. The return on investment can be calculated in several ways, as described in Sec. 6.2.

The return on investment of a brownfield project must be higher than that on a property without environmental contamination. Higher perceived risk leads to the lender's requirement for a greater equity component of financing. Equity return can range widely and often requires 50 percent or more of the project profits. Higher equity requirements, added to the bank's requirements, can make project profitability difficult. Additionally, the marketing of a previously contaminated site or one that has lingering contamination issues will require below-market pricing, reducing potential profit on the project. A very conservative review of ROI must be made because of these problems. It is important for all parties involved, including government agencies, to understand that these higher ROI requirements are necessary for the project to go forward and for the developer to take on the risks involved.

Attorneys are treated as expert advisors and clients take their attorneys' word on potential risks. For their part, attorneys depend on environmental engineers to determine potential risks. Lack of experience and a conservative view of these transactions incline attorneys to give very conservative advice, and an overly exaggerated view of potential risk, to their clients. One attorney told this author that he gets no pat on the back when he suggests that a risk is low and the deal goes well; but he can be sued if he underestimates the risks. Legal and environmental advice can determine the deal. It is the author's recommendation that all persons involved in these deals hear the advice of each of the advisors (environmental engineer and attorney) for themselves and ask questions until they understand the real and perceived risks, because they have to make the decisions themselves and not delegate the decisions to others.

During due diligence there are several parts of the process that need to be reviewed carefully to get to the bottom line of return on investment. The survey of a property can uncover subtle information regarding the site and its redevelopment potential—for example, data on easements, encroachments, and underground services. A survey will often be needed to obtain accurate renovation budgets. Easements and encroachments may need to be addressed prior to closing because there is negotiating leverage with the owner before closing while there is none afterward. If the easements or encroachments are extensive enough and cannot be satisfactorily resolved, this might be a deal breaker. For example, some properties have decades-old railroad easements that crisscross the property. This could eliminate the possibility of adding structures on the property, thus making the redevelopment of the site not feasible.

Title insurance and a review of the title policy provides the assurance that there are no additional liens or encumbrances on the title. This is one of the most important parts of due diligence. Brownfields are problem properties with contamination and often insufficient funds to achieve cleanup prior to sale. There may be additional "clouds" on the title such as liens, unpaid taxes, mortgages, or other claims on the property. This is why fully understanding the title issues is so important to the final deal.

Zoning of the property can be one of the most critical issues in determining the final value and potential marketing of the completed site. If the site is within the Planned Manufacturing Districts or some other established manufacturing areas, the zoning could be considered stable. If the site is in an area that is being converted from industrial to commercial or residential property, then the site potential redevelopment may have higher value as residential and that may be the best use for the site. This situation also may reduce the potential redevelopment for industrial use because of the proximity of residential neighbors. The zoning may not be stable. There are opportunities to convert properties to a different zoning designation, and it is often best to redevelop a property to a use that is compatible with the other properties in the area; this may mean requesting a zoning change from industrial to residential, which often happens. Occasionally there may be changes from residential to industrial zoning that are mandated by the city council if an area has been designated for conversion to a manufacturing district.

Projects can be much easier to manage and less risky if there is a specific end user for the redevelopment. If the project proceeds on a speculative basis, the carrying costs during marketing may be too large to merit continuing on the project. Projects like these often must be redeveloped to suit a specific type of end user, if not a specific company, in order to minimize marketing risks. This end use must meet zoning and neighborhood requirements.

Once a property is defined as a Superfund site, it will always have a stigma to overcome. This has few advantages and many problems. The positives are that a prospective purchaser agreement is possible on some sites. This will provide significant protections from suits and cleanup costs. Unfortunately, regardless of any agreement and the actual site cleanliness, the stigma created by the Superfund designation is significant. Many potential users will shy away, if not run away, from these sites. Sites that have imminent danger to health and the environment may be cleaned up by EPA's Superfund Emergency Response arm, but this often results only in cleanup of the superficial surface problems and leaves the underground contamination. The agreements provided as a result of the limited cleanup will severely limit the end use and development of the site. Great care needs to be taken when creating the agreement and development plan to ensure that the process does not trigger extremely large environmental liabilities and also the requirement to repay the Superfund cleanup costs.

The environmental timeline can be either very short (e.g., tank removal) or extremely long (e.g., groundwater cleanup). A realistic discussion of the potential marketability of the project if there is a long-term cleanup must be had at the time this decision is made. If the end user will not take title or even lease the property until the site is clean, all remedial activities have been completed, and an NFR letter is received, then the cleanup timeline may destroy the project's schedule, potential profitability, and financing.

Finally, a decision must be made. After all the factors are weighed, is the project worth doing? The answer to that question has been the purpose of all the estimates, planning, and due diligence.

For biographical information on Robert Rafson, see Sec. 5.1.

7

The Private
Developer—
Remediation

The project for redevelopment goes ahead.

In Sec. 7.1, Robert Rafson discusses how the developer deals with the contractors.

In Sec. 7.2, Carey S. Rosemarin and Steven M. Siros bring in the lawyer concerning the contamination issues.

In Sec. 7.3, Andrew Warren treats another aspect of the legal issues concerning the prospective purchaser agreement, which limits liability.

In Sec. 7.4, Ernest Di Monte deals with accounting issues and tax effects of the project costs.

In Sec. 7.5, Dinah Szander discusses environmental insurance, which is a newer approach to dealing with uncertainties. This section considers the need for caution in dealing with the insurance industry. It follows years of litigation for nonpayment of claims and years of unwillingness to insure pollution risks. Now, new insurance vehicles and commitments are becoming an increasing and useful approach, and new types of coverage are being developed.

In Sec. 7.6, the remediation work now has to be done. The soil and water contamination issues must be corrected, the regulations satisfied, and all work properly sampled, tested, analyzed, interpreted, and monitored. This is the time of project management. Everything has led up to this work, and simultaneously the developer will be performing the demolition and reconstruction. Harold Rafson provides an overview of the issues involved.

7.1   Redevelopment Costs

Robert Rafson

Redevelopment costs of a brownfield project include standard site preparation, construction and landscaping, and environmental cleanup. This discussion focuses on how the brownfield environmental cleanup affects both project timing and costs.

Architectural plans are often the first thing the developer works to complete. These plans provide the basis for the project to move forward. The plans will provide the information the subcontractors need to bid on their portion of work, and allow the general contractor to lay out a plan and schedule for the project and determine the details that allow *critical path analysis*—the process of determining each item that will, if delayed, cause the next step in the project to be delayed. The critical path starts with the creation of architectural plans and then the permitting process. The architectural plans must coordinate with the cleanup, which will likely be affected by soil conditions and restrictions on uses of the site.

Environmental plans (Phase III) are often the next critical path item. The development of those plans, like that for architectural plans, need to be managed to ensure that the design of the cleanup will fit the time, money, and cleanup objectives of the project. Phase III is described in detail in the environmental remediation section, but from the developer's point of view the environmental plan is critical to the project. Depending on the plan, the cleanup may require area, materials, and time that would interfere with the remainder of the project, thus making it a key consideration in the critical path of the project.

There are other issues related to the environmental plans that need to be explored at this point. There is no obligation for the state to be involved in the process of remediation except if there is a desire for a no further remediation (NFR) letter by means of the site remediation program and removal of underground storage tanks. The NFR letter states that once the property is cleaned up to the agreed-upon plan objectives, which comply with the state TACO, then for this use no further remediation is necessary. This letter is additional proof for lenders and buyers that the site is clean. Additionally, since the state has a memorandum of understanding with the federal EPA, the NFR letter protects the site from additional cleanup requirements imposed by USEPA.

Underground storage tank removal must be done in the presence of the state fire marshal. This is done to ensure that the tank is safely removed and that no visible leakage is present. In Chicago, the City Department of Environment manages the tank removal process and must also be present when a tank is removed within the city limits. If the tank removal and cleanup necessary is put through the IEPA leaking underground storage tank (LUST) program, a state NFR letter can be obtained.

The process of gaining state acceptance of the cleanup plan can be time consuming, depending on the complexity of the cleanup. If the cleanup is a tank pull, the state review is extremely short, but if groundwater or alternative technologies are included, the plan may take a while to be accepted. One of the most important functions of the negotiations for the acceptance of the cleanup plan is to keep the plan within the timing and expenses anticipated in the planning phase. It is easy for these state reviews to request additional or more expensive testing, lower cleanup objectives, or more reporting. Risks of additional testing and differing cleanup objectives are much less of a factor than they were before the adoption of TACO, which more clearly defines cleanup objectives.

The state voluntary cleanup program (also called a site remediation program) also has disadvantages. The state wants to ensure that once a cleanup begins, it will be complete. There are two options when a project enters the program. The applicant can apply with a $500 fee or advance partial payment of the anticipated costs. The retainer will be returned once the cleanup is completed and accepted by the state. That means that the developer must come up with the cleanup costs plus advance partial payment plus engineering costs, all of which may be equity money. That makes the cleanup of sites more costly in the voluntary cleanup program. There also will be much more oversight and the possibility of changes in the states' opinion of the extent of the cleanup. For these reasons, many cleanups proceed to successfully remediate the property without assistance or oversight by the state. However, the lack of the NFR letter (approval by the state) will make a property harder to sell or lease. In some projects the NFR letter is a contractual requirement.

Most developers will, however, enter into the site remediation program to receive the NFR letter because it is a powerful tool for obtaining financing and smoothes the sale of the property. Usually these advantages outweigh the delays and expense.

Site use is an important part of determining project costs. The site use may require specific redevelopment that could add to the redevelopment costs. The use could also significantly change the cleanup objectives and the determination of the cleanup methods. Therefore, if the end use or likely use is known, the evaluation of redevelopment can be much more accurately determined.

Coordination of cleanup with property use may be critical to deciding what to do, or may even impose significant limitation on the possible end uses of the property. If, for instance, the use requires that the building be surrounded by paved parking or loading/unloading areas, and some or all of the contamination is contained in that area, the cleanup dilemma might be solved because the paving can be the engineered barrier needed. On the other hand, there are cleanups that have nothing to do with the end use. Underground storage tanks must be abandoned, removed, or upgraded to the required standards by December 22, 1998, or be liable for fines. Underground storage tanks also add a great amount of stigma to the property, and many developers remove these tanks as a matter of course to remove the stigma even if they are not required to do so by law.

Asbestos removal, underground tank removal, and disposal of hazardous materials are part of the demolition of almost any building. Virtually all demolition contractors in Chicago are asbestos removal firms, and most have experience in tank removal and disposal of hazardous materials since these are standard procedures in the demolition of a site. These costs are also very predictable, since these parts of the cleanup are easily determinable.

Setting up for long-term remediation is the most risky part of any remediation project. It is difficult to determine both the remediation cost and duration of cleanup required. The choice of method can ease the determination, but there are many geological factors that make the cleanup, especially when in situ methods are involved, hard to estimate. The risk of cost overruns from these types of cleanups is significant. Many developers will add 50 percent contingency to the environmental engineer's estimate to cover these potential overruns.

Bioremediation, soil venting, soil washing, and other remediation methods may also affect the construction schedule and determine the redevelopment timetable. These remediations may be in the path of the new development and thus need to be completed before any significant building can start. This being the case, the faster the remediation method, the more attractive it is. Multiyear cleanups add a level of uncertainty that the market for this property will remain strong. There are additional risks involved in longer-term cleanups.

Inner city cleanups also involve problems of vandalism, theft, and other access issues that must also be weighed in the determination of cleanup methods and redevelopment paths.

Last, the costs of cleanups have significantly changed with the implementation of risk-based cleanup objectives (TACO in Illinois). Risk-based cleanups mean, for the developer, that the cleanup objectives can be specifically determined and the hurdle can be defined. Risk-based analysis can also open opportunities for natural attenuation, engineered barriers, and legislative barriers to limit cleanup costs and requirements. This is discussed in great detail in Sec. 6.4, but for the developer and for determining project schedule and costs it is the most important development in the recent past. Now a developer can significantly reduce the risk based on negotiated objects as it was in the past. Cost overruns since TACO are rare, and when they occur, the costs are small in comparison to those incurred before TACO regulations were adopted.

For biographical information on Robert Rafson, see Sec. 5.1.

7.2 Legal Devices Used to Deal with Contamination

Carey S. Rosemarin and Steven M. Siros

The significant adverse impact of the Comprehensive Response Compensation Liability, 42 U.S.C. § 9601 *et seq.* (CERCLA or Superfund), on the transfer of real estate and businesses is well known. The concept, embodied by CERCLA, that a party might be held liable merely on the basis of its status as an owner of contaminated property, combined with the magnitude of environmental remediation costs, has dampened the enthusiasm of many would-be property purchasers. Yet, while an incalculable number of transactions have dissolved on the basis of environmental problems, others have proceeded to closing. The difference lies in the fact that in the latter transactions the parties have found acceptable ways to allocate the liability. This chapter discusses the process by which parties have divided the potential liabilities, and the legal devices parties have drawn upon to facilitate transfers of contaminated property.

It is helpful to separate the process into two parts—defining the legal risk (of which environmental risk is a subpart) and allocating it—recognizing that the parts overlap. The legal risk is defined by acquiring information about those aspects of the property that will imply costs for the owner, or upon which liability can be premised. In that sense, buying contaminated property is little different than buying other items known to harbor potentially hidden costs. For example, the purchase of a used car presents risk because a car is a complex machine and in many instances little is known about its history. Some of the risk can be defined by having a mechanic inspect the car prior to purchase. A more germane example is the purchase of a building constructed in the early 1950s. It is suspected that the structure, full of pipe chases and deteriorating insulation, contains asbestos, which will have to be addressed prior to renovation of the building. Until an asbestos survey is conducted, there exists a risk that the costs of dealing with the asbestos will exceed the value of the building. By conducting the asbestos survey, the prospective purchaser buys information to define the risk.

In some cases, defining the risk ends the exercise, either because the information collected confirms that the environmental concerns are minimal, or because it shows that the degree of remediation required will spoil the economics of the deal. However, most cases lie somewhere between these two extremes. The parties have come to the bargaining table because the transaction presented perceived benefits to each, and often those perceived rewards provide the incentives to pursue resolutions to the environmental issues. It is at this point that the second part of the process begins, and legal devices may play a crucial role. Returning to the automobile example, suppose the shopper is a single parent who can ill afford to have a car retained at the repair shop, but who also cannot afford a new car. A risk-averse purchaser in this situation might find comfort in an extended warranty covering certain possible repairs. The warranty contract does not make the possible mechanical problems disappear, but it makes the risks of unwanted costs tolerable. Analogous devices may be used to effect the sale in the case of the 1950s-vintage building. For instance, the parties may agree that the seller will be responsible for removing asbestos only in the areas the buyer plans to renovate. Such mechanisms are obviously more complicated, but are limited only by the creativity of the parties.

The following paragraphs discuss legal provisions that guide parties in collecting relevant information and thus maximize the efficiency of their information-gathering dollars. The remainder of this section focuses on the use of that information. We will discuss how certain legal devices may be applied to reduce the legal risk presented by environmental conditions.

Gathering Information

As noted earlier, transfers of contaminated property did not cease upon the enactment of Superfund. Rather, because of the potentially high costs that can be imposed upon an owner of contaminated property, environmental concerns quickly found their place among the numerous business factors considered in making the decision of whether to purchase the asset. Accordingly, in the course of a prospective purchaser's due diligence (the gathering of information to determine whether the transaction under consideration is a sound business investment), information on the environmental condition of the property is routinely collected.[1]

Phase I and Phase II
Environmental Assessments

The starting point is a Phase I environmental assessment.[2] The purpose of a Phase I inspection is to identify any physical conditions on or around the property that may form the basis of the imposition of liability on the new owner under environmental laws. The Phase I inspection is used to determine whether a more in-depth (and more costly) investigation, the Phase II assessment, should be performed. There exists no singular definition of a Phase I inspection, although general agreement about the fundamental concept has developed over the years. The term is generally understood to refer to the collection and synthesis of information derived from a review of publicly available sources, and an on-site visual inspection. A number of useful checklists have been published that identify numerous factors that a Phase I assessment should contain.[3] The purpose of this chapter is not to add to the numerous articles explaining the contents of a Phase I inspection, although the following explanation should provide a general sense of how a Phase I inspection should be conducted and used.[4] Phase I inspections are discussed further in Secs. 5.6 and 6.5.

Among the publicly available sources that are universally considered in a Phase I assessment is the chain of title, which is reviewed to determine whether the property was occupied by persons whose activities may have caused pollutants to be deposited on the property. The fact that "Ace Metal Plating" may have owned the property between 1948 and 1962 would likely cause a prudent purchaser to conduct further investigation, but ownership by "Harry's Family Shoes," a retail establishment, may not. Ownership records may not be sufficiently revealing, however. For example, historically a common practice in Illinois was to place the ownership of real property in land trusts.[5] In such cases, the owner of record may be a bank, acting in its capacity as trustee, and the beneficial owner is not identified in the office of the recorder of deeds. Old aerial photographs, historic newspaper articles, and other public sources may provide more information about the historic uses of the property. Interviews with long-time residents of the area are also commonly used for this purpose.

Invasive sampling, such as is accomplished by soil borings or installing monitoring wells to sample and analyze groundwater, is relatively expensive, and therefore is usually not included within the scope of a Phase I assessment. These activities are most often reserved for the Phase II study. But the Phase I assessment, if well researched and well written, should indicate whether invasive sampling is necessary. It should provide a reasoned answer to the question of whether the prospective purchaser has a reasonable basis to be concerned about the existence of hazardous substances on the property that are likely to cause future legal problems. Depending on the magnitude and implication of the potential problem, the purchaser can make a reasoned decision as to whether to expend the additional resources for subsequent invasive sampling. The design of the Phase II study is unique to the problem at hand, and may include a number of separate trips back to the property to collect samples for laboratory analysis. Aside from observing proper health and safety procedures and using accepted sampling and analytical techniques, there is no standard procedure. Within the aforementioned parameters, cost is the dominant factor in how the Phase II portion of the assessment proceeds.

Innocent Purchaser Defense

For years after the passage of CERCLA in 1980, the manner in which Phase I and Phase II studies were conducted varied widely, as did their quality.[6] Some degree of uniformity was injected into the practice during CERCLA's second decade, after the amendment to the statute in 1986. CERCLA contains several narrowly interpreted defenses to liability (i.e., prescribed sets of facts that, if proved by the defendant, defeat the plaintiff's cause of action), one of which essentially states that an owner may not be held liable if he or she can demonstrate that the contamination was caused by a third party with whom the owner had no *contractual relationship*.[7] The main effect of the contractual relationship provision was to prevent owners of industrial facilities from evading liability by stating that the contamination was caused by their tenants, suppliers, independent contractors, or others with whom they regularly dealt. However, another effect of that provision was to prevent current owners from evading liability by claiming that contamination was caused by former owners who may no longer be viable. Thus, owners who sought to defend against CERCLA court actions on the basis that a third party (the former owner) caused the contamination were usually not successful because they had contractual relationships with the former owners (i.e., the purchase and sale agreement).[8]

In 1986, Congress enacted the Superfund Amendments and Reauthorization Act (SARA).[9] SARA ostensibly provided some relief for property owners by including a definition of contractual relationships.[10] The definition was particularly complex, but its key provision required that an owner would be deemed *not* to be in a contractual relationship with the former owner (and thus eligible to take advantage of the defense) if the present owner could show that he or she "had no reason to know" of the presence of the hazardous substances even though he or she undertook "all appropriate inquiry into the previous ownership and uses of the property consistent with good commercial or customary practice in an effort to minimize liability." The defense in 42 U.S.C. § 9607(b)(3), containing the contractual provision language, is commonly known as the *third-party defense*, and its invocation by current property owners has come to be known as the *innocent purchaser defense*.[11] Most efforts to invoke the innocent purchaser defense have not been successful, and this has been true both before and after the passage of SARA.[12]

The respected American Society for Testing and Materials (ASTM) has devised a standard method of environmental assessments that is premised on this statutory provision.[13] This ASTM method established a good commercial and customary practice for conducting environmental site assessments in order to facilitate the use of CERCLA's innocent landowner defense. The ASTM method includes guidance on Phase I assessments, records review, site reconnaissance, interviews with owners and occupants, and report preparation.[14]

The Illinois Environmental Protection Act contains a number of provisions that mirror the federal CERCLA statute.[15] Illinois also has a third-party defense, which is virtually identical to the federal version.[16] And, like CERCLA, the state statute also provides that there will be no contractual relationship for purposes of the third-party defense if the owner/defendant can show that he or she "had no reason to know" of the existence of the hazardous substances. However, the Illinois statute parts company with its federal counterpart by stating in detail what is necessary to prove that the owner had no reason to know.[17] Specifically, it states that the defendant owner must have conducted a Phase I audit that did not indicate the presence of contamination—and it prescribes the elements of a Phase I audit. The statute states in relevant part:[18]

> . . . the term "Phase I Environmental Audit" means an investigation of real property, conducted by environmental professionals, to discover the presence or likely presence of a release or a substantial threat of a release of a hazardous substance or pesticide at, on, to, or from real property, and whether a release or a substantial threat of a release of a hazardous substance or pesticide has occurred or may occur at, on, to, or from the real property.

The statute goes on to require that a valid Phase I study include reviews of the chain of title for the prior 50 years, aerial photographs, environmental liens, reasonably obtainable government documents, and "business records" for the previous 50 years, as well as a "visual site inspection."[19]

The statute notes that if the results of the Phase I audit suggest a reasonable basis to believe that contamination exists, but the purchaser did not proceed to a Phase II audit, which involves soil and groundwater testing, then the exemption from the contractual relationship definition will not be available, and thus the innocent purchaser version of the third-party defense will not be an option.[20] This provision is consistent with the purpose of a Phase I assessment. As noted previously, the purpose of the Phase I assessment, whether in the context of the statutory third-party defense or the potential purchase of a brownfield, is to indicate whether additional problems exist, that is, to define the risk. A Phase II assessment may confirm or disprove the theories developed as a result of the Phase I assessment, but ultimately the purchaser reaches the point at which he or she will cease spending money to pay for information, and either walk away or proceed to the next stage of the transaction—allocating the risk.

Allocating the Risk

After the parties have acquired some sense of the environmental issues presented by the site, they enter a new stage of the negotiations in which they further identify the risk, and then divide and reduce it. These functions are not distinct, and blend in the course of the parties' efforts to forge a deal. Nonetheless, the parties usually attempt to establish a basic agreement fairly early in the process as to who will be responsible for which environmental conditions. This tends to be a rather informal process shaped by the particular needs of each of the parties. After the agreement is struck, various legal devices are used to facilitate the deal.

Division of Liability

To some extent, the risk identification that was the object of the Phase I and Phase II environmental assessments spills over into the negotiations concerning division of liability. This occurs because legal risk is not solely a function of the presence of contaminants on the property, but of the relative likelihood that the owner will be required to spend money to remediate the contamination.[21] The distinction referred to here is that between a person's status as one of the parties identified in Section 107 of CERCLA, 42 U.S.C. § 9607 (i.e., a present owner of contaminated property), and as a person who has an affirmative duty to remediate. In other words, ownership of contaminated property may make a person susceptible to a cleanup order (from the government, or from a court in cases initiated by private parties), but until such an order is issued (absent spills or certain other events), an owner may not have a duty to remediate the contamination. Thus, after the environmental conditions have been identified, the parties continue to identify the risk by identifying those events that are likely to trigger a need to clean up.

Consider the example of a developer purchasing a parcel known to contain lead in soils in the southwest quadrant, but not in the groundwater. If the developer's plans call for the placement of a parking lot in the contaminated area, the contamination may precipitate little marginal cost, and the seller, who may be responsible for the disposal of the lead, may agree to retain the liability as long as the purchaser maintains the parking lot. On the other hand, the developer's plans may not be crystallized, and a review of government records may show that the state environmental enforcement authority has some concerns about the lead, in light of the location of a playground within 200 feet of the area. In these situations, the legal risk is increased because potential triggers of liability have been identified.

Notwithstanding the fluidity of certain factors influencing the degree of legal risk, the deal demands that the liabilities be allocated. That division assumes a wide variety of forms. In theory, the most logical starting point is a proposal that the seller remain liable for all contamination on the property up to the closing date, and the buyer accept liability for all contamination placed on the property after the closing date.

Although such agreements are not uncommon, various forces driving the deal may not allow such simplicity. Ultimately, the market may require one of the parties to accept more of the liability than this starting point might suggest. If the demand for property in a particular location is high, it may not be unreasonable for the seller to require that the new owner accept responsibility for existing contamination. Conversely, if the demand for the property is low, to effect the transaction, the seller may have to agree not merely to remain liable for the contamination but to affirmatively clean it up before closing.

Numerous other forces enter the mix. For example, the seller may not be willing to remain responsible for all contamination on the property as of the closing date, especially if the contamination was deposited on the property by one of the seller's predecessors in title or a former tenant. Or, for any number of reasons, the seller may have a need to cut all ties with the property and be insulated from all liability relating to it after the closing date. The buyer may be willing to purchase a brownfield, but may require that it be clean at the time of purchase. The buyer may also require a seller to be available as a potential source of cleanup funds long after the closing date. Although such factors may kill the deal, a number of legal devices may be employed to work a compromise and reduce the risk assumed by each party to an acceptable level.

Indemnity Agreements

The most widely used arrangement is an *indemnity*, which is a promise by one party to remain responsible for certain liabilities. It may be limited in its scope of coverage, time, and funding. Again, reflecting the market and other forces that shape the transaction, the indemnity may run from the seller to the buyer, or vice versa. Courts have widely honored private indemnity agreements providing for future CERCLA liabilities.[22] However, they carefully scrutinize the language in an effort to determine the intent of the parties at the time the agreement was entered into.[23] Therefore, the indemnity language must be carefully drafted and is often the subject of intense negotiation.

An equally important consideration in determining that an indemnity sufficiently reduces the risk at hand is the realistic protection it may afford, and the practical difficulty the indemnitee may encounter in seeking to enforce the indemnity agreement. Conceptually, the indemnity can be thought of as an insurance policy issued by a private party. (Increasingly, parties are using insurance policies to allocate the risk of environmental costs, as discussed further in Sec. 7.5.) However, as a practical matter, it is very different. Litigation is often necessary to force a party to honor its alleged indemnity obligations. Parties should consider a number of factors before concluding that the indemnity agreement satisfies the need to reduce the risk.

The primary question that must be asked is whether the indemnitor is likely to have the financial ability to honor the indemnity, or whether the indemnitor will even be in existence.[24] If the indemnitee harbors any doubt about these matters, it should confront them in the course of the negotiations. The consequence of not doing so could be the loss of the risk reduction sought by the indemnity. Thus, just as indemnities are familiar provisions in contracts providing for the sale of contaminated property, so are provisions that secure them.

For instance, in the example just given, the parties may have agreed that the seller is to indemnify the buyer for all remediation costs "resulting from an order from any duly authorized government authority to remediate the lead in the soil on the southwest quadrant." If the seller or its business plans are not well known to the buyer, the latter may be wise to demand that a reasonable amount of money based on an estimate of the cost of the anticipated remediation be deducted from the purchase price and held in escrow for a certain amount of time. Alternatively, in the event that the seller has a parent corporation that is more creditworthy than the subsidiary, it may be possible for the parent to serve as a guarantor. In turn, the seller may want some assurances of its own. In the event that contamination for which the seller is responsible is to be left on the property, the seller's interests are advanced by selling to a buyer that has no intentions of expanding the improvements on the property or otherwise disturbing the affected soil. Therefore, the seller may condition its indemnification obligations on the buyer's commitment to leave the area undisturbed.

Cleanups and Cleanup Standards

Under the best of circumstances, it is clear from the preceding discussion that some degree of legal risk will remain, even with the most carefully drafted indemnity. Accordingly, parties with a need for more certainty may decide that the best course of action is to remediate on their own, with or without the involvement of a governmental entity. In making this decision, the parties remove the risk by excising the physical problem. This course of action is often taken when the problem is relatively small, such as with a limited amount of accessible asbestos. In such instances, the seller (before closing) or the buyer (after closing) may cause the asbestos to be removed and for all intents and purposes eliminate the problem without government intervention. The removal of underground storage tanks is also a subject on which the parties are often able to agree. However, in most cases these projects require some governmental supervision because federal regulations require that the project be supervised by the appropriate state agency.[25]

In transactions involving properties with larger and more complex contamination patterns, greater effort may be required to reduce the risk to acceptable levels, but in recent years the law has responded to meet this need. Parties have always had the option of devising and implementing their own cleanup plans, and of attempting to remove the legal risk along with the physical risk. However, historically, such actions were viewed as being fraught with the risk that a government agency would intervene at a later date and require additional or different remedial work to be undertaken. This risk existed because of the lack of universally accepted cleanup standards. Indeed, this condition, combined with the significant powers of the federal and state governments to force current owners to clean up, was a significant factor in creating the brownfields problem that is the subject of this book.[26] This aspect of the practice of environmental law has changed significantly, and may be the most important development of this area of the law in the 1990s.

USEPA has adopted soil screening levels (SSLs), which are concentrations of various contaminants in soil below which USEPA is not likely to apply its enforcement resources.[27] The levels are very conservative, and are premised on residential use of property. Perhaps a more important development was the publication by ASTM of its procedures for risk-based corrective action (RBCA, also referred to as *Rebecca*).[28] The fundamental concept advanced by RBCA is that remedial efforts should be applied to contamination where it presents a risk to human health or the environment. Concomitantly, RBCA allows for the removal of the risk by means other than remediation. For example, contamination on the subject property may exist at a level that is unacceptable if the property is to be used as a residence, but tolerable if the property is to be used for industrial purposes. In such cases, RBCA allows that the contaminants need not be removed if some constraint, such as a deed restriction, is instituted that assures that the property will not be used for residential purposes. A number of states have adopted some form of the RBCA process.[29]

Many states have also adopted voluntary cleanup programs that in turn incorporate some form of the RBCA process. Under state voluntary cleanup programs, owners of contaminated property may clean up their properties under the auspices of the state government in the absence of an enforcement action. Upon completion of the cleanup, the owner receives some recognition by the state that the site has been properly remediated. Such programs offer parties to a potential transaction an alternative means to acquire the certainty sought by each and get the deal done.

Illinois' Site Remediation Program is a good example.[30] Under the program, an applicant submits a cleanup plan premised on cleanup levels derived under the TACO system. The cleanup plan can be implemented upon approval from the state and the remediation project is reviewed by the state upon completion. If the state acknowledges that the project is complete, it issues a no further remediation letter. The importance of the no further remediation letter is shown by the language of the statute:[31]

> The Agency's issuance of the No Further Remediation Letter signifies a *release* [emphasis added] from further responsibilities under this Act in performing the approved remedial action and shall be considered prima facie evidence that the site does not constitute a threat to human health and the environment and does not require further remediation under this Act, so long as the site is utilized in accordance with the terms of the No Further Remediation Letter.

Furthermore, the statute provides that the letter applies not only to the current owner of the site, but also to a transferee of the owner of the site and financial institutions that foreclose on the property, among others.[32]

The Illinois Site Remediation Program appears to be having a positive effect on transfers of contaminated property in Illinois. Undoubtedly, a letter from the state environmental authority that states that the site has been cleaned up to acceptable levels and that releases the owner and the transferee from further remediation liability provides a powerful incentive to enroll in the program. It also provides a useful structure for parties to obtain the certainty they require in the context of a transaction. With information derived during the course of the environmental assessment, the parties can premise their deal on the completion of remediation under the program. They can agree that the property will be transferred subsequent to the receipt of the no further remediation letter. This device does not necessarily replace the indemnification provisions, but it reduces the risk inherent in any indemnity.[33]

Voluntary cleanup programs may be criticized for providing a sense of false security because the owner remains liable as a matter of federal law under CERCLA. Therefore, the concern has been expressed that an owner could clean up under a state voluntary cleanup program yet remain vulnerable to a federal enforcement action.[34] Technically, that may be true, but in practice the exposure may not be significant. A number of states have entered into memoranda of understanding with USEPA in which the federal government essentially states that it has reviewed the state's voluntary cleanup program and finds it acceptable, and that cleanups that proceed within the confines of such programs are not likely to be the subject of federal enforcement actions.[35] USEPA does not release the owners of such sites from future liability. However, as a practical matter, USEPA has not shown a great propensity to devote its enforcement resources to sites that have already been the subjects of the state's regulatory attention.

Prospective Purchaser Agreements

Perhaps a more substantive test of the federal government's concern about a parcel of contaminated property is its willingness to enter into a prospective purchaser agreement, the last legal device to be discussed in this chapter. Prospective purchaser agreements are written agreements between a prospective purchaser of contaminated property and the federal government. (They are executed by both USEPA and the U.S. Department of Justice.) The agreement provides that in exchange for the prospective purchaser conducting some specified remediation on the property and/or reimbursing the federal government for all or part of the remediation costs it has expended on the property, the federal government will not seek to hold the prospective purchaser liable for further costs after it purchases the property. (The commitment is embodied in a covenant not to sue, which is similar to a release). The agreement also provides the prospective purchaser with contribution protection, a provision that may be equally valuable. An *action for contribution* is a suit by one liable party against another liable party to recover the amount that the plaintiff has paid in excess of its fair share. Absent contribution protection, a purchaser of contaminated property, as the owner (and thus potentially liable under the statute), would be vulnerable to suits by other persons who may be liable (perhaps because they arranged for the disposal of hazardous substances on the property) and who were forced to pay for the government's cleanup costs.

A prospective purchaser agreement with the federal government can provide the certainty needed to acquire contaminated property. However, such agreements may not be readily available. USEPA has issued a policy statement that outlines the conditions under which it will agree to enter into prospective purchaser agreements, and sets forth a model agreement.[36] USEPA will enter into such agreements where the agreement will result in (1) a substantial direct benefit to the government in terms of the cleanup or (2) a substantial indirect benefit to the community leading to the economic revitalization of a site. However, USEPA has stated that it will not enter into a prospective purchaser agreement if it is not otherwise planning to conduct a cleanup action on the site, or if it has not already expended funds to remediate the site.

Conclusion

As this section demonstrates, a variety of legal devices are available to parties dealing with environmental contamination of brownfields sites. So long as the legal risk of the project is properly understood, potential environmental liability related to brownfields sites can be accounted for and dealt with. In order to understand the legal risks, information is the most critical component, followed by an allocation of legal risk.

References and Notes

1. No known statutory provisions in state or federal environmental laws prohibit parties from purchasing property (even industrial property) without first conducting an environmental assessment. Virtually all institutional lenders will require an environmental assessment, but an aggressive buyer with sufficient cash may choose to forge ahead for any number of reasons. For example, such a person may find the risk of losing the deal as a result of the time necessary to complete an environmental assessment greater than the risk of potential environmental liabilities.

2. Such reviews are often referred to as *environmental inspections, reviews,* or *audits.* However, the term *environmental audit* has a dual meaning. In the present context it is used to refer to an investigation whose purpose is to determine whether the property is contaminated (i.e., whether the purchaser of the property is likely to incur liability under CERCLA based on that party's future status as an owner). The term is also used to refer to an assessment not only of existing contamination, but of management practices involving hazardous substances, to determine whether the practices are in compliance with applicable regulations.

3. See the Bureau of National Affairs (BNA's) Friedman, Frank, "Practical Guide to Environmental Management." In: *Environmental Due Diligence Guide,* Chapter 51, 111, 289–356. (1995)

4. For a more in-depth review of Phase I inspections, see Motiuk, Leo, *Environmental Due Diligence,* Practicing Law Institute (1997); Sander, Ram, "The Importance of Due Diligence in Commercial Transactions: Avoiding CERCLA Liability." Ford. Envt'l L.J. (Spring 1996)

5. See 765 ILCS 405/10.01 *et. seq.*

6. See, e.g., *Levy v. Versar*, 882 F. Supp. 736 (N.D. Ill. 1995) (consultants could be liable as PRPs where actions of consultants contributed to contamination)

7. 42 U.S.C. § 9607(b) reads in its entirety:

   There shall be no liability under subsection (a) of this section for a person otherwise liable who can establish by a preponderance of the evidence that the release or threat of release of a hazardous substance and the damages resulting therefrom were caused solely by—

   (1) an act of God;

   (2) an act of war;

   (3) an act or omission of a third party other than an employee or agent of the defendant, or than one whose act or omission occurs in connection with a *contractual relationship,* [emphasis added] existing directly or indirectly, with the defendant (except where the sole contractual arrangement arises from a published tariff and acceptance for carriage by a common carrier by rail), if the defendant establishes by a preponderance of the evidence that (a) he exercised due care with respect to the hazardous substance concerned, taking into consideration the characteristics of such hazardous substance, in light of all relevant facts and circumstances, and (b) he took precautions against foreseeable acts or omissions of any such third party and the consequences that could foreseeably result from such acts or omissions; or

   (4) any combination of the foregoing paragraphs.

8. See *United States v. Carolawn,* 21 Env't. Rep. (BNA) 2124, 2129 (D.S.C. 1984) (denying summary judgment for property owner under 107(b) where title held for less than one hour)

9. Pub. L. No. 99-499, Oct. 17, 1986

10. 42 U.S.C. § 9601(35)(A)

11. 42 U.S.C. § 9607(35)(A), 9607(B)

12. In order to take advantage of the innocent purchaser defense, the purchaser must meet additional requirements, including the exercise of due care with regard to the hazardous substances. See, e.g., *Foster v. United States,* 922 F. Supp. 642 (D.D.C. 1996) (court rejects third-party defense because owner failed to abate contamination, failed to notify governmental agencies, and failed to restrict access to the property); *Idylwoods Assoc. v. Made Capital, Inc.,* 915 F. Supp. 421 (W.D.N.Y 1997) (court rejected third-party defense because current owner failed to prevent illegal dumping and failed to secure the site); but see, *New York v. Lashins Arcade Co.,* 856 F. Supp. 153 (S.D.N.Y. 1994, aff'd, 91 F.3d 353 (2d Cir. 1996) (second circuit affirmed lower court's grant of summary judgment to current owner based on third-party defense)

13. "ASTM Standard Practice for Environmental Site Assessments: Phase I Environmental Site Assessment Process," E1527-93; "ASTM Standard Practice for Environmental Site Assessments: Transaction Screen Process," E1528-93.

14. "ASTM Standard Practice for Environmental Site Assessments: Phase I Environmental Site Assessment Process," E1527-93; "ASTM Standard Practice for Environmental Site Assessments: Transaction Screen Process," E1528-93.

15. See, e.g. 415 ILCS 5/22.2(f) (provides for owner/operator liability for environment contamination), but see 415 ILCS 5/58.9 (prohibits action against property owner for more than property owner's fair share of the liability)

16. 415 ILCS 5/22.2(j)(1)(C)

17. The enactment of 415 ILCS 5/58.9, providing for proportionate share liability, may have lessened the importance of the innocent purchaser defense as a matter of state law, but the discussion in the text is nonetheless instructive for purposes of analyzing the contents of environmental assessments.

18. 415 ILCS, 5/22.2(j)(6)(E)(v). The statue also defines the term *environmental professional* as "an individual (other than a practicing attorney) who, through academic training, occupational experience, and reputation (such as engineers, industrial hygienists, or geologists) can objectively conduct one or more aspects of an Environmental Audit. . . ." 415 ILCS 5/22.2(j)(6)(E)(iii)

19. 415 ILCS, 5/22.2(j)(6)(E)(v)

20. 415 ILCS 5/22.2(j)(6)(E)(ii)(II)

21. As is discussed minimally later, and in further depth in Sec. 4.3, this distinction is the basis for the Illinois Tiered Approach to Corrective Action Objectives regulations. See, e.g., 35 Ill. Admin. Code, Part 742.

22. See *Aluminum Co of America v. Beazer East, Inc.*, 124 F.3d 551 (3d Cir. 1997) (clear, unambiguous language of 1954 agreement encompassed future CERCLA liabilities)

23. *Beazer East, Inc. v. Mead Corp.*, 34 F.3d 206, 215 (3d Cir. 1994) (where the indemnity agreement fails to specifically mention CERCLA liability, courts will carefully scrutinize the agreement to determine if the agreement "evince[s] the parties' broad intent" to encompass and allocate CERCLA liability).

24. Corporations may dissolve for numerous reasons, and the ability to sue thereafter may be limited. See 805 ILCS 5/12/80 (actions barred against corporation five years after corporation's dissolution)

25. See 40 C.F.R. § 280.66 (requiring approval by relevant state or federal authority). The Illinois counterpart of this regulation is found in 35 Ill. Adm. Code § 732.400. Additionally, the Illinois Environmental Protection Act provides for the issuance of a no further remediation letter, which will provide an official statement from IEPA that it is satisfied with the status of the site, and no additional cleanup will be required. (415 ILCS 5/57.7)

26. An analogous problem pervaded the enforcement arena, and still does, to a large extent. Since the advent of CERCLA, respondents or defendants in enforcement actions have bemoaned the fact that they have little power to resist allegedly excessive cleanup demands by governmental authorities, because of the vast amount of discretion that statutes have vested in environmental agencies. See, e.g. 42 U.S.C. § 9613(h)

27. See "USEPA Soil Screening Guidance" (Jan 17, 1995); see also 61 Fed. Reg. 27349 (May 31, 1996)

28. See "Emergency Guidelines for Risk-Based Corrective Action Applied at Petroleum Release Sites," ASTM E1739-95

29. See e.g., 35 Ill. Code, Part 742 (setting forth Illinois' Tiered Approach to Corrective Action Objectives). Illinois' TACO system is the subject of Sec. 4.3 of this book. See also 30 TAC § 334.203 (setting forth the Texas System of risk-based criteria for establishing SSLS)

30. 415 ILCS 5/158 *et seq.*

31. 415 ILCS 5/58.10(a)

32. 415 ILCS §§ 5/58.10(d)(8); 5/58.10(d)(10)

33. Lenders appear to favor the program, and require their borrowers to obtain a no further remediation letter. However, as tempting as the letter may be, some may find the cost of participation in a state voluntary cleanup program, in terms of time and red tape, unbearable. Persons who have the ability to proceed without lenders (or who can persuade their lenders that the letter is not necessary) may choose to forgo the voluntary cleanup program but to take advantage of the state's RBCA system if one has been incorporated into the state's regulations. If the state intends to generally apply cleanup values delivered pursuant to its interpretation of the RBCA procedure, then arguably cleanup to those levels, even absent supervision by the state, will provide some protection. In other words, if the cleanup is essentially the same as that which the state would require, then the basis for a future cleanup action has been removed. Of course, an owner would have to be prepared to demonstrate to the state at some future time that the cleanup was equivalent to that which the state would have required.

34. See *Brownfields Forum*, PLI (1997) (noting the risk of federal enforcement even if full compliance with state voluntary cleanup program has been achieved).

35. See also, e.g., Superfund Memorandum of Agreement Between the Illinois Environmental Protection Agency and the United States Environmental Protection Agency, Region V, April 6, 1995

36. "Announcement and Publication of Guidance on Agreements with Prospective Purchasers of Contaminated Property and Model Prospective Purchaser Agreement," 60 Fed. Reg. 34792. (July 3, 1995)

For biographical information on Carey S. Rosemarin and Steven M. Siros, see Sec. 1.2.

7.3 Federal Prospective Purchaser Agreements

Andrew Warren

A prospective purchaser agreement (PPA) facilitates redevelopment of brownfields subject to action under the Comprehensive Environmental Response, Compensation and Liability Act (CERCLA or Superfund) by the U.S. Environmental Protection Agency (EPA). A PPA resolves the buyer's potential Superfund liability as an owner or operator of contaminated property. In fact, such agreements are the only legally enforceable mechanism, other than settlement of an EPA enforcement action, that addresses the Superfund liability of a buyer and subsequent owners of the property.[1]

A PPA is an administrative settlement agreement between EPA, with participation by the U.S. Department of Justice, and the buyer prior to purchase of the property. In return for consideration (a legal requirement for value in the form of payment or other commitments) from the buyer, the agreement provides the buyer with a commitment from the United States not to sue the buyer under CERCLA, and other environmental statutes,[2] for contamination present at the property. In addition, the agreement provides protection for the buyer against private party claims under CERCLA for costs incurred in connection with contamination at the property. Finally, the buyer can transfer the benefits of a PPA to subsequent owners of the property.

PPAs were initially a rarely used cost recovery option utilized in EPA enforcement actions against parties with limited assets. EPA first authorized the use of PPAs in 1989 in a guidance document governing settlements with landowners.[3] One of the guidance criteria limited EPA's use of PPAs to those circumstances where EPA received a substantial benefit that was not otherwise available. Therefore, EPA typically utilized PPAs as a settlement mechanism of last resort when EPA could not obtain adequate recovery from an owner defendant.

In 1995, as part of its brownfields initiative (see discussion of federal brownfields initiatives in Sec. 4.2), EPA revised the PPA guidance. Under the 1995 guidance,[4] EPA began to use PPAs as a mechanism to return brownfields to productive use. In particular, the guidance expanded the criteria for the consideration received pursuant to a PPA beyond the former narrow use as a cost recovery tool. EPA can now consider the beneficial side effects of redeveloping brownfields as a basis for entering into the agreement.

Since issuance of the 1995 guidance, EPA has significantly increased its use of PPAs. Between 1989 and 1995, EPA entered into only 16 PPAs. Through April 1999, EPA entered approximately 95 PPAs.

To obtain a PPA, the buyer must demonstrate satisfaction of the following criteria:

1. EPA must have taken action or anticipate taking action at the property under Superfund.

2. In return for entering into the agreement, EPA must receive a substantial benefit either in the form of a direct benefit (cash payment or cleanup) or as an indirect public benefit in combination with a reduced direct benefit to EPA.

3. The operation of the new site development, with the exercise of due care, cannot aggravate or contribute to the existing contamination or interfere with EPA's response action.

4. The operation of the new site development cannot pose health risks to the community and those persons likely to be present at the site.

5. The prospective purchaser must be financially viable.

The second criterion—the benefit EPA receives from the prospective purchaser—often presents a problem during negotiations. EPA must obtain "adequate" consideration in return for the covenant not to sue included in the PPA. The guidance allows EPA to accept cash payment, performance of a cleanup at the property, or some combination thereof. In addition, EPA is expressly authorized to accept an indirect public benefit in the form of job creation, development of abandoned or blighted property, or creation of recreation areas as consideration.

The 1995 guidance provides few guidelines, however, about the methodology EPA should employ for calculating consideration. Instead, the guidance lists the following relevant factors: the amount of costs EPA incurred at the site; the estimated future costs; the potential cost recovery from other responsible parties; the purchase price of the property; the market value of the property; the value of an EPA CERCLA lien; and the potential that the prospective purchaser may obtain a windfall by buying property cleaned up with unreimbursed federal funds. Given the number of variables, EPA and the prospective buyer may reach differing conclusions on what constitutes reasonable consideration.

Prospective purchasers face extensive negotiations with EPA. EPA utilizes a model document, included with the 1995 guidance, as its starting point. The model document contains several provisions that a buyer should attempt to address during negotiations. For example, the covenant not to sue covers claims associated with existing contamination, which in the model PPA includes contamination present or under the site. This definition fails to recognize the mobility of contamination and subjects the buyer to potential liability for off-site migration. The definition also fails to conform to CERCLA's definition of a "facility," which includes all areas where hazardous substances come to be located.[5] The model PPA contains language regarding transfers to subsequent owners that gives EPA authority to grant or deny the transfer at its sole discretion. Finally, the model PPA provides an extremely broad and irrevocable right of access to EPA.

A buyer should also be prepared for a slow negotiation process that bears no relation to the timing of a typical private real estate transaction. Since the promise not to sue the buyer comes from the United States, EPA must seek the assistance and approval of the U.S. Department of Justice. EPA's internal review of a proposed PPA includes a technical and legal review conducted at both the regional and national level. In most cases, a proposed PPA must also undergo a 30-day public comment period by publication in the *Federal Register*. Consequently, the entire negotiation process can take as long as a year. For an illustration of the PPA process, see the Autodeposition Site case study (Sec. 9.2).

References and Notes

1. Other settlement options, such as de minimis settlements, only are available to potentially responsible parties under CERCLA (i.e., owners, operators, transporters, or generators). A buyer does not fall into any of the categories until it takes title to or control of the property.

2. EPA can also promise not to sue a buyer under Section 7002 of the Resource Conservation and Recovery Act, 42 U.S.C. § 6902.

3. *Guidance on Landowner Liability under Section 107(a) of CERCLA, De Minimis Settlements under Section 122(g)(B) of CERCLA, and Settlements with Prospective Purchasers of Contaminated Property,* OSWER Directive No. 9835.9 and 54 *Fed. Reg.* 34235. (Aug. 18, 1989)

4. *Guidance on Agreements with Prospective Purchasers of Contaminated Property,* dated May 24, 1995, 60 *Fed. Reg.* 34792. (July 3, 1995)

5. Section 101(9) of CERCLA, 42 U.S.C. § 9601(9)

For biographical information on Andrew Warren, see Sec. 5.3.

7.4   Tax Treatment of Environmental Costs

Ernest R. Di Monte

On August 5, 1997, Congress passed into law Section 198 of the Internal Revenue Code. Section 198 in effect allows a tax incentive to be taken by businesses or individuals that invest in properties contaminated with hazardous substances and that remove the dangerous contaminants. This incentive, given by Congress, is in response to the Clinton administration's stance on environmental issues.

Section 198 allows businesses or individuals to expense the costs associated with the cleanup of the contaminated land. Alternatively, the business or individual may add the costs of remediation to the basis of the property. However, the first option offers an incentive that allows for a quicker recapture of the expenditures. This law is in effect until the end of the fiscal year 2000.

Environmental costs associated with the acquisition of the property may be fees such as land survey costs, engineering costs, legal fees, or consulting costs. These costs must be "qualified environmental remediation expenditures." Qualified environmental remediation expenditures are defined in Paragraph L-6159 of the Internal Revenue Code as any expenditures otherwise chargeable to a capital account and paid or incurred in connection with the abatement or control of "hazardous substances" at a "qualified contaminated site." Hazardous substances are described in Paragraph L-6164 of the Internal Revenue Code, and qualified contaminated sites are described in Paragraph L-6160.

Taxpayers may elect to expense these costs or may add them to the basis of the property, which is what is normally done with costs to improve such property. The election to expense these costs in the year in which they are associated is optional. These costs must also be incurred after August 5, 1997 but not after December 31, 2000.

Environmental costs associated with the actual abatement of the hazardous substances may be expensed, with the exception of property acquired with a character subject to allowance for depreciation. Any equipment bought for the abatement process will be subject to capitalization of the asset and cannot be expensed with regards to the cleanup of the site. For example, a truck purchased for the purpose of removing hazardous substances from the site would be subject to depreciation over the useful life of the truck.

Costs that can be expensed during the cleanup of the land are contractor fees, disposal costs, and almost any expenditure associated with the cleanup, with the exception of capitalized property. These costs would normally be subject to being added to the basis of the land. These costs must also be within the time frame of August 5, 1997 to December 31, 2000.

The term *hazardous substance*, for purposes of defining qualified environmental remediation expenditures, must be any substance that is a hazardous substance as defined in Sec. 101(14) of the Comprehensive Environmental Response and Liability Act of 1980 (CERCLA), and any substance designated as a hazardous substance under Sec. 102 of CERCLA, which is described in previous chapters. These guidelines must be adhered to, or the Internal Revenue Service may not allow these incurred costs to be expensed.

A *qualified contaminated site* is any area held by the taxpayer for use in a trade or business or for the production of income, or property included in inventory in the taxpayer's hands that is within a targeted area and at or on which there has been a release (or threat of release) or disposal of any hazardous substance. The taxpayer must also receive a statement that the area meets certain requirements from the appropriate agency of the state in which the site is located. The requirements include that the site must be within a targeted area and there must have been a release, or threat of release, of hazardous substances. Prior to initiating any work on the targeted area, the taxpayer should ensure that the intended area meets the aforesaid requirements, because without these requirements the Internal Revenue Service (IRS) will not allow the incurred costs to be considered a qualified environmental remediation expenditure. The Environmental Protection Agency of the state can provide such a list of targeted areas and can also provide the statement required by the IRS.

Electing to expense the costs of the environmental cleanup is highly recommended. By electing to expense the costs of the cleanup, the taxpayer is able to more quickly recover the monies spent on the abatement. The taxpayer is better off expensing the environmental cleanup costs, because the expense is recaptured faster that way than when the cost is added to the basis of the land.

There are also tax incentives provided by the state of Illinois for the cleanup of environmental wastelands. On July 21, 1997, Governor Jim Edgar signed into law Senate Bill 93 to help expedite the return of brownfields to worthwhile use. A two-pronged incentive was created by Senate Bill 93. The first is a new environmental remediation tax credit. Developers will be eligible for a state income tax credit equal to 25 percent of the remediation costs, but not exceeding $150,000 per site, when they complete approved environmental cleanups. The second incentive is the Illinois Brownfields Redevelopment Grant Program. Municipalities will be able to receive up to $120,000 for identifying and evaluating sites that have cleanup potential. The law authorizes $1.2 million in grants for each of the next five years to be transferred from the current Response Action Contractor Indemnification Program. The director of the Illinois Environmental Protection Agency, Mary Gade, commented that "This legislation is indicative of our ongoing effort to work in partnership with our private and local stakeholders to clean up contaminated sites faster, cheaper and better while protecting public health and the environment."

To help businesses and individuals interested in redeveloping properties contaminated with hazardous substances, the Illinois Environmental Protection Agency has developed a Web site containing useful information regarding which properties are targeted areas. The address of the Web site is www.epa.state.il.us/land/seids. (SEIDS stands for Site Environmental Information Data System.)

For biographical information on Ernest Di Monte, see Sec. 1.3.

7.5 Environmental Insurance in Brownfields Redevelopment

Dinah L. Szander

In the 1980s and early 1990s, uncertainties arising from suspected or detected environmental contamination stalled mergers and acquisitions, deprived buyers of access to regular commercial lending sources, and motivated deep-pocket sellers to mothball their tainted surplus properties. When transactions involving contaminated properties did close, they were usually characterized by parties of fairly comparable size, no need for a mortgage, and requirements for attorneys' fees and executive time out of proportion to the quantity of risk transferred. Nor did the legal expense and the disruption of executive time end at the closing; it continued for several years, as it often fell to the lawyers and the executives to oversee the retained remediation projects and adjust the indemnity claims between the transacting parties.

More often than not, the buyers who offered to take the surplus property off the hands of the corporate owners were poorly capitalized and looking for seller financing of the cleanup and redevelopment. Many, if not most, owners considered this an unacceptable liability management strategy. The disincentives to deal were numerous. Lending was outside the business plans of most corporate owners. The prospect of foreclosing on the property in an even worse condition was unpalatable. In those days before outsourcing, the owners may have been staffed up to manage any cleanup to completion at least as well as the prospective buyers could. Finally, the owners were skeptical of deed restrictions as an adequate method of controlling the future uses to which the property might be put by limited-asset buyers. As a result, almost every one of the pre-1994 offers to purchase adversely impacted property did not make it past the due diligence stage. Those that got that far inevitably generated more sampling data, which might have to be reported to the government. Then the deal died, often due to the prospective buyers' lack of financing and real estate redevelopment expertise. A corporate inertia set in, in which even prime pieces of surplus impacted real estate located in areas with an established infrastructure and a ready supply of willing labor were mothballed in order to avoid risks deemed too uncertain for transfer.

Laying off this uncertainty to stabilize the economics of a particular reuse plan against a large risk pool is a primary function of environmental insurance in brownfields redevelopment today. The essential purpose of this section, therefore, is to dispel uncertainty by providing practical information that can be used as a resource in transactions involving contaminated property.

Insurance cannot reduce the basic business risks associated with a real estate transaction, and it cannot turn a poor piece of real estate into a prime one. However, it can facilitate the closing of a good deal. The use of insurance may reduce transaction costs as well as alleviate seller and lender uncertainties. The seller may look to a viable third party (insurance company) as a source for funding environmental liabilities rather than relying only on the viability of the brownfield buyer's indemnity and covenant to remediate. Typically, the lender is additionally named as an insured. Insurance may substitute for traditional indemnification provisions and/or escrow agreements, or may be structured to be triggered if the buyer's indemnity payments are uncollectable, or if the limit on the payments has been exceeded, or to fill the gaps in the parties' indemnities. Environmental insurance may be used to mitigate balance sheet reserves and reports to the Securities and Exchange Commission (SEC).

Environmental insurance is a specialty line of the liability insurance market, though some of the applicable concepts are derived from the fixed-location property insurance market. It is not general liability (GL) insurance, the source of the coverage litigation so many deep-pocket owners nationwide filed after their historical GL carriers denied an intention to cover pollution liability in occurrence policies written prior to 1986. Such GL policies written to this day largely exclude pollution, and the specialty environmental market distances itself from the ongoing historical GL dispute.

There are approximately half a dozen specialty environmental insurers participating in the discussion of brownfields insurance issues. Because of the need for flexibility to customize integrated programs, the market is restricted to the surplus lines portion of the insurance business. As a result, the policies will not have been reviewed by insurance regulators and are not required to meet some standards applicable to the so-called *admitted* insurance contracts. The insurance must be sold through a licensed surplus lines broker, and these policies are typically taxed directly and at a higher rate than admitted policies.

The crafting of the integrated package of multiple coverages suitable for a brownfield transaction is time intensive and requires highly specialized underwriting and claims-adjusting skill sets. The policy language should be crafted to avoid disputes over the intent of coverage, and in each transaction a fair amount of time should be spent arriving at common expectations between the policyholder and the insurer concerning future claims-adjusting scenarios. At the same time, the underwriting evaluation must incorporate more functions than most insurance policies, due to the unique project risks, technical risks, and regulatory risks associated with brownfields redevelopment. This evaluation requires significant investment in research and interpretation of quantitative information dealing with environmental loss experience, as well as judgment determinations as to the viability of the project, the ability of the developer to manage the project to completion within anticipated budgets, and the confidence of the underwriter in the business integrity of the insurance applicant. Price is relative, but, in view of this relatively high underwriting expense and the degree of certainty provided to all insureds in terms of cleanup cost amounts, project delays, and unknown environmental liabilities that could otherwise derail a promising brownfield redevelopment, the premium may be surprisingly affordable.

In addition to the unique risks just discussed, the integrated policy covers the environmental risks of unknown liability that are customarily allocated between transactional parties. The first category is cleanup liability arising under CERCLA or its state progeny, or common law causes of action for trespass and nuisance. The second category is third-party claims of toxic tort. The third category is off-site property damage, which if indirect such as diminution in value (legal results vary widely by jurisdiction) or business interruption, is typically excluded from transactional indemnity agreements. The second and third categories (collectively, the so-called third-party risk), are not a part of government brownfields programs, and this residual risk is commonly cited as the reason these programs have not motivated deep-pocket owners to take their surplus properties out of mothballs. The fourth risk category is natural resource damage liability, a little understood and still evolving area of law. The discovery of each of the four categories of environmental risk may be included in the integrated policy.

Quantifying highly technical environmental and liability risks requires specialized training and is a skill set most noninsurance corporations choose to outsource. Therefore, if insurance expertise is not employed, outside attorneys, assisted by environmental consultants, labor mightily to arrive at a degree of certainty with which they are comfortable, under severe time pressures legitimately imposed by financing and seller interests in the deal. Those in the business of assuming and managing environmental risk, on the other hand, typically require a lower degree of certainty before allocating or assuming environmental risk, and they quantify risk in a fraction of the time required by lawyers. For one thing, insurers have a larger database of losses and trends in environmental liabilities than any single law or environmental consulting firm could possibly compile using its limited client base as the only available information source.

The cost to defend environmentally related third-party claims can be daunting. Litigation is often protracted, occupying management time and resources for years. As explained earlier, absent insurance, the third-party risk creates a major barrier to brownfields investment and lending. However, developers also should be concerned that natural resource damages can be claimed even at properties that have been cleaned up and settled under CERCLA's other provisions.

The relative speed of the insurer due diligence (also known as underwriting), as well as the confidence level afforded to the end product (the quantification of the risk), is appealing to all parties in the often time-strapped brownfield transaction. With an insurance company involved in the due diligence, the transacting parties are afforded ample time to adjust the purchase price, the postclosing cleanup covenants, or the indemnities to reflect the quoted premium (which will include not only the expected losses but also claim adjustment expense, defense costs, and a margin for underwriting profit).

A number of coverage parts are considered in any brownfield project. The first is coverage for required or voluntary cleanup program cleanup liability resulting from newly discovered contamination or the reopening of previously remediated conditions due to changes in law. The second is third-party risk and natural resource damages, including defense costs. Both may be covered by the real estate environmental liability (REEL) coverage part, which covers not only preexisting unknown pollution events but also those that may commence in the future. (If the future use to which the property will be put is industrial instead of commercial or residential, this coverage part will be called *environmental impairment liability,* reflecting the greater risk of (and hence the higher premium for) future releases from ongoing operations. For purposes of this article, this coverage part will be referred to as *the REEL* in either event.) An important exclusion is coverage for costs associated with the presence of contamination identified before the policy's inception; this coverage to protect against unforeseen escalation in the cost to correct conditions that are a known part of the project is provided by the remediation stop loss coverage part. The REEL incepts when the known cleanup project is completed.

Other coverages that may be included in the integrated policy are (1) finite risk programs to provide funding for cleanups or financial assurance to address closure and postclosure liability funding through payment of equal premiums over a designated time period; (2) asbestos coverage designed to protect building owners against claims resulting from releases at a covered location subject to an operations and maintenance program; (3) nonowned disposal site coverage respecting specified waste disposal and treatment sites used after policy inception; and (4) owner-controlled environmental contractor insurance applicable to the brownfield cleanup project.

An example of the cleanup cost coverage of the REEL follows. Assume that the completed brownfield project included a cleanup of pentachlorophenol (PCP) from a historic wood treatment operation. During the

REEL policy term, there is a discovery of additional PCP contamination or a change in the PCP cleanup standard that lowers the allowable level left behind in the brownfield cleanup. Subject to policy limits, deductibles, and other terms and conditions, the new cleanup costs would be covered, both on and off site. In addition, if a requirement to clean up contamination on the covered location originating from an off-site source were discovered during the policy period, those cleanup costs would also be covered. Nor is there a requirement that cleanup be directed or ordered; costs incurred in state-sanctioned voluntary cleanup programs are covered.

Unlike the REEL, the remediation stop loss coverage does not cover third-party risk and natural resource damages or defense costs related thereto. This coverage assures that brownfields project cleanup costs above a self-insured retention level (SIR) will be capped, subject to the available policy limit. It indemnifies the insured for financial losses that arise when the anticipated cost of a remediation project is exceeded. Based upon a scheduled remediation project defined through an environmental site investigation, this type of coverage addresses unanticipated cost over-runs. Coverage is provided above the applicable SIR, which equals the projected cost of cleanup plus a buffer. For example, the developer of a project with expected remediation costs of $2 million may carry a $500,000 buffer (SIR) and purchase $4 million in stop loss coverage. If actual cleanup costs for the project are $4 million, then when the costs of the project run over the $2 million anticipated costs plus the $500,000 SIR, the remediation stop loss coverage would respond by paying the unanticipated $1.5 million.

The asbestos liability coverage is designed to protect building owners against claims resulting from releases of asbestos at covered locations. For instance, building owners with operation and maintenance programs for asbestos-containing materials can purchase coverage to protect against bodily injury and property damage claims resulting from release of the asbestos on an occurrence basis.

Nonowned disposal site coverage protects against exposure at specified disposal sites owned by others. Exposure at such sites could lead to liability stemming from Superfund cost recovery actions. Coverages are very specific and can be limited.

Environmental remediation may itself generate environmental liabilities. An aquifer may be cross-contaminated or an invitee to the project may suffer bodily injury. These are activity-oriented exposures, not fixed-location exposures. The developer will require the remediation contractors to provide evidence of blanket annual aggregate coverage for all work performed during the project. In some cases, owners and prime contractors may not have adequate protection for their potential exposures, because the policy limit for the contractor's blanket coverage is shared with other projects, or the contractor may not carry insurance for environmental risks (called *contractor pollution liability coverage*), which are excluded from the GL coverage. Owner-controlled insurance coverage fills this gap in the loss exposure with consistent coverage for all contractors on the environmental remediation project. This coverage may offer substantial benefits through the provision of standardized forms and relief from the administrative burdens associated with coordinating additional-insured coverage.

Knowledge of contingent environmental liability is required to be recorded as a liability and disclosed to public shareholders, which tends to impair the property owner's ability to attract investment or otherwise maximize use of capital. A combination of finite risk policies (a funding mechanism for a known loss) and stop loss policies has reportedly been used by publicly traded companies in conjunction with title transfer in brownfields redevelopment to offset and spin off certain environmental liabilities from these companies' balance sheets onto the balance sheets of the brownfield redeveloper. AICPA Statement of Position 96-2, as well as SEC Bulletin No. 92, should be considered, and appropriate legal and accounting professionals should be consulted, before such a use of environmental insurance is undertaken.

In conclusion, environmental insurance adds significant value in brownfields redevelopment. Environmental insurance allows overall project costs to be estimated more accurately by minimizing an area of risk or by quantifying costs associated with identified risks. This insurance also facilitates the closing of the transaction itself, beginning with giving the deep-pocket seller the degree of confidence necessary to take mothballed real estate out of storage, and the lender the comfort it needs to finance the redevelopment. With the deal launched, the integrated policy can cut transaction costs (attorney and executive time) in negotiating and documenting indemnities or escrows. This cost saving results from insurance being a faster and more credible (as well as more objective from the standpoint of the transacting parties) gauge of the quantity of risk to be transferred or retained. In some deals, it can even substitute for the indemnity or escrow. In any case, it removes uncertainty by allowing the transacting parties and the lender to look to a highly creditworthy financial institution for protection from unknown cleanup costs, from third-party risk and natural resource damage liability (as defined), including defense costs, or from cost overruns in the brownfield remediation project. Finally, there may be advantageous balance sheet and SEC disclosure impacts if environmental insurance is obtained to address such contingencies. All that having been said, this coverage product is a significant contract in its own right, to be afforded appropriate due diligence, negotiation, and drafting resources to assure that the policyholder and the insurer reach common ground on what is intended to be covered and how claims will be adjusted. Because the payment of covered claims is the primary reason for buying insurance coverage, it is critical that the parties have confidence in the insurer's expertise in this core function and its commitment to pay claims fairly. Insurance companies are not created equal, and insurer resistance to paying legitimate claims, lack of specialized skills in this regard, or insolvency could adversely impact the public image of the deep-pocket seller as well as the brownfield redeveloper, and distract both parties from their own core businesses.

Dinah L. Szander, J.D., joined Zurich U.S. Specialties in 1998 as risk management executive and counsel. Specialties is a strategic business unit of a global financial services company with over $375 billion in assets under management and 68,000 employees. Ms. Szander has 6 years of experience at two large law firms and 14 years of senior management legal experience at a Fortune 200 company. She graduated summa cum laude and Phi Beta Kappa, and chairs the environmental insurance subcommittee of the environmental transactions committee of the American Bar Association.

Throughout her career, Ms. Szander has provided risk management and transactional consultation to corporate clients. She came to Specialties from Landels Ripley & Diamond, LLP, of San Francisco, which she joined as a partner in 1996. At Landels for two years, Ms. Szander spearheaded the firm's effort to expand its share of the brownfields transaction and merger and acquisition (M&A) legal markets. Her other four years of private practice were at Bronson, Bronson & McKinnon, of San Francisco, from 1978 to 1982.

In 1996, she joined Landels from McKesson Corporation, a $20 billion pharmaceutical company. At McKesson, from 1982 to 1984, she managed toxic tort litigation. From 1984 to 1987, she functioned as the general counsel of the chemical line of business. After the chemical business was sold, she returned to real estate/M&A legal work, in addition to managing McKesson's retained environmental liabilities. In 1994, she assumed the further responsibility of directing the corporate compliance function reporting to the audit committee of the board of directors.

Specialties is headquartered in New York. Ms. Szander maintains an office in San Francisco as well.

7.6   Technical Remediation

Harold J. Rafson

Decisions

We come now to the subject of *remediation,* which is the correction of the contamination on the site and the reduction of contamination to acceptable levels as stated by the regulations. Decisions have to be made concerning remediation.

When considering a remediation method, there are three issues to keep in mind:

1. Effectiveness
2. Cost
3. Time

Any method must be effective, or else you will not fulfill your obligations to the regulators (and your neighbors and society). Different technologies can be variously effective at different costs, and can be greatly affected by time considerations. For example, if you wish to burn or volatilize a contaminant from soil in an incinerator, but don't get the temperature high enough, then there can be residuals. You will have saved money on heating costs but lost effectiveness. Or, for example, phytoremediation—that is, natural remediation achieved by plants or trees—is almost certainly a cheaper alternative, but it may take years to achieve acceptable remediation levels, and the treatment is likely to be inconsistent. This may limit the use of the property during this period, and therefore in many cases will be an unacceptable option. Therefore, in the following discussion of remediation methods, all three of these factors must be reviewed. Further, there are other relevant issues to be considered before we get to the technology of remediation.

Strategy for Remediation

It is not possible to give a strategy that will be applicable to all the possible variations of projects that come to the minds of readers. We will chose a straightforward example, where there is a building with some asbestos and lead paint contamination and all of the soil contamination is outside of the plant in two locations, one at an underground storage tank (UST) close to the plant and one further back on vacant land behind the plant. The developer wants to close on the deal and get to the point of obtaining a no further remediation letter, to allow sale or redevelopment of the site.

The developer's strategy may be as follows:

1. *Building*—rapidly clean up the building so it can be occupied as soon as possible. This will include removing asbestos insulation where exposed, sealing asbestos insulation where necessary, removing asbestos floor tiles, and removing or sealing lead-based paint. The developer no doubt will also have to clean out miscellaneous materials from the plant, fix the windows and the roof, and do some landscape work. When this process is done, which may take a few months, the building will be ready to be occupied.

2. *USTs*—rapidly remove the UST, and any contaminated soil surrounding the tank, if necessary. This can be accomplished within one month after receiving the proper permits.

3. *Back lot contamination*—with time, the back lot contamination can be corrected, let us say by soil venting. Within a year or so the venting will be complete, and with time for testing and review and approval by the regulators the NFR letter will be finally obtained. At this point, this section of the property can be sold to the building owner, if a satisfactory arrangement has been made.

In this scenario, the developer has been able to close the deal, make a profit, and, in time, clean up the loose ends of the project without having too much money tied up for a long period. The owner has been satisfied in obtaining the desired building rapidly and in good order.

This is only one strategy, and it is up to the developer to creatively consider the problems of the site and the needs of the client to make the deal work.

Remediation Technologies

Background

Before beginning to discuss remediation technology, it is important that the reader have a concept clearly in mind: the Law of Conservation of Matter. This states simply that matter is neither created nor destroyed. Therefore, all you can do is change something—you cannot make something magically disappear. A volatile compound such as perchlorethylene (dry cleaning fluid) can be volatilized, which takes it out of the soil and puts it into the air. Now what to do with it? It can be absorbed on carbon: but then what do we do with the perchlorethylene-laden carbon—take it to a hazardous landfill elsewhere (that's a physical location change) or incinerate it (that's a chemical change)? Remember that the individual elements of the perchlorethylene do not disappear; the carbon will be changed to carbon dioxide or carbon monoxide (depending on the operating conditions of combustion), and the chlorine may be changed to hydrochloric acid. What do we do with the hydrochloric acid? Possibly scrub the incinerator exhaust with a caustic scrubber, where the hydrochloric acid reacts to form sodium or calcium chloride. This pattern of reasoning can be continued. The point is that the laws of nature are not superseded. Compounds do not disappear, they are only changed, and the outcomes of any action must be considered carefully.

Compounds and Concentrations

Before considering removing contaminants, you have to understand the contaminants you intend to remove. There are certain fundamentals about the characteristics of a compound—its volatility, its water solubility, its biodegradability, its reaction kinetics, and so on—that will ultimately determine the best remediation process to use. We would try to volatilize gasoline from soil or groundwater; but we would not try this approach with heavy metal contamination, which is not volatile. Treatment of soils contaminated with heavy metal is very different from treatment of hydrocarbon contamination.

Following the identification of the compounds that are contaminants, review is made of the technologies, and the technologies most likely to be successful are selected.

The developer hires an environmental engineer for advice about what to do. The advice is heavily influenced by the type and concentration of the compound to be remediated.

Concentration also has a significant effect on both the need to remediate and the selection of remediation technology. If the concentration is around the soil saturation limit, the soil must be remediated; if lower, more options are available. With incineration, a contaminant present at 10 ppm concentration will cost about 10 times as much to remove as the same quantity of contaminant present at 100 ppm. A technology can be cost effective at a higher concentration and cost prohibitive at a lower concentration.

Substrates

In a separate section on geology (Sec. 6.7) there is a more detailed discussion of the issues concerning the characters of the soils and their impact on remediation choices. But solely from the viewpoint of remediation, it is obvious that compounds are absorbed and adsorbed to different degrees by different soils. A grain of sand is like a small impervious rock, and a hydrocarbon will adhere to the surface and can be revolatilized from or treated on the surface. With a porous rock, or a highly organic soil (loam), the hydrocarbon will be absorbed into the structure, and may be tightly bound. Removal of a contaminant will be much more difficult.

The likelihood of the use of in situ remediation methods is primarily determined by the soil type and permeability.

Remediation Technologies

There are a great many different technologies that are used for soil remediation, and each proprietary system is hailed by its supplier as something unique and superior. However, the remediation technologies can be grouped into classes, primarily based on the major mechanism used for removal. This grouping can be as follows:

- Biological treatment
- Chemical treatment
- Physical treatment
- Thermal treatment
- Stabilization, solidification, and encapsulation

We can also consider a different classification system: those methods that are most frequently found to be useful, and those less used, (though, in many cases, these are well suited for specialized situations). We will group these into technologies as follows:

Frequently used technologies:
    Solidification/stabilization
    Pump and treat
    Off site incineration
    Natural attenuation

Less frequently used technologies (innovative):
    In "Innovative Treatment Technologies—Annual Status Report,"[1] November 1996, the following brief descriptions are given of various innovative technologies.

Source Control Technologies
    EX SITU BIOREMEDIATION uses microorganisms to degrade organic contaminants in excavated soil, soil, sludge, and solids. The microorganisms break down the contaminants by using them as a food source. The end products typically are CO_2 and H_2O. Ex situ bioremediation includes slurry phase bioremediation, in which the soils are mixed in water to form a slurry, and solid-phase bioremediation, in which the soils are placed in a cell or building and tilled with added water and nutrients. Land farming and composting are types of solid-phase bioremediation.

In application of IN SITU SOIL BIOREMEDIATION, an oxygen source and sometimes nutrients are pumped under pressure into the soil through wells, or they are spread on the surface for infiltration to the contaminated material. Bioventing is a common form of in situ bioremediation. Bioventing utilizes extraction wells to circulate air with or without pumping air into the ground.

The CONTAINED RECOVERY OF OILY WASTES (CROW™) process displaces oily wastes with steam and hot water. The contaminated oils are swept into a more permeable area and are pumped out of the soil.

In CYANIDE OXIDATION organic cyanides are oxidized to less hazardous compounds through chemical reactions.

DECHLORINATION is a chemical reaction which removes or replaces chlorine atoms contained in hazardous compounds, rendering them less hazardous.

For IN SITU FLUSHING, large volumes of water, at times supplemented with treatment compounds, are introduced into soil or waste, to flush hazardous contaminants from a site. Injected water must be isolated effectively within the aquifer and recovered.

With HOT AIR INJECTION, heated air is injected and circulated through the subsurface. The heated air volatilizes volatile organic compounds so they can be extracted and captured for further treatment or recycling.

PHYSICAL SEPARATION removes contaminants from a medium in order to reduce the volume of material requiring treatment.

PLASMA HIGH TEMPERATURE METALS RECOVERY is a thermal treatment process that purges contaminants from solids and soils as metal fumes and organic vapors. The organic vapors can be burned as fuel and the metal fumes can be recovered and recycled.

SOIL VAPOR EXTRACTION (SVE) removes volatile organic compounds from the soil in situ through the use of vapor extraction wells, sometimes combined with air injection wells, to strip and flush the contaminants into the air stream for further treatment.

SOIL WASHING is used for two purposes. First, the mechanical action and water (sometimes with additives) physically remove the contaminants from the soil particles. Second, agitation of the soil particles allows the more highly contaminated fine particles to separate from the larger ones, thus reducing the volume of material requiring further treatment.

SOLVENT EXTRACTION operates on the principle that, in the correct solvent, organic contaminants can be solubilized preferentially and removed from the waste. The solvent used will vary, depending on waste type.

For THERMAL DESORPTION, the waste is heated in a controlled environment to cause organic compounds to volatilize. The operating temperature for thermal desorption is usually less than 1,000°F (550°C). The volatilized contaminants usually require further control or treatment.

VITRIFICATION melts contaminated soil at temperatures of approximately 3,000°F (1,600°C). Metals are encapsulated in the glass-like structure of the solidified silicate compounds. Organics may be treated by combustion.

Groundwater Treatment Technologies

AIR SPARGING involves injecting air or oxygen into the aquifer to strip or flush volatile contaminants as the air bubbles up through the groundwater and is captured by a vapor extraction system. The entire system acts as an in situ air stripper. Stripped or volatilized contaminants usually will be removed through soil vapor extraction wells and usually require further treatment.

Air sparging often is combined with IN SITU GROUND-WATER BIOREMEDIATION, in which nutrients or an oxygen source (such as air) are pumped under pressure into the aquifer through wells to enhance biodegradation of contaminants in the groundwater.

DUAL-PHASE EXTRACTION removes contaminants simultaneously from both the saturated and the unsaturated zone soils in situ. The new technology applies soil vapor extraction techniques to contaminants trapped in saturated zone soils, which are more difficult to extract than those in the unsaturated zone. In some instances, this result may be achieved by sparging the groundwater section of a well that penetrates the groundwater table. Other methods also may be employed.

IN SITU OXIDATION oxidizes contaminants that are dissolved in groundwater, converting them into insoluble compounds.

PASSIVE TREATMENT WALLS act like chemical treatment zones. Contaminated groundwater comes into contact with the wall, which is permeable, and a chemical reaction takes place. Limestone treatment zones increase the pH, which effectively immobilizes dissolved metals in the saturated zone. Another type of passive treatment wall contains iron filings that dechlorinate compounds.

SURFACTANT FLUSHING of non-aqueous phase liquids (NAPL) increases the solubility and mobility of contaminants in water, so that the NAPL can be biodegraded more easily in the aquifer or recovered for treatment aboveground by a pump-and-treat system.

A summary of contamination source control technologies used at Superfund sites shows a preponderance of the use of established technologies. A few newer technologies such as soil vapor extraction and thermal desorption have found increasing use. A review of data also shows that, while the use of innovative technologies had been planned in numerous projects, when these projects entered the design phase, plans were changed to use more established technologies. In the eighth edition report there is a summary table that discusses the changes to plans that occur at the design stage. In almost all cases the innovative technologies about which less is known are dropped in preference to the known technologies. Though in some cases there may be opportunities for savings through the use of innovative technologies, the brownfield developer is much concerned about uncertainty and timing. As a result, it can be expected that the developer will, in almost all cases, use established technologies.

Selection of Technology

Let us describe various technologies that have been used for soil and groundwater remediation and then the selection of technologies for an application.

The list for the developer is short.

Nontreatment technologies:

Landfilling and capping

Treatment technologies:

Thermal desorption

Soil vapor extraction

Pump and treat

Solidification/stabilization

Bioremediation

Natural attenuation

This does not mean that there may not be cases where imaginative and innovative technologies will find application to extraordinary benefit. But the developer is not oriented toward research and development, and is dealing with regulators who are interested in proven results. Unfortunately, this mind-set results in slowing the development of new technologies. If, indeed, there are half a million brownfields sites, society could benefit from these cases and learn to improve remediation technologies. However, as far as this book is concerned, the developer (who is often not technologically sophisticated or adventuresome) is encouraged to stick with tried-and-true technologies.

The preceding list of treatment technologies can be further rewritten to illustrate application:

Soil (Including Effect of Soil on Groundwater):   *Groundwater:*

Incineration   Pump and treat

Thermal desorption

Soil vapor extraction

Solidification/stabilization
Bioremediation
Natural attenuation

Hopefully groundwater contamination will not become an issue, because pump and treat can turn out to be a long-term affair.

Let us look at the case of soil contamination. We can subdivide the technologies according to whether the contaminants to be located are more or less volatile:

More Volatile:
Incineration
Natural attenuation
Bioremediation
Soil vapor extraction
Thermal desorption
By Concentration:

Less Volatile:
Solidifaction/stabilization
Incineration
Natural attenuation
Bioremediation
Thermal desorption

Higher Concentration:
Solidfaction/stabilization
Incineration

Lower Concentration:
Bioremediation
Natural attenuation
Soil vapor extraction

By Time to Accomplish:

Short Time:
Solidfaction/stabilization
Incineration
Thermal desorption

Long Time:
Natural attenuation
Soil vapor extraction
Bioremediation

By Cost:

Higher Cost:
Incineration
Thermal desorption

Lower Cost:
Solidfaction/stabilization
Natural attenuation
Soil vapor extraction
Bioremediation

Setting these various factors against your particular need, the engineer will arrive at a first screening geared toward selecting a remediation technology that will work well for you.

Description of Frequently Used Technologies

Solidification/stabilization. Solidification/stabilization may be defined as addition of, or encasement in, a medium to prevent the transport of hazardous or toxic contaminants within the soil. Commonly, this is either done by mixing the soil with concrete to reduce the leaching potential of the soil or by using capping materials to reduce water flow through and access to the contaminated soil. Capping is done where groundwater impact is unlikely. There are thermal treatments to convert sand to glass (used for radioactive waste).

Pump and Treat. Pump and treat refers graphically to a process used to treat groundwater. The costs and effectiveness depend on the contaminant type and concentration, the soil permeability, the treatment method selected, and the outlet concentration that is acceptable. Once the groundwater has been brought to the surface, there are many types of treatment methods that are available, and the selection depends upon the stated factors. Among the treatment methods are all those typical of wastewater treatment (filtration, stripping, adsorption, biological degradation either in aerobic or anaeobic conditions, and others as well as combinations of treatment methods), and also specialized chemical treatments suited to the specific contaminants.

The treated groundwater may or may not be reintroduced into the soil, based upon site-specific circumstances. When groundwater is reintroduced either back into the aquifer or onto the surface, the procedure would be considered soil flushing. This is useful for some types of contaminants and soil types to speed cleaning.

If the groundwater contamination is in an aquifer, it is likely that reduction of the contaminant levels may take a long time. However, the pump and treat operation frequently can go forward at the same time as beneficial use of the property, if that property is not adding to the existing contamination and if the procedure does not result in additional exposure of citizens.

Incineration—off Site or on Site. The idea behind incineration (or thermal oxidation) is to oxidize a compound to a harmless form. The commonly presented example is a carbon molecule being oxidized to carbon dioxide. However, the contamination is never just carbon alone, but organic compounds that usually include molecules of chlorine, sulfur, nitrogen, and other elements. The oxidation of these compounds may form harmful by-products and organic acids, and other compounds are generated that may require subsequent treatment. There is also the possibility of the addition of contaminants from the fuel used for incineration.

The incineration process requires the compound to be heated to at least 1500°F. The compound may be present in parts per million, but all million parts must be raised to this high temperature. Equipment is designed to recover heat, but that is only partially successful, and incineration is a high-cost method of remediation unless the quantities to be treated are small.

Whether incineration is to be performed off site or on site is an economic evaluation and is determined primarily by the amount of material to be treated and shipping costs.

Natural Attenuation.   Nature has a way of cleaning itself; that is how the human race has survived so long. However, increasing population and the industrial revolution have altered this balance. We cannot always rely on natural attenuation, and have had to develop more specific and faster methods, as discussed previously. But natural attenuation works, if you have the time. A wide variety of biota are extant in soils, and, when these soils are contaminated, if adequate moisture, temperature and foods are present, bacteria or plants will perform (on most compounds) a methodical degradation. Sometimes the process can be assisted by aerating the soil and adding moisture, foods, selected cultures of bacteria, or other additives. If natural attenuation is applicable, where contamination is not transferred to the population by any means and the property can be idled, then this is a very attractive option. The only requirement is that a monitoring program be in place that will identify the decreasing contamination levels and that contamination does not migrate.

Often the contaminated area has taken many years to accumulate contaminants and we now face those areas expecting immediate solutions. The slow process of natural attenuation, if it is imaginatively considered in a remediation plan, may be extremely beneficial.

As an example, often an industrial site is significantly larger than called for by current production requirements. In redevelopment, it may be possible to remove gross contamination without going after a larger area of lower contamination that is still somewat above acceptable levels. This larger area can be treated at a different pace, and may heal itself in several years. If there is no pressure to use this land, and no possibility that workers will be exposed, such an area can be idled and monitored, with more aggressive measures taken only if the land must be used before natural attenuation is complete. Such plans must be carried out in agreement with regulators, and there must be recognition of the decreased value of this land during this period.

Recently, oxygen-containing compounds have been added to groundwater to increase biological activity. This helps to speed the breakdown of many digestible compounds and has found increasing acceptance by the regulatory community.

Time.   In many cases, time to accomplish remediation is the most critical factor in technology selection. The saying "Time is money" does not begin to describe the importance of delay in this business. A developer is interested in satisfying the needs of a client. Often, a client is looking for a facility into which it can expand economically. There always is the choice between a brownfield and a greenfield site. While a brownfield site may offer many benefits (location, cost, labor pool, etc.), these fade into insignificance if a client has to wait two or more years before moving into the location. In such cases, a more expensive cleanup technology that will get the job done in one year might be considered by a developer—otherwise there may be no deal.

The developer generally wants to get in and out of a project quickly. A long remediation job ties up the developer's money and personnel. A developer is prone to lean toward faster technology than will a company doing its own project more patiently.

Remediation Plan

Remediation plans consist of three basic elements: first, an analysis of the existing data and exploration of the process by which the cleanup will be performed; second, a health and safety plan to demonstrate how the cleanup can be done safely, protecting the workforce and the community at large; third, a testing plan to prove that the cleanup was effective. The testing may include long-term monitoring if pump and treat or natural attenuation or biological treatments are used.

These plans should include milestones defined by the regulators. When a particular level of cleanliness is met, and the site is acceptably clean, remediation and monitoring can end. This stipulation in the plan will save late negotiations with regulators as cleanup progresses.

Project Management

The project management of a remediation project is no different than that for any other construction project, with the additional requirement that accurate records are to be maintained of all testing and all disposition of contaminated materials, and that continuing communication is to be maintained with the relevant regulators to apprise them of the progress of the project.

Included in the project management will be the overseeing of the following issues with contractors:

- Technical ability and experience to ensure proper implementation, safe work, and testing to prove that the site is clean. Rapport should be maintained with regulators to ensure NFR letters at completion. Bonding to ensure completion of the project is desirable.

- Regular, prompt notification of the developer by the project manager of any extraordinary issues, and other notification as agreed. Reports will be provided of work accomplished for comparison to billings the developer will receive from contractors.

Regulatory Meetings

The engineer will prepare reports to confirm that the remediation has been successfully completed, and, with the developer, will meet regularly with the regulatory authorities to ensure that cleanup and testing schedules are met.

Monitoring

Monitoring after remediation is completed may be required until the contaminants are dispersed or are biologically decomposed below cleanup objectives. The monitoring will need to be negotiated as part of the site remediation plan. The testing will be specific to the contaminants of concern and site conditions.

Unexpected Events

All remediation plans have been based upon a limited amount of data. A certain number of holes have been drilled and samples drawn. Only certain analyses have been performed. Therefore, any remediation plan is based upon what amounts to only a best guess, using available data, as to what will actually be found when a site is dug up. Is it better or worse than predicted? It can go either way. But the major concern is the uncovering of something totally unexpected. It is these uncertainties that have plagued remediation efforts in the past—and have led to the stigma attached to these efforts by lenders and investors.

But all events should be looked upon as to be expected, since initial information was never perfect. Contingencies have to be provided for to account for these cases. When the situation is considered in this light, risk considerations become a more normal part of doing business. The original investigations should be good enough to avoid disasters—but some level of underestimation can be tolerated.

Nevertheless, this is not the usual kind of business; uncertainties are greater. Contractual obligations must be met. The developer may end up in a situation where the projected profits are eroded by unexpected events, which means that on another project the developer will have to estimate profits higher, since overall the developer does deserve a reasonable profit for his or her efforts. It is a normal practice in doing business that as uncertainties are present, higher contingencies and higher allowances for profits to cover the risk must be estimated. Elsewhere in Sec. 6.2 an example is given that is estimated to be profitable, but any number of small changes (and reasonable events) could evaporate the profit easily. The developer is therefore faced with reality. If he or she estimates too low, he or she can easily lose all profit and have spent unproductive time and capital on a project. If he or she estimates too high, he or she will not get the work. It is a troubling balance to the developer, and creative thinking by developers and clients is required to allow projects to go forward on a reasonable basis.

But, because of more certain cleanup objectives and better testing and analytical methods, cleanup estimates have become more predictable and cost overruns, which plagued the environmental remediation business in the past, are not nearly as bad as they once were.

The Illinois TACO Process

The Illinois Tiered Approach to Corrective Action Objectives (TACO: 35 Ill. Admin. Code Part 742 R97-12A)[2] has been described in more detail in Sec. 4.3. This section's objective is to follow through the technical aspects of that process (Illinois Pollution Control Board, Jan 5, 1997).

We will first give a short summary and a guide to how the process works (illustrated by several flow charts, which are included in App. 1).

A Tier 1 analysis requires the applicant to compare levels of contaminants of concern at the remediation site to predetermined remediation objectives.

Flow chart AA—provides the overview for soil remediation

Flow chart AB—provides the overview for groundwater

Flow chart BA—provides the flow chart for Tier 1

A Tier 2 analysis uses the equations set forth in the rules to develop alternative remediation objectives for contaminants of concern using site-specific information.

Flow chart CA—provides the Tier 2 flow chart for soil

Flow chart CB—provides the Tier 2 flow chart for groundwater

A Tier 3 analysis allows the applicant to develop remediation objectives using alternate parameters not found in Tier 1 or Tier 2.

Let us follow through some examples to illustrate the use of the rules.

Example 1

Let us take a simple case where the soil behind a plant has been contaminated with only one chemical—trichlorethylene. A developer wishes to take this land and build townhouses on it (residential use). To what level must the trichlorethylene be reduced before it is acceptable to the Illinois EPA?

Looking at flow chart BA, we see that we have to go to the Regulation Appendix B, Tables A and E of the rules. Table A for trichlorethylene states:

Exposure Route—Specific Values for Soils	Ingestion—mg/kg	58
	Inhalation	5
Soil Component for Groundwater Exposure Routes		
	Class I—mg/kg	0.0
	Class II—mg/kg	0.3

and that the acceptable detection limit (ADL) is less than remediation levels stated.

The property is 5 acres, of which only ¼ acre shows any contamination with TCE (at a level of 12 ppm). This is true for a stratum of soil between 6 feet and 10 feet deep. This means that we pass the ingestion limit, but fail the inhalation limit. There is no use of the groundwater for drinking purposes, so the groundwater limits are not relevant. What should be done to lower the trichlorethylene concentration level from 12 to 5 ppm? Should we excavate and land farm? Shall we soil vent? Should we use biological methods? Should we employ phytoremediation? A feasibility study is in order. For simplicity, let us add that time invested is very costly, so that the developer rules against phytoremediation, biological methods, and soil ventilation. Soil farming is selected, and the process goes forward and is completed in two months.

This illustration demonstrates the ability to quickly set, in Illinois, goals that EPA must live with and that are doable by the developer.

Let us say that the remediation cost is higher than the developer cares to accept. Does he or she have any other options? We can look at Tier 2 (refer to flow chart CA) or Tier 3. Both provide the opportunity for alternative remediation goals for the specific site, which may prove to be advantageous. But in this example of a small site, the developer can count on additional effort and costs to get regulatory approval.

References:

1. U.S. Environmental Protection Agency, "Innovative Treatment Technologies—Annual Status Report" (8th ed.). (November 1996)
2. Illinois Pollution Control Board, Tiered Approach to Corrective Action Objectives (TACO): 35 ILL ADM, Code Part 742, R97-12A. (June 5, 1997)

For biographical information on Harold Rafson, see Sec. 1.1.

8

The Private Developer—Closure

In Sec. 8.1, Robert Rafson discusses closure. The developer now counts either profits or losses, sells the property, and fulfills all obligations to all the parties involved—prior owners, new owners, regulators, and contractors.

In Sec. 8.2, Ernest Di Monte deals with the accounting aspects of the sale.

8.1 Closure

Robert Rafson

Contracts for the sale of a cleaned-up site may be more complicated and delicate to negotiate than those for almost any other type of real estate sale. There may be requirements for indemnification of the buyer and other stipulations by the buyer that may erode the developer's profit.

Indemnification of the buyer is one of the most difficult requirements for a developer to accept. Often the buyer wants complete and unconditional release from liabilities that arise from the environmental condition of the site. It is difficult for the developer to assume these open-ended liabilities and to protect the buyer from problems either known or unknown. This kind of indemnification is additionally difficult if either legislative or engineered barriers are used to separate humans from exposure. There still could be lawsuits over the contamination and its effect on the neighboring property, the value of the property, or perceived health problems. These types of indemnification should not be underestimated. There is a fairly large amount of case law to indicate that the population does pursue owners and operators of properties for these issues, and therefore any developer must think about these issues before beginning the redevelopment of a contaminated property. When the end of the redevelopment comes, the buyer will likely ask for these indemnifications, and the developer must have planned for this.

"The happiest day of a property owner's life is the day the property is sold" is an old adage. Know that properties—even clean ones—that have environmental stigma are difficult to sell. Buyers are frightened of contamination and often react irrationally to that fear, especially when large amounts of money are at risk. Buyers will require many additional reviews and contract addenda, and will worry about each and every detail of the negotiation. They also will expect that the price should be lower than the market value because they feel they are taking an additional risk (which may be true). There are also buyers who will not take on any real or perceived environmental risk; these people should never consider brownfields redevelopment properties in the first place.

The comfort letter has also become a factor in the sale and redevelopment of these properties. A potential buyer with questions as to the environmental risk is more comfortable with the state and federal sign-off than with the assurances of developers or environmental engineers. It is true that NFR letters (or other comfort letters) provide legal protections against being pursued for additional cleanup, but they do not protect the owner against unknown contamination or provide financial protections against that cleanup. Professional liability insurance of the environmental engineer provides the most direct protection against loss due to cleanup costs, unknown contamination, and assessing the potential for problems. This, however, is not the view of the banking community or many potential buyers, who will always take the state or federal EPA sign-off over any other person's knowledge. Therefore many developers in Illinois will choose to enter the property into the voluntary site remediation program with the goal of providing an NFR letter to the prospective buyer.

There is a cost to entering into such an agreement and program. No developer wants to depend on an act of government to determine the timeline of a project. In this case, however, the NFR letter is so critical to the perceptions of the buyer, and often the lender, that there is no getting around the need for the letter. This has economic implications for the project as a whole. It is ironic that NFR letters, consent letters, and PPAs have become an integral part of the redevelopment of brownfields. The legislators who wrote these regulations aimed to keep EPA and the government from being involved with property transfer. However, these assurances have proven to provide the assurance that EPA will leave the prospective purchaser alone, as would have been the case had the developer built on virgin land.

The voluntary site remediation program requires a lengthy and detailed plan. This remediation plan, once accepted by the regulators, provides the structure for the cleanup and provides positive milestones for work. This has significant advantages in the development of the timeline and budget for the project. Unfortunately, NFR letters have become such a commodity that the state reviewing process has gotten bogged down. There often are extensive, expensive, and time-consuming requirements for site assessment (Phase II).

When the remediation is done, the state has been satisfied, the NFR letter request has been sent in, and the NFR letter is granted, what remains is the presentation to the banks. Both the developer's bank and the buyer's bank must be satisfied with the environmental engineer's work. It is important to know that the environmental engineer's report will be acceptable to all the parties involved. Further, the consulting engineer must state that his or her report is for the use of the developer, bank, and buyer (if the property is for sale), as this will allow the consultants' professional liability insurance to protect all the parties. It is important to make sure that this insurance is of sufficient size for the project and will cover any reasonable problem that may arise.

Environmental insurance provides additional protections against cleanup costs or cost overruns, as well as toxic torts and other third-party lawsuits. Even though these insurance policies are in their infancy, many property transfers are completed using environmental insurance to lend comfort to those taking on the liabilities that exist with a property.

There have been some interesting suits over the coverage provided by environmental engineers. These suits claim that, regardless of the limitations stated in the report, the property was purchased on the consultants' suggestions and its value was estimated on that opinion; therefore, if the opinion was wrong, the consultant is liable for misleading the customer (the property buyer) and could be pursued for the difference in the property's value (or cost of cleanup). This provided some downside protection, though suing an insurance company for this type of loss can be expensive and time consuming. It is best if you have good consultants with good credentials, whose opinions all parties will ultimately trust.

If a site requires continuing monitoring or cleanup after redevelopment, the site may be difficult to sell until the cleanup is completed. Leasing, however, is a very possible option as long as the monitoring or cleanup does not interfere with or cause a hazard to the company leasing the site. Long-term monitoring costs of the property can be funded as part of the purchase agreement, possibly through an escrow account.

If the buyer is sophisticated enough and the liabilities can be defined well enough, then a transfer of those liabilities can be included as part of the contract. Unfortunately, most of the time the transfer of liabilities happens as a matter of law without the notice of the parties. Presently, the law in Illinois is that responsibility for contamination lies with the party that caused the contamination; but often that entity or person is no longer around or financially viable to pay for the cleanup, and thus the cost reverts to the present owner.

Ultimately, the transfer of liabilities and other site responsibilities is the key issue in any brownfield redevelopment. These transfers can take place either at the beginning of the process, with the purchase of the property, or at the end, with the sale or lease of the property. In either case, it is important to get good advice to estimate the real risk, but it is up to the purchaser of the property to decide whether that risk is acceptable. Too often, the owners of businesses relinquish the determination of what is an acceptable risk to their environmental consultants or attorneys. This conservative position makes it difficult for the purchase to be consummated.

For this reason, many large industrial real estate firms purchase only virgin land on which to build new industrial buildings. This trend cannot be sustained forever. Existing brownfields sites must be redeveloped, for all the reasons stated in this book. Fortunately, the purchase, cleanup, and redevelopment of brownfields properties are becoming more frequent—but are still less than common.

It is an unfortunate realization that the value of a brownfield property, even after cleanup, inherently reflects the stigma related to the property. This will make the marketing and reuse of the site more difficult, but for those who are willing to take risks, there are great opportunities to expand operations or purchase new properties priced below market value, due to the stigma retained by the property.

Ultimately, business decisions come down to the bottom line: is the risk outweighed by the benefit of discounted property price? For businesses that could use land and/or buildings priced below market value, there is opportunity to balance the risk with the acquisition of cheap space, which helps the bottom line.

For biographical information on Robert Rafson, see Sec. 5.1.

8.2   Tax Treatment/ Recognition of Environmental Liabilities

Ernest Di Monte

Sale of the Property

The sale of property that has been cleansed of its hazardous substances offers no tax incentives, as written in Sec. 198 of the Internal Revenue Code. The gain from the sale of the newly created "greenfield" is treated as ordinary income to the extent of past remediation expenditures expensed instead of capitalized. If the sale of the property occurs within a short period of time (i.e., 24 months) after the completion of the abatement, it will negate the incentive that Congress has given to taxpayers who buy contaminated properties and clean them up. This statement does not imply that there is not substantial gain to be made by abating property contaminated with hazardous substances. It simply means that the incentive is not for investors looking to clean up property and then sell it within a 24-month period, but is best used by taxpayers looking to acquire property for long-term use, such as building a factory or warehouse on the land.

According to Sec. 198(e) of the Internal Revenue Code:

1. the deduction allowed by this section for such expenditure shall be treated as a deduction for depreciation, and

2. such property shall be treated as section 1245 property solely for purposes of applying section 1245 to such deduction.

Section 198(e) simply states that the deduction taken for the environmental expenditures on the property that would otherwise have been added to the basis of the land under normal practices will have to be considered as accumulated depreciation.

Recognition of Environmental Remediation Liabilities

In January 1993, the American Institute of Certified Public Accountants (AICPA) held an environmental issues discussion. The main objectives of the discussion were to examine practice problems in applying generally accepted accounting principles to environment-related financial statement assertions; to detect environmental issues that may need authoritative accounting and auditing guidance; and to make inroads toward the development of guidance on applying existing accounting and auditing standards to environment-related matters. Out of this discussion came Statement of Position 96-1 from AICPA.

Statement of Position 96-1 is provided to improve and narrow authoritative literature of existing principles as applied by entities to the specific circumstance of environmental liabilities. This may include the recognizing, measuring, and disclosing of environmental remediation liabilities in the financial statements. For purposes of this book we will discuss the aspects of recognition, measurement, and disclosure of environmental remediation liabilities.

Recognition involves determining *when* amounts should be disclosed in financial statements for the purpose of reporting environmental liabilities. *Measurement* has to do with the *amount* to be reported in the financial statements. According to Financial Accounting Standards Board (FASB) Statement of Financial Accounting Standards No. 5, *Accounting for Contingencies,* the accrual of a liability is required if (1) information available prior to issuance of the financial statements indicates that it is probable that an asset has been impaired or a liability has been incurred at the date of the financial statements, and (2) the amount of the loss can be reasonably estimated.

Once an entity has determined that it will probably incur costs for the remediation of an environmental liability, the entity should estimate the liability it figures to bear. This estimate should be based on available information. The entity should include its allocable share of the liability for a specific site. Many sites that are contaminated have been contaminated by a few different entities—for example, a waste disposal site that has been used as a dumping ground for many companies. This estimate should then be recognized as a liability on the balance sheet of the entity and as a charge to income. The liability should also be disclosed in notes to the financial statements, which should describe the liability along with any additional appropriate information.

Estimates formed in the early stages of remediation can vary significantly. Many times early estimates require major revision. Many factors are essential to forming cost estimates, such as the extent and type of hazardous substances at a given site and the technologies that can be applied to abatement of the site. Also to be considered when developing estimates are the number of potentially responsible parties and the financial ability of those parties to pay their share of the environmental cleanup costs.

In disclosing notes to the financial statements, entities are encouraged to report the nature and a brief description of the environmental remediation liability. Included in the description may be the estimated time frame of disbursements for expenses, the estimated time frame for probable recoveries (i.e., insurance proceeds and other responsible party recoveries), accounting principles used, and other pertinent information. Disclosures to the financial statements themselves include the amount of the liability to be incurred. This amount should only be incurred on the financial statements if the liability is probable. If the liability is reasonably possible, it should be disclosed in the notes to the financial statements but not accrued on the financial statements. Also to be included on the financial statements are any future receivables, such as receivables from other responsible parties, recoveries from insurers, or recoveries from prior owners, that are related to the environmental remediation liability. Remediation expenses should be disclosed as a charge against operating income, since the costs are considered a part of the normal operation of a company.

For biographical information on Ernest Di Monte, see Sec. 1.3.

Case Studies of Private Developers

Sections 9.1 and 9.2 are case studies. Even projects that go forward reasonably well have problems, and these case studies, like the one in Sec. 4.4, tell the stories chronologically and review the results. Each redevelopment project is a learning experience.

9.1 Case Study— D. C. Franche

Robert Rafson

On Christmas Day 1994, Darlene Franche, the owner-operator of D. C. Franche Paint Company, closed the doors of the company, delivered the keys of the building at 1401 W. Wabansia Street in Chicago (see Fig. 9-1) to the bank, and left for Florida. The paint factory was left with over 35,000 gallons of oil-based paints and solvents (Fig. 9-2).

During the next five months, developer Robert Rafson and Steve Safran, president of Safran Metal Corp., located at an adjoining property, worked jointly as the Wabansia Corp. to make a deal with the bank, which held the note on the property, to purchase the note and foreclose to gain title to the property. Unfortunately, three days before the planned purchase of the note, Superfund showed up to do the cleanup.

A fire had broken out at the other shuttered plant of D. C. Franche, located in Davenport, Iowa, and the Davenport Fire Department called the Chicago Fire Department. The Chicago Fire Department contacted the Chicago Department of Environment, which in turn contacted the emergency response arm of Superfund. The emergency group received fund-

Figure 9-1. Abandoned paint company (former D. C. Franche Paint site), 1401 W. Wabansia, Chicago, IL.

Figure 9-2. Sampling of unknown waste.

ing and permission to proceed with the cleanup of the site, much to the surprise of the bank and the would-be developers. By November 1995, the emergency removal action (see Figs. 9-3 and 9-4) was completed. At that time, no underground storage tanks or contamination were discovered. Interviews with former employees indicated that all the tanks had been removed in about 1989. Unfortunately, no records of that removal survived the failure of the business.

The developer met with the EPA on-site coordinator, the Superfund attorney, the Chicago Department of Environment coordinator, and others to determine possible courses of action. Once a site is designated a Superfund site, the cleanup must progress according to Superfund cleanup protocols. This makes the cleanup of the products on the site much more expensive due to the testing, handling, and reporting requirements. There appeared to be only three options for the developer: the first was to simply walk away. The Superfund cleanup costs were anticipated to be about $300,000, and, with the mortgage, back taxes, and physical problems of the abandoned buildings, the site was more of a liability than an asset. The second option was to complete the Superfund cleanup privately. But the anticipated cost savings would not make up the gap in liabilities, thus making this option untenable. Third, prospective purchaser agreement could be negotiated with Superfund. Those agreements were promul-

Figure 9-3. Pump-out of paint drums for off-site disposal.

Figure 9-4. Packing hazardous materials for off-site destruction.

gated late in 1995 or early in 1996, but the draft language was already available.

A few similar agreements had been negotiated by USEPA, but no formal process had been developed at the time. The PPA process was not only fairly straightforward, but was also encouraged by an initiative by President Clinton, Vice-President Gore and the Chief of EPA, Carol Browner, to put Superfund sites back into use. In the past, EPA would require that all of the cost of cleanup be recovered, which would have made the deal impossible. The new direction was to seek compensation for the cleanup costs for a specific project in economically reasonable terms. Economic compensation can consider social benefits such as jobs or a public use. The developer prepared a redevelopment plan including estimated costs and partial EPA cost reimbursement.

It took 15 months to negotiate this agreement, which was the first prospective purchaser agreement in Region 5 and the first transferable agreement nationwide. On May 17, 1996, Wabansia Corp. was granted the PPA with USEPA. This agreement protects Wabansia Corp. from the existing conditions as they relate to the environmental condition of the site, as well as providing a covenant not to sue for cleanup costs. Once the PPA was granted, Wabansia Corp. proceeded to purchase the bank's notes and to foreclose.

On August 13, 1996, Wabansia Corp. purchased 1401 W. Wabansia and gained title. The redevelopment was completed and fully occupied almost two years later (see Figs. 9-5 and 9-6), having suffered delays due to the challenges presented by financing a former Superfund site.

These delays added to the real costs of redevelopment in terms of both added costs and decreased income. The property has been divided into sections providing warehousing spaces for Safran Metal Corp. and space for an adjoining machine shop to expand; the main office building has been remodeled to provide modern office facilities for Greenfield Partners, Ltd., a brownfields redevelopment company.

A PPA does not end the story when it comes to the redevelopment of a Superfund site. The stigma of being a former Superfund site is far-reaching. The PPA, though critical to the redevelopment of the site, presents additional difficulties in the financing, leasing, or sale of the property. There are additional notification requirements, and the purchaser of the site must agree to the terms of the PPA to enjoy its protection. Educating a potential buyer as to the benefits of the PPA and residual responsibilities make sale or leasing extremely difficult. These difficulties also greatly affect the sale price and value for financing of the site. The problems of a contaminated site exist for any former Superfund site, regardless of the PPA or actual site cleanliness. Banks or lending institu-

Figure 9-5. Repairing and tuckpointing brickwork.

Figure 9-6. Completed facade of 1401 W. Wabansia, Chicago, IL.

tions are unwilling to take title to a Superfund site or contaminated property even with the significant protection provided by the recent changes in CERCLA.

For biographical information on Robert Rafson, see Sec. 5.1.

9.2   Case Study—The Autodeposition Site PPA

Andrew Warren and Robert Rafson

In 1997, EPA Region 5 entered into a prospective purchaser agreement (PPA) for the Autodeposition site in Chicago, along with Greenfield Partners, Ltd. The site, a former metal plating and coating facility (Fig. 9-7), was subject to a 1995 Superfund removal action. EPA's removal action addressed the immediate threat posed by hazardous substances associated with the plating operations at the facility. After conducting the removal action, EPA perfected a Superfund lien[1] on the site property, securing its incurred response costs.

Soon after the EPA action, the abandoned buildings on the property caught fire and burned (Fig. 9-8). The city of Chicago demolished the

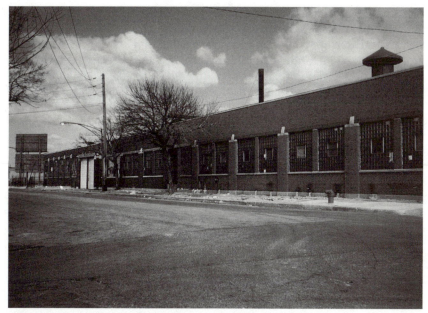

Figure 9-7. Abandoned plating plant with significant deterioration: Autodeposition site, 1518 W. Hubbard, Chicago, IL.

Figure 9-8. During negotiations with EPA, the middle section of the building burned. The building was demolished by the city of Chicago.

remaining building structures and removed all rubble and debris. Thus, at the end of EPA's action, the site consisted of a concrete slab with mild residual subsurface contamination and a $400,000 federal lien in place, as well as city claims for the demolition, unpaid county real estate taxes, and a mortgage in default, totaling approximately $850,000. Claims for demolition were $85,000. Normally, EPA would take no further action and the site would languish as a low-priority issue for the state to address at some point in the future.

Fortunately, Greenfield Partners, Ltd. approached EPA about entering into a PPA to facilitate redevelopment of the property. The buyer proposed to remediate certain remaining items of concern to EPA (EPA's removal action did not address underground storage tanks, contaminated brick and slab, an asbestos-insulated boiler, and residual soil contamination) and construct a loading dock at the site for use by an adjacent manufacturing facility. In return for performance of this work, the buyer sought a PPA and removal of the EPA lien.

The parties eventually negotiated a PPA with the following terms: the buyer agreed to perform work with an estimated value of $140,000; the buyer agreed to remove the fuel oil from the two underground storage tanks (Fig. 9-9), remove the asbestos insulation from the boiler, cap the site with a vapor barrier and concrete (Figs. 9-10 and 9-11), and construct a

Figure 9-9. Two 20,000-gallon underground fuel oil tanks were pumped out.

Figure 9-10. A vapor barrier was installed as part of an engineered barrier.

Figure 9-11. A reinforced concrete cap was installed.

Figure 9-12. Newly constructed 51,000-square-foot warehouse.

loading dock at the facility with facilitated reuse by a neighboring industry. EPA agreed to release the Superfund lien from the site property, and
the United States granted the buyer a covenant not to sue and contribution
protection for any claims associated with existing contamination at the
site. In addition, the protections granted to the buyer by the PPA were
explicitly extended to the buyer's lessees or assignees, upon notice to EPA.

Thus, the PPA satisfied all EPA criteria under applicable guidance and
facilitated expansion of a neighboring industry. That expansion (see Fig.
9-12) helped retain neighboring business within Chicago, preserving 305
jobs, and led to the creation of 30 additional jobs.

Note

1. Section 107(1) of CERCLA, 42 U.S.C. §9607(1), authorizes the imposition of a
   lien against property subject to an EPA response action as a means to secure
   recovery of EPA's costs.

*For biographical information on Andrew Warren, see Sec. 5.3. For bio-
graphical information on Robert Rafson, see Sec. 5.1.*

10
Conclusion

This is a rapidly developing field; here Harold Rafson adds some of his reflections and hopes for the future.

Final Thoughts

Harold J. Rafson

This is a very exciting time in the saga of brownfields redevelopment. We believe that we have bottomed out, and that the redevelopment curve is now clearly on the upswing. We are beyond the time when liabilities and unrealistic remediation standards virtually stopped redevelopment. There is a long way to go, and this is a period when experimentation and change are required. We would like to comment on some issues discussed in this book, and on some peripheral issues as well.

After having read the seller's concerns in Chap. 3, it is clear that there are major issues still to be resolved. There are beginnings toward those solutions. Insurance is one such solution, but the coverage and cost of such insurance vehicles are still to evolve. The atmosphere of distrust between insured and insurer must also be overcome. Another approach is joint ventures between the seller and developer, engineer, or contractors.

In Chap. 3, reference is made to "sleeping dogs" that are not really asleep. In the coming years, those dogs (liabilities, financial recognition of liabilities, and costs of cleanups) will have to be dealt with. Only when this is done will there be great progress toward resolving brownfields issues nationally.

As has been made clear from the historical and other sections, brownfields are just a part of the cost of urban blight, suburban sprawl, and technological changes in transport and manufacturing. They are also a part of the changing profile of American business, as there is less manufacturing and more service business. A large number of those changes are inevitable. Just as we, as a society, have learned to recycle more of our waste products, so we must learn to recycle more of our land, buildings, and infrastructure. We cannot continue as a society intent upon disposing of what is considered outmoded. Further progress must be made in programs to encourage such facility recycling—in laws, incentive programs, lending vehicles, and elimination of stigma. In fact, current attitudes toward stigma can be turned on their heads—and just as it is now virtuous to recycle paper and other materials, the same attitude can be achieved toward retaining and improving good old buildings.

In all brownfields discussions, another aspect that the authors believe is not frequently considered is research. It was necessary to do some research to make paper, plastics, metals, and municipal waste into usable, economical recyclable by-products. In the same way, it will be necessary to consider research in the field of brownfields revitalization. It is clear that research plays a role in improving methods of environmental remediation. One example is the use of natural attenuation, biological methods, and phytoremediation. All of these processes are slower, more natural, and less costly. More information on how to use these longer-term processes and their effectiveness is needed.

Another aspect of technological change that requires further knowledge is the design of manufacturing processes. In the old days, gravity flow was a major design concept. Then, in the days of the assembly line and specialized worker function, long, flat buildings were designed. These also were bracketed by warehouses for incoming parts and outgoing product. (Or the layout was U-shaped). These buildings were considered desirable because of their flat openness. We are now in an age of just in time warehousing, and computerized and mechanized warehouses that have other design possibilities. The manufacturing process has shifted its focus from great quantities of the same product to endless variations of specialized designs—we are changing to the age of computer-integrated manufacturing (CIM). The best design layouts for such technologies are not necessarily linear. All these changes require a rethinking of industrial design. It does not require a great leap of imagination to see how some of the sturdy older buildings would be easily applicable to new manufacturing concepts. Again, architects and engineers must be enlisted to revitalize building designs for new manufacturing methods, to study optimum design possibilities, and to avoid the temptation of designing the same one-story manufacturing/warehousing space again and again.

It is the viewpoint of the authors that at some point the nation is going to have to decrease its wastefulness of energy, which is costly and degenerative to the environment. As with recycling, society will change its perspective and decreasing energy waste will be looked upon as meritorious. Then urban sprawl will be limited.

The authors spoke with a major redeveloper, who stated he would "never buy land that has been used for anything besides farming." He would not "mess with" any property that has a possibility of being contaminated. While this developer has decided that this viewpoint is good for business, it is costly to society. We believe that abandoned and contaminated land must be recycled, and that, if this costs more than building on virgin land, society must find ways to support such redevelopment. The benefits will include better use of existing infrastructure, cleaner air and water, and revitalized cities.

When we started to write about brownfields, we realized that studies show there are over 500,000 brownfields sites nationally. It is quite possible that many "sleeping dogs" have been left uncounted. We believe it is only through an exchange of information through networks, journals, media, and government officials that opinions can be changed to result in more constructive efforts by society. The brownfield is a result rather than a cause, and our efforts should not only be directed toward correcting the blight of brownfields, but toward limiting those causes that create them.

We would like to continue to contribute toward these beneficial changes, and we believe that we can be of help by creating links with people working to solve brownfields issues. People can make changes working together. Information can be shared, and support can be provided. The authors offer their willingness to help. If you believe that an exchange of information or contact will be helpful, please contact us.

Harold J. Rafson
42 Indian Tree Drive, Highland Park, IL 60035
hrafson@worldnet.att.net
Telephone: 848-433-3026
Fax: 847-433-3073

Robert Rafson
1401 W. Wabansia, Chicago, IL 60622
rafson@idt.net
Telephone and fax: 773-384-3841

For biographical information on Harold Rafson, see Sec. 1.1.

Tiered Approach to Corrective Action Objectives (TACO)

ILLINOIS POLLUTION CONTROL BOARD
June 5, 1997

IN THE MATTER OF:)
)
TIERED APPROACH TO CORRECTIVE   )   R97-12 (A)
ACTION OBJECTIVES (TACO): 35 ILL.   )   (Rulemaking - Land)
ADM. CODE PART 742   )

Adopted Rule. Final Order.

OPINION AND ORDER OF THE BOARD (by M. McFawn and J. Yi):

The Board adopts today as final, rules which create a tiered approach to establishing corrective action, *i.e.*, remediation objectives, based on risks to human health and the environment, allowing consideration of the proposed land use at a subject site. These rules are located at a new part, 35 Ill. Adm. Code 742, entitled the Tiered Approach to Corrective Action Objectives, and have therefore become known as the TACO rules. Part 742 is unusual because it does not regulate activities at a site or mandate fixed clean up standards. Rather, the TACO rules at Part 742 provide the acceptable methodologies for determining site-specific, risk-based remediation objectives; while the programs to which TACO is applied govern the scope and extent of the site investigation preceding the application of TACO, as well as the no further remediation determination made by the Illinois Environmental Protection Agency (Agency) after the TACO derived remediation objectives are achieved. The TACO rules are to be applied to all types of remediation programs under the Illinois Environmental Protection Act (Act) (415 ILCS 5/1 *et seq.* (1994)), including the Site Remediation Program adopted today as a new Part 740, and the Underground Storage Tank rules found at Part 732 and the Resource Conservation and Recovery Act programs.

The TACO methodology is premised upon the statutory mandates in the Site Remediation legislation, P.A. 89-431, which was signed and became effective December 15, 1995, and later amended by P.A. 89-443, effective July 1, 1996. The Site Remediation legislation, also known as the Brownfield legislation, added Title XVII to the Act. Title XVII is intended to achieve five objectives. Those objectives are to: 1) establish a risk-based system of remediation based on the protection of human health and the environment relative to present and future use of the land; 2) assure that the land use for which remedial action was undertaken will not be modified without consideration of the adequacy of such remedial action for the new land use; 3) provide incentives for the private sector to undertake remedial action; 4) establish expeditious alternatives for the review of site investigation and remedial activities, including a privatized review process; and 5) assure that the resources of the Hazardous Waste

The Board gratefully acknowledges the efforts of the entire staff throughout this rulemaking, and in particular the concerted efforts of Kevin Desharnais as the hearing officer and attorney-assistant; Charles Feinen and Amy Muran-Felton, attorney-assistants; Anand Rao and Elizabeth Ann of the Technical Unit, and Kemelyau Pittman. Their help greatly assisted the Board in deciding and managing the complexities this rulemaking entailed.

Fund are used in a manner that is protective of human health and the environment relative to present and future uses of the site and surrounding area. The TACO rules address the first two of these objectives; the remaining three are the focus of the Site Remediation Program at Part 740.

PROCEDURAL HISTORY

On September 16, 1996 the Agency filed a proposal to add the TACO rules as a new Part 742 to the Board's rules. The Board accepted this matter for hearing on September 19, 1996. On November 7, 1996, the Board sent this matter to First Notice without commenting on the merits of the proposal. Subsequently, on December 6, 1996, the proposal was published in the *Illinois Register* (20 Ill. Reg. 15429.)

Development of the Proposal. Section 58.11 of the Act, adopted as part of the Site Remediation Program legislation, created the Site Remediation Advisory Committee (SRAC) to advise the Agency in developing the mandated TACO and the Site Remediation Program regulatory proposals. The SRAC consists of one member from each of the following organizations: the Illinois State Chamber of Commerce, the Illinois Manufacturers Association, the Chemical Industry Council of Illinois, the Consulting Engineers Council of Illinois, the Illinois Bankers Association, the Community Bankers Association of Illinois, and the National Solid Waste Management Association. In addition, representatives from the Illinois Petroleum Council, the Illinois Petroleum Marketers Association, and the City of Chicago participated. The Agency met with the SRAC, or subgroups thereof, ten times between March 14, 1996 and August 30, 1996, to discuss both the TACO rules and the rules for the Part 740 Site Remediation Program. The TACO rules proposed by the Agency and adopted for First Notice represented the consensus reached by the SRAC and the Agency on the TACO rules. (See Exh. 1 at 11.) Two sets of hearings were held in this matter during the First Notice period. The first set of hearings, held on December 2 and 3, 1996, in Chicago, and on December 10, 1996 in Springfield, was reserved for the Agency's presentation of its proposal and questions for Agency witnesses. The second set of hearings, held on January 15 and 16, 1997 in Springfield, was for the purpose of addressing remaining questions for the Agency, allowing the presentation of testimony by other interested participants, and allowing questions directed to those testifying.

Subsequent to those hearings and after the close of the public comment period, on April 17, 1997, the Board sent the proposal to Second Notice, pursuant to the Administrative Procedure Act (5 ILCS 100/1-1 *et seq.* (1994)), for consideration by the Joint Committee on Administrative Rules. The Board opinion accompanying the second notice order explains in detail how the TACO methodology was developed and how the rules are to be applied in conjunction with other Board rules governing site remediation. At that time the Board also bifurcated this rulemaking and adopted a separate opinion and order creating a Docket B, wherein the Board proposed for First Notice new rules concerning a single issue. The Board found it necessary to do so because the Agency had requested that the Board adopt a "mixture" rule, *i.e.*; a rule which requires that an applicant consider the cumulative effect of similar-acting contaminants at a site when developing the appropriate remediation objectives. Shortly thereafter, the Secretary of State informed the Board that it would not accept Docket B for First Notice publication because the rules proposed therein were amendments to Part 742 which was not yet adopted as final. Consequently, on May 1, 1997, the Board vacated its April 17, 1997 opinion and order, and replaced it with an opinion and order adopting the mixture rule under Docket B as proposed rules only.

On May 20, 1997, the Joint Committee on Administrative Rules voted no objection to the new Part 742, as proposed under Docket A. Today, the Board adopts Part 742 as final rules to become effective on July 1, 1997. The July 1, 1997 effective date coincides with the effective dates of the Site Remediation Program rules also finalized today as a new Part 740: In the Matter of: *Site Remediation Program and Groundwater Quality*, docketed as R97-11; and In the Matter of : *Leaking Underground Storage Tanks*, docketed as R97-10, adopted by the Board on March 6, 1997 amending the existing Part 732 which govern remediation of underground storage tanks. (Like the TACO rules, the Site Remediation Program was mandated by P.A. 89-431, while the leaking underground storage tank amendments were mandated by P.A. 89-457, effective May 22, 1996.)

Docket B: For the most part, the Agency's request for a mixture rule was developed in a series of filings subsequent to the public hearings in this matter, and with minimum justification in support of such rules. In its initial rulemaking proposal, the Agency had only requested a mixture rule under Tier 2 for noncarcinogenic chemicals. In its filings during the public comment period, but after the close of hearings, the Agency requested that the Board also adopt a mixture rule applicable to the development of groundwater remediation objectives under Tier 1 for both carcinogenic and noncarcinogenic chemicals, and further requested that the Tier 2 rule be applicable to carcinogenic chemicals in groundwater. The record before the Board at the time of Second Notice was insufficient for the Board to adopt the entire mixture rule ultimately requested by the Agency. However, the justification provided in support of expanding the rule's applicability did indicate that absent such a rule, remediation objectives determined using TACO may not be protective of human health at sites with multiple, similar-acting chemicals. Therefore, the Board found it necessary to clearly examine the mixture rule proposed by the Agency to determine to what extent it is necessary to insure that the remediation objectives developed under TACO are protective of human health in all circumstances. Docket B was opened for that purpose. Docket B will proceed through regular rulemaking, albeit on an expedited schedule.

OVERVIEW OF THE TACO PROCESS

The TACO rules establish procedures for developing remediation objectives for soil and groundwater at remediation sites based on risks to human health, taking into account the existing pathways for human exposure and current and future use of the remediation site. The methodology consists of a three tiered approach for establishing remediation objectives. The tiers can operate fully independent of each other, and it is not necessary to perform a Tier 1 analysis before performing a Tier 2 or Tier 3 analysis, or to perform a Tier 2 analysis before performing a Tier 3 analysis. Each successive tier allows the person conducting a remedial investigation pursuant to the Act (hereinafter referred to as the "applicant") to rely on more site-specific information, and requires a concomitant increase in the level of site-specific investigation and analysis under Part 742.

As a prerequisite to using the tiered approach to establish remediation objectives, the applicant must determine the contaminants of concern at the site. This is done by conducting a site investigation under the applicable remediation program; such investigation is not part of the TACO process. Again, the programs with which TACO is to be used include the Underground Storage Tank program at Part 732, the Site Remediation Program proposed at Part 740, and the RCRA Part B Permits and Closure Plans at Parts 724 and 725. As mentioned at the outset, these programs govern the activities at the site which address the contamination, including the scope of the site investigation and ultimately the no further remediation determination made by the Agency. (Hereinafter in the opinion, these programs are referred to as the "governing programs.") The specific requirements of the governing program control how TACO is applied to determine the applicable remediation objective. After identifying the contaminants of concern, the applicant can use the TACO process to establish remediation objectives. Each tier of the TACO process requires the applicant to consider up to four potential exposure routes for each contaminant of concern: 1) the inhalation exposure route; 2) the soil ingestion route; 3) the dermal contact exposure route[1]; and 4) the groundwater ingestion route. The groundwater ingestion route is further subdivided into two components: 1) the migration to groundwater, or soil component, which must be investigated to establish a soil remediation objective; and 2) the direct ingestion of groundwater, or groundwater component, which must be investigated to establish a groundwater remediation objective. (Hereinafter each component of the groundwater ingestion route is referred to as the "soil component" or the "groundwater component"). Alternatively, as described in greater detail below, the applicant can: 1) demonstrate that a particular exposure route is not available for a contaminant of concern, and thereby exclude further consideration of that exposure route for that contaminant, or 2) rely on area background concentrations in establishing remediation objectives or to demonstrate that further remediation is not warranted.

A Tier 1 analysis requires the applicant to compare levels of contaminants of concern at the remediation site to pre-determined remediation objectives. The pre-determined remediation objectives are listed in the rules at Appendix B, Tables A through E. Separate remediation objectives are established for properties designated for residential use and for industrial/commercial use. The residential levels are the most stringent and are considered protective for all uses. The industrial/commercial levels are less stringent and must be accompanied by an institutional control, such as a deed restriction, in order to assure that the site is used only for industrial/commercial purposes. Additionally, if the site is to be

---

[1]The dermal contact exposure route need only be considered if the applicant elects to use the Tier 2 Risk Based Corrective Action (RBCA) equations set forth in Appendix C, Table C, or a Tier 3 formal risk assessment, to establish remediation objectives.

remediated to industrial/commercial levels, the applicant must assure that the remediation levels established for construction workers are also achieved. If any contaminants of concern at a remediation site are found to exceed the applicable pre-determined levels, the applicant is required to remediate the contamination until the remediation objectives are achieved, or alternatively, to develop site-specific remediation objectives using a Tier 2 or Tier 3 analysis. Under Tier 1, if multiple noncarcinogenic chemicals with similar-acting properties are present in the groundwater, their cumulative effect must be evaluated as part of the development of remediation objectives. This is the Tier 1 component of the mixture rule adopted pursuant to the Agency's public comments during First Notice, and currently under further consideration under Docket B.

A Tier 2 analysis uses equations set forth in the rules to develop alternative remediation objectives for contaminants of concern using site-specific information. The equations used to develop site-specific remediation objectives are from the United States Environmental Protection Agency's (USEPA) Soil Screening Levels Guidance (SSL) and the American Society of Testing and Material's (ASTM) Risk Based Corrective Action (RBCA). The equations are set forth in the proposed rules at Appendix C, Tables A and C, respectively. If any contaminants of concern are found to exceed the remediation objectives developed using the Tier 2 equations, the applicant is required to remediate the contamination until the objectives are achieved or to develop alternative objectives using a Tier 3 analysis. The mixture rule for noncarcinogens is also applicable under Tier 2. Unlike a Tier 1 analysis, however, it is applicable when developing both soil and groundwater remediation objectives. This component of the Tier 2 mixture rule, as well as a mixture rule for carcinogens in groundwater, are bother under further consideration in Docket B.

A Tier 3 analysis allows the applicant to develop remediation objectives using alternative parameters not found in Tier 1 or Tier 2. It allows the applicant great flexibility in developing remediation objectives appropriate for a particular site based upon site-specific information rather than relying on general categories of information. The options available under Tier 3 include: use of modified parameters in the Tier 2 equations; use of alternative models; conducting a site-specific risk assessment; assessment of impractical remediation; and variation of the target risk level. If any contaminants of concern are found to exceed the remediation objectives developed using the Tier 3 analysis, the applicant is required to remediate the contamination until the objectives are achieved. At this time, the mixture rule is not specifically applicable to a Tier 3 analysis. However, a mixture rule for carcinogens and noncarcinogens is under consideration for soil and groundwater remediation objectives in Docket B.

Outside of the individual tiers of analysis, there are two alternative means for addressing the presence of contamination: exclusion of pathways and reliance on area background. The first option, exclusion of pathways, is based on the premise that an exposure pathway must exist for contamination to present a threat to human health. If it can be shown that a pathway does not exist for any contaminants of concern, the applicant need not address that exposure pathway for those contaminants. The methods for evaluating and excluding exposure routes are set forth at Subpart C. The second option, reliance on area background, is based on Section 58.5(b)(1) of the Act, which provides that applicants shall not be required to remediate contaminants of concern to levels that are less than area background levels. If it can be shown that a contaminant of concern is present at levels that do not exceed area background levels for the site, the applicant need not further address that contaminant. Under appropriate circumstances, the applicant can also use background levels as remediation objectives. The methods for determining area background concentrations are set forth in Subpart D.

The applicant can use any combination of tiers if multiple contaminants of concern are present at a site. Remediation objectives established under any tier are considered equally health protective for a particular land use. Upon completion of remedial activities which achieve the established remediation objectives, the applicant is entitled to a no further remediation determination in accordance with the terms of the governing program. The TACO rules do not provide for the no further remediation determination; they provide only the process for determining site-specific remediation objectives based upon risk. The no further remediation determination is made at the conclusion of the process by the Agency pursuant to the governing program. For example, the Agency's no further remediation determination in the Site Remediation Program is effected through a No Further Remediation Letter. The same instrument is used in the Underground Storage Tank Program.

The following section contains a more detailed summary of the components of the rules. A detailed summary of the major issues raised concerning various components of the TACO rules, is set forth in the third section of the Second Notice opinion adopted on April 17, 1997. (See R97-12(A), Opinion and Order of April 17, 1997, beginning at page 33.)

SUMMARY OF THE SUBPARTS A THROUGH K

Subpart A: Introduction

This subpart contains sections concerning intent and purpose, applicability, overview and key elements of the tiered approach, and the requirements for site characterization. Section 742.100, entitled "Intent and Purpose," states that Part 742, the TACO process, contains the procedures for use in evaluating risks to human health posed by environmental conditions, and procedures for use in developing objectives for remediation which assure that risks are at acceptable levels. Furthermore, Section 742.100(b) states that the procedures are intended to provide adequate protection of human health and the environment based on risks to human health posed by environmental conditions while incorporating site-related information.

Section 742.105 sets forth the situations in which the rules are intended to apply. The applicant may use the Part 742 procedures to the extent allowed by state and federal law. The procedures must be used in accordance with the requirements of the program pursuant to which the remediation is being conducted. Section 742.105 specifically references the Underground Storage Tank program, the proposed Site Remediation Program, and the RCRA Part B Permits and Closure Plans. The use of Part 742 is subject to the limitation that it cannot be used where there is an imminent and substantial endangerment to human health and the environment. Section 742.105 also makes clear that groundwater remediation objectives established pursuant to the TACO process can exceed the groundwater quality standards set forth at 35 Ill. Adm. Code Part 620. This exception is based upon a statutory provision; the record does not otherwise support such a rule. Section 58.5 of the Act authorizes the use of groundwater remediation objectives for contaminants of concern that are greater than the groundwater quality standards established by the Board at 35 Ill. Adm. Code 620 pursuant to the Illinois Groundwater Protection Act. (415 ILCS 55/1 *et seq.*) The Board has made clarifying changes to Section 742.105 to notify the applicant that remediation objectives greater than the Part 620 standards may be developed only under Tier 3. Under Docket B, the Board is considering whether such remediation objectives may also be determined under Tiers 1 and 2.

The Agency also proposed a rule to the effect that a no further remediation determination constitutes *prima facie* evidence that the contaminants of concern addressed at a site do not cause or tend to cause water pollution pursuant to Section 12(a) of the Act, or create a water pollution hazard pursuant to Section 12(d) of the Act. Such a statement would be particularly critical at sites remediated to groundwater objectives greater than the State's groundwater quality standards, because those groundwater quality standards were adopted as being the minimum levels protective of human health and the environment pursuant to the Illinois Groundwater Protection Act. (*Id.*) As explained above, Section 58.5 of the Act allows an applicant to propose, and the Agency to approve pursuant to Tier 3 of TACO, remediation objectives greater than the State's groundwater quality standards. Therefore, once such a remediation objective is achieved, the Agency's no further remediation determination in effect deems the levels of contamination remaining at the site as protective of human health. Yet, the programs used in conjunction with TACO govern the scope and extent of the legal protection provided by a No Further Remediation Letter or any other type of no further remediation determination made by the Agency. Therefore, the Board did not adopt under TACO the rule proposed by the Agency addressing the effect of a no further remediation determination.

Section 742.110 contains an overview of the tiered approach, which is similar to the summary of the rules set forth above. We will not repeat that discussion here. The applicant is well advised to consult the illustrations in the Appendices for assistance in understanding the TACO process generally, and any particular provisions. The Illustrations provide road maps and "decision trees" which further clarify the TACO process. Generally, Illustrations A and B of Appendix A provide decision trees for developing soil and groundwater remediation objectives, respectively. Illustration A of Appendix B provides such a road map for Tier 1, and Illustrations A and B of Appendix C provides the same for Tier 2. However, none of the illustrations include the effect of the mixture rule adopted today, or being considered under Docket B.

Section 742.115 addresses the "Key Elements of the Tiered Approach." It sets forth the exposure routes that must be evaluated, the factors that must be considered in determining the remediation objectives for contaminants of concern, and the potential land use classifications for the site. Section 742.115(a) sets forth the potential exposure routes that must be addressed at a remediation site, specifically: inhalation, soil ingestion, groundwater ingestion, and dermal contact with soil. The groundwater ingestion route is further divided into two components: the soil component and the groundwater component. The dermal contact exposure route need only be addressed if the applicant develops remediation objectives using RBCA equations set forth in Appendix C, Table C, or through a formal risk assessment under Tier 3. For each contaminant of concern, the applicant must develop remediation objectives for each applicable exposure pathway or demonstrate that a pathway has been excluded from consideration.

Section 742.115(b) sets forth the factors that must be considered when identifying the contaminants of concern at the remediation site. These factors include: 1) the materials and wastes managed at the site; 2) the extent of the no further remediation determination which the applicant is seeking from the Agency under the governing program, *e.g.*, under the Site Remediation Program, a comprehensive or focused No Further Remediation Letter; and 3) the general requirements applicable under the governing program. In the Site Remediation Program or a Section 4(y) voluntary cleanup, the applicant determines the scope of the remediation and the contaminants of concern that will be addressed. In other programs, the scope of the remediation and contaminants of concern will be dictated by the governing program's requirements. At the conclusion of the process, the Agency will make a determination about whether further remediation is necessary under the governing program. If the Agency determines that no further remediation is necessary, the scope of the determination will extend only to the scope of the remediation performed.

Section 742.115(c) sets forth the possible land use classifications under the TACO process. The rules allow the proposed land use to be characterized as one of the following: residential, conservation, agricultural, or industrial/commercial. The land use classification determines the expected exposure scenario at the site, which is a principal factor in establishing health protective remediation objectives.

Tier 1 sets forth separate remediation objectives for residential and industrial/commercial uses. Similarly, Tier 2 has separate equations for developing remediation objectives for residential and industrial/commercial uses which reflect the different exposure rates expected for each land use. The remediation objectives for industrial/commercial property are premised upon a lower exposure rate, and therefore are less stringent than those established for residential property. Accordingly, an industrial/commercial designation under any tier requires an accompanying institutional control to assure that the land use is appropriately restricted to that property classification. Furthermore, because the rules do not reflect consideration of the appropriate exposure expected for a conservation or agricultural land use designation, these designations require a Tier 3 demonstration based on individual site use characteristics to assure appropriate protection of human health and the environment.

Section 742.120 sets forth the requirement that the applicant perform a site characterization prior to developing remediation objectives pursuant to TACO in order to establish the extent and concentrations of contamination at the site. The site investigation must be conducted in accordance with the requirements of the governing program. The TACO rules therefore do not set forth a separate site investigation procedure.

Subpart B: General

Subpart B of the rules contains general sections including definitions, a severability clause, and incorporations by reference. Additionally, Subpart B sets forth procedures for determining the soil attenuation capacity, the soil saturation limit, and demonstrating compliance with remediation objectives. These procedures apply across the entire TACO process. Finally, Subpart B contains the general rule that submittals to the Agency and subsequent review and approval by the Agency are to be done in accordance with the governing program's rules.

Definitions. Section 742.200 sets forth the definitions of terms used in these proposed rules. Most of these are self-explanatory, however several warrant further discussion. The definition of the term "carcinogen" repeats the statutory language from Section 58.2 of the Act. This definition requires that contaminants that fall into any of the following categories be considered carcinogens: 1) Category A1 or A2 carcinogens, as defined by the American Conference of Governmental Industrial Hygienists; 2) Category 1 or 2A/2B carcinogens, as defined by the World Health Organization's International Agency for Research on Cancer; 3) a "Human Carcinogen" or "Anticipated Human Carcinogen," as defined by the United States Department of Health and Human Service National Toxicological Program; or 4) a Category A or B1/B2 carcinogen as defined by the United States Environmental Protection Agency in its Integrated Risk Information System (IRIS), or a final rule issued in a Federal Register notice by the USEPA. Because the USEPA Soil Screening Level Guidance, which the Agency relied upon in developing the Tier 1 Tables for soil remediation objectives, includes Category C carcinogens within its definition of "carcinogens," the Agency recalculated the soil remediation objectives for those contaminants classified as Category C carcinogens by USEPA. The Board agrees that this properly reflects the statutory intent.

At the Board's request, the Illinois Department of Transportation (IDOT) introduced into the record the definitions of the terms "highway," "highway authority," and "right of way" from the Illinois Highway Code. (See Dec. 10, 1996 Transcript at 114 - 115.) The Agency proposed that these definitions be included in the proposed rules in its Errata Sheet No. 2. The Board believes that including these definitions in the rules clarifies these terms and ensures consistency in their application and, therefore, adopts these definitions.

The definition of "residential property" proposed by the Agency paraphrased the statutory language for Section 58.2 of the Act. The Agency's proposed definition reads: "Residential Property" means any real property that is used for habitation by individuals or properties where children have the opportunity for exposure to contaminants through ingestion or inhalation at educational facilities, health care facilities, child care facilities or playgrounds. To more closely reflect the statutory intent, the Board adopts a slightly modified definition for Part 742, as well as in Part 740 in the R97-11 rulemaking. The definition as adopted reads:

> "Residential Property" MEANS ANY REAL PROPERTY THAT IS USED FOR HABITATION BY INDIVIDUALS, or where children have the opportunity for exposure to contaminants through soil ingestion or inhalation at educational facilities, health care facilities, child care facilities, or outdoor recreational areas.

Soil Attenuation Capacity and Saturation Limit. Section 742.215 requires that the concentrations of organic contaminants remaining in the soil not exceed the attenuation capacity of the soil and sets forth the method for determining soil attenuation capacity. The requirement that the soil attenuation capacity not be exceeded is intended to insure the integrity of the soil remediation objectives established under the tiered approach, since the models which are used to derive the soil remediation objectives do not account for the existence of free product. Contaminant transport models generally assume equilibrium between contaminants that adhere to soil particles and contaminants that dissolve in water in the soil pores. This assumption is violated if the soil attenuation capacity is exceeded; then the models cannot accurately predict the behavior and movement of contaminants. (See Exh. 4 at 3-5.) Furthermore, John Sherrill of the Agency testified that the requirement that the soil attenuation capacity not be exceeded will achieve three objectives. First, it will ensure that there will be no migration of mobile free products. Second, it will protect against potentially unacceptable health risks from accidental exposure to contamination left in place which might occur if an engineered barrier or institutional control is breached. Finally, it will provide a ceiling on the level of exposure from high contaminant concentrations from multiple organic contaminants. (Dec. 2, 1996 Transcript at 151-152.)

Similarly, because the models which are used to derive the soil remediation objectives do not account for the existence of free product, Section 742.220 provides two circumstances under which remediation objectives cannot exceed the soil saturation limit (C_{sat}). The soil saturation limit is defined in Section 742.200 as "the contaminant concentration at which soil pore air and pore water are saturated with the chemical and the adsorptive limits of the soil particles have been reached." Pursuant to Section 742.220, the applicant must ensure that the soil saturation limit is not exceeded when establishing a Tier 2 or Tier 3 remediation objective for the inhalation exposure route for an organic contaminant with a melting point below $30° C$, or when establishing a Tier 2 or Tier 3 remediation objective for the soil component of the groundwater ingestion exposure route for any organic contaminant.

Section 742.220 establishes three methods for determining the soil saturation limit. These methods are: 1) use of the chemical-specific default values set forth in Appendix A, Table A; 2) use of a value derived from Equation S29 in Appendix C, Table A; or 3) use of a value derived from another method approved by the Agency.

Compliance with Remediation Objectives. Section 742.225 sets forth the method for demonstrating compliance with remediation objectives. For groundwater remediation objectives, compliance with remediation objectives is demonstrated by comparing discrete samples to the applicable groundwater remediation objective. The location of groundwater sampling points is determined in accordance with the requirements of the governing program pursuant to which remediation is being conducted.

Similarly, compliance with soil remediation objectives can be demonstrated by comparing discrete samples of contaminant concentrations to the applicable soil remediation objective, unless the applicant elects to composite or average soil samples in accordance with subsections (c) and (d) of this section, as explained below. Again, the number of locations is determined by the governing program.

Subsection (c) of Section 742.225 sets forth the requirements and limitations applicable to compositing or averaging soil samples for the soil component of the groundwater ingestion exposure route. For contaminants other than volatile organic compounds (VOCs), discrete samples from the same boring may be composited or averaged. For VOCs, discrete samples from the same boring may be averaged but compositing of samples is not allowed. This is because compositing would tend to allow VOCs to volatilize and escape and the sampling would thus underestimate the presence of volatile contaminants of concern in the soil. A minimum of two sampling locations for every 0.5 acres of contaminated area is required, with discrete samples at each location taken at every two feet of depth, beginning six inches below the ground surface, and continuing through the zone of contamination. Samples may not be taken from below the water table.

Subsection (d) of Section 742.225 sets forth the requirements and limitations applicable to compositing or averaging soil samples for the inhalation or ingestion exposure routes. The compositing and averaging requirements will be established on a site-specific basis, based upon a sampling protocol approved by the Agency.

Pursuant to Section 742.225(e), for the purposes of calculating averages under Section 742.225, if no more than 50 percent of the samples are reported as non-detect or below detection limits, such results must be included in the sampling results as one-half of the reported detection limit for the contaminant. If more than 50 percent are reported as non-detect, the applicant must obtain Agency approval for an alternate procedure which is statistically valid for determining the average.

Section 742.230 returns to the more general format of Subpart B. An omnibus provision, Section 742.230 addresses Agency review and approval of submittals. This section makes clear that the applicant must submit documents and requests in accordance with the governing program under which the remediation is being addressed and the Agency will review and approve the same in accordance with the governing program.

Subpart C: Exposure Route Evaluations

Subpart C sets forth the requirements and methodologies for determining and evaluating the following exposure routes: inhalation, soil ingestion, and groundwater ingestion. The rules allow the applicant to exclude from consideration contaminants of concern for one or more exposure routes if the applicant demonstrates that the identified exposure route is not available for that contaminant. The principle underlying the pathway exclusion is different from that underlying the development of numeric remediation objectives. It is premised on the concept that an exposure route must exist which enables a contaminant to reach a receptor for the contaminant to present a threat to human health. Thus, Agency witness Mr. Gary King testified that pathway exclusion is based on effective source control, coupled with site conditions and an appropriate institutional control that effectively prohibits human exposure through a given pathway. (Exh. 2 at 4.) The rules set forth five general criteria, as well as exposure route-specific criteria, which must be satisfied to exclude an exposure route for a particular contaminant of concern.

General Criteria for Exclusion of Pathways. There are five general criteria which must be satisfied before any exposure route may be excluded from consideration. These criteria are intended to insure that the contamination left in place when the pathway is excluded will not present a threat to human health. The first two criteria, set forth in Sections 742.305(a) and (b), require that the soil attenuation capacity and the soil saturation limit capacity not be exceeded, as set forth in Section 742.215 and 742.220, respectively. These criteria are intended to insure that there is no free product present and to insure that the behavior of the contaminants can be accurately modeled.

Criteria three through five, set forth in Sections 742.305(c) through (e), require the applicant to insure that the contaminated soil which remains in place will not exhibit the hazardous characteristics of reactivity, corrosivity, or toxicity. Section 742.305(c) requires that any soil that contains contaminants of concern cannot exhibit any of the characteristics for reactivity for hazardous waste, as determined under 35 Ill. Adm. Code 721.123. Section 742.305(d) requires that any soil that contains contaminants of concern cannot exhibit a pH less than or equal to 2.0, or greater than or equal to 12.5. Finally, Section 742.305(e) requires that any soil that contains one or more of the inorganic chemicals arsenic, barium, cadmium, chromium, lead, mercury, selenium, silver, or the salts of any of these chemicals cannot exhibit characteristics of toxicity. The toxicity determination is made pursuant to the methods set forth at 35 Ill. Adm. Code 721.124.

Specific Criteria for Exclusion of Pathways. Sections 742.310, 742.315 and 742.320 set forth the specific criteria which must be satisfied to exclude from consideration each particular pathway, *i.e.*, inhalation, soil ingestion, and groundwater ingestion respectively. These criteria must be satisfied in order to exclude the applicable exposure route, in addition to the five general criteria set forth in Section 742.305.

Section 742.310 sets forth additional criteria which must be satisfied in order to exclude the inhalation exposure route for contaminants of concern. In order to exclude this exposure route, the concentration of any contaminant of concern within ten feet of the land surface or any man-made pathway cannot exceed the Tier 1 remediation objective for the inhalation exposure route. Alternatively, the applicant can install an engineered barrier, as set forth in Subpart K, which is approved by the Agency. The applicant must also obtain an institutional control, in accordance with the requirements of Subpart J, which ensures compliance with these requirements and ensures the safety of construction workers.

Section 742.315 sets forth additional criteria which must be satisfied in order to exclude the soil ingestion exposure route for contaminants of concern. To exclude this pathway, the applicant must demonstrate that the concentration of any contaminant of concern within three feet of the land surface does not exceed the applicable Tier 1 remediation objective or that an engineered barrier has been installed, in accordance with the requirements of Subpart K and approved by the Agency. Furthermore, the applicant must obtain an institutional control which ensures that these requirements are met and ensures the safety of construction workers.

Section 742.320 sets forth the additional criteria which must be satisfied in order to exclude the groundwater ingestion exposure route from consideration. These criteria include location and groundwater quality demonstrations, as well as a requirement that an institutional control, *i.e.* an ordinance adopted by the local government, be in place. Taken together, these criteria are intended to ensure that potable drinking water supplies will not be impacted by contamination left in place.

Specifically, the applicant must demonstrate that corrective action measures have been completed to remove any free product. The applicant must also demonstrate that the source of the release is not located within the minimum or designated maximum setback zone or within a regulated recharge area of a potable water supply well. The applicant must also demonstrate that the concentration of any contaminant of concern in the groundwater within the minimum or designated maximum setback zone of an existing water supply well meets the applicable Tier 1 groundwater remediation objective. Finally, the applicant must demonstrate that the concentration of any contaminant of concern in groundwater which discharges to a surface water will meet the applicable surface water quality standards under 35 Ill. Adm. Code 302. These last two demonstrations must be made using Equation R26 in Appendix C, Table C. In order to exclude the direct ingestion of groundwater pathway, in addition to satisfying the location and groundwater quality demonstrations, the applicant must demonstrate, in accordance with Subpart J: Institutional Controls, that the unit of local government has adopted an ordinance that effectively prohibits the installation or use of groundwater as a potable supply of water. Such an ordinance must be in effect for any area within 2500 feet of the source of the release. Additionally, the unit of local government must enter into a Memorandum of Understanding (MOU) with the Agency if the local ordinance used as an institutional control does not prohibit the local government from installing and using a potable water supply well. The MOU must commit the local government to: 1) keep a registry of sites within its boundaries that have received no further remediation determinations; 2) consider whether groundwater contamination from those sites may be present at potential public well sites; and 3) take appropriate protective measures if wells are sited near such locations. (PC 15 at 5-6; Errata Sheet No. 2.)

Subpart D: Determining Area Background

This subpart sets forth the procedures for determining area background concentrations for contaminants of concern. As set forth in Section 58.5(b) of the Act, applicants shall not be required to remediate contaminants of concern to levels that are less than area background levels, subject to two statutory exceptions which have been included in the rules at Section 742.415. First, if the contaminant concentration is equal to or less than the area background level, yet it exceeds the Tier 1 residential level, the property cannot be converted to residential use unless that remediation level or an alternative developed under Tier 2 or Tier 3 is achieved. (Section 58.5(b)(2) of the Act.) Second, if the Agency determines in writing that the area background level poses an acute threat to human health or the environment in consideration of the post-remedial land use, appropriate risk-based remediation levels must be developed. (Section 58.5(b)(3) of the Act.) If neither of these exceptions applies, and the applicant can demonstrate that contaminant concentrations are at area background levels, no further remediation is required. Another use of area background levels is for the applicant to demonstrate that area background levels should be used as the remediation objectives for contaminants of concern. This alternative is limited to industrial/commercial properties only, and accordingly requires the use of an institutional control.

Determination of Area Background for Soils. Section 742.405 sets forth the method for determining area background for soils. Subsection (a) sets forth the sampling requirements. Section 742.405(b) sets forth the two options available to the applicant for determining the area background level for inorganics. The first option is referred to as the statewide area background approach. This approach relies upon data previously compiled by the Agency concerning area background concentrations throughout the State, which is set forth in Appendix A, Table G. Under the statewide area background approach, the applicant must set the upper limit of the area background concentration for the site at the value of the concentrations of inorganic chemicals in background soils listed in Appendix A, Table G. The applicant's second option is to use another method which is statistically valid for the characteristics of the data set and which has been approved by the Agency.

Determination of Area Background for Groundwater. Section 742.410 sets forth the method for determining area background concentrations for groundwater. Subsection (a) sets forth the sampling requirements and is intended to ensure that the sampling points are of sufficient quantity and appropriately located so as to be representative of actual background concentrations. Section 742.410(b) sets forth the two options available to the applicant for determining background levels for groundwater. The first option is referred to as the "Prescriptive Approach"; the second option is the use of another statistically valid approach which is appropriate for the data set and approved by the Agency.

Under the Prescriptive Approach, the upper limit of the area background concentration for the site is set at the Upper Tolerance Limit (UTL) for sample sets of ten samples or more, or at the maximum value of the sample set for sets of less than ten samples. The Prescriptive Approach establishes the method for determining the UTL of a normally distributed sample. If the sample set contains less than fifty (50) samples, the applicant must use the Shapiro-Wilke Test of Normality to determine whether the sample set is normally distributed. The Prescriptive Approach can only be used if the samples are determined to be normally distributed.

The Prescriptive Approach cannot be used if more than 15 percent of the groundwater sampling results for any chemical are less than the appropriate detection limit for that chemical. If 15 percent or less are less than the appropriate detection limit, a concentration equal to one-half the detection limit must be used for that chemical in the calculations. Additionally, the Prescriptive Approach cannot be used for determining area background for pH. For these exceptions and in any case, Section 742.410(b) concludes with the provision that another statistically valid approach may be used on a site-specific basis if approved by the Agency.

Pursuant to Section 742.415, area background concentrations can be used in two ways. First, an area background concentration can be used to support a request to exclude a chemical as a contaminant of concern from further remediation due to its presence as a result of background conditions. Second, an area background concentration can be used as the remediation objective for a contaminant of concern. For either of these to occur the applicant must submit the request to the Agency. Again, however, pursuant to Section 58.5(b)(3) of the Act, area background cannot be used in either manner if the Agency determines, in writing, that the background level poses an acute threat to human health or the environment taking into consideration the post-remedial land use of the site.

Subpart E: Tier 1 Evaluation

A Tier 1 evaluation compares the concentrations of contaminants of concern to established baseline remediation objectives which are set forth in Appendix B, Tables A through E. The Tier 1 objectives are numerical chemical concentrations that represent a level of contamination at or below which there are no human health concerns for the designated land use. The Tier 1 objectives for individual chemical contaminants do not exceed an excess cancer risk of 1 in 1,000,000 for carcinogens (also referred to as 1×10^{-6}), or a hazard quotient[2] of 1 for noncarcinogens. The pre-established remediation objectives under Tier 1 are

---

[2]"Hazard Quotient" is defined as the ratio of a single substance exposure level over a specified time period to a reference dose for that substance derived from a similar exposure period. The reference dose, which is derived for noncarcinogens as an acceptable daily chemical exposure, is that dose at which no harmful consequences occur. A hazard quotient greater than 1

based upon the SSL and the screening levels therein are designed to insure that contaminants of concern individually will not present a greater risk. However, in some instances where multiple contaminants are present at a site, the cumulative effect of similar-acting chemicals may cause the target risk of 1 x 10^{-6} or the hazard quotient of 1 to be exceeded. To correct this, the rules adopted today include a mixture rule which provides that remediation objectives developed under Tier 1 for the groundwater ingestion exposure route must take into account the cumulative effect of noncarcinogens affecting the same target organs at a site. This rule, as well as a mixture rule for carcinogens in groundwater, are the subject of further consideration in Docket B.

In order to allow consideration of the proposed land use for the site, different objectives are set forth for different receptor populations: residential, industrial/commercial, and construction workers. Where the remediation objectives are based on an industrial/commercial property use, institutional controls must be adopted in accordance with Subpart J to ensure that the land use is appropriately restricted. The applicant need not further evaluate an exposure route if all contaminants of concern are below Tier 1 values for that exposure route, with the one exception for groundwater remediation objectives.

The Tier 1 remediation objectives are set forth in Appendix B, Tables A through E. These tables set forth remediation objectives for 117 chemicals. The tables are generally divided into two groups: those applicable to soil remediation objectives, and those applicable to groundwater remediation objectives. The groundwater component and the soil component of the groundwater ingestion route are further divided into objectives for Class I or Class II groundwater.

Tier 1 Soil Remediation Objectives. Under Tier 1, the applicant must consider two different direct exposure routes for soil when establishing remediation objectives pursuant to the TACO approach: the inhalation exposure route and the ingestion exposure route. Additionally, the applicant must consider the soil component of the groundwater ingestion route. Because these objectives are considered sufficiently protective, the applicant need not examine the dermal contact exposure route under Tier 1. The Tier 1 soil remediation objectives are set forth in Appendix B, Tables A, B, C, and D. The mixture rule adopted today for similar-acting chemicals is not applicable to these remediation objectives.

Appendix B, Table A sets forth the soil remediation objectives based upon residential property use, for the soil ingestion exposure route, the inhalation exposure route, and the soil component of the groundwater ingestion exposure route. Where appropriate, Table A also sets forth the Acceptable Detection Limit (ADL).[3] Because the Tier 1 residential levels are based

---

signifies a potential adverse effect, since a value greater than one occurs when a chemical exposure is measured to be greater than the reference dose.

[3]ADL is defined at Section 742.200 to mean the detectable concentration of a substance which is equal to the lowest appropriate Practical Quantitation Limit (PDL). PDL is defined as the lowest concentration that can be reliably measured within specified limits of precision and

upon protection in a residential exposure scenario, they are considered sufficiently protective that it is not necessary to establish separate remediation objectives for construction workers.

Appendix B, Table B sets forth the Tier 1 soil remediation objectives based upon industrial/commercial property use. As for the residential remediation objectives in Table A, separate remediation objectives are established for the soil ingestion exposure route, the inhalation exposure route, and the soil component of the groundwater ingestion exposure route. For the soil ingestion exposure route and the inhalation exposure route, separate remediation objectives are established for two receptor populations: the industrial/commercial population and the construction worker population. For the soil component of the groundwater ingestion exposure route, separate remediation objectives are established for Class I and Class II groundwaters.

The Tier 1 objectives for the soil ingestion and inhalation pathways were derived using the SSL with modifications as necessary to comply with Illinois law. The SSL was developed by USEPA for use in the Superfund program as a mechanism for screening out sites which do not require further study or action. The screening levels in the SSL are soil concentrations at or below which there is no concern in the Superfund program that some type of further action is required. They were developed based on a conceptual site model of a one-half acre site, with contamination extending to the water table, upon which a future residence with a private well would be built. (December 2, 1996 Transcript at pages 52-54; Exh. 5 at 11.)

Again, the screening levels in the SSL are designed to insure that the contaminants of concern individually will not present a greater than 1 in a million excess cancer risk for carcinogens or have a hazard quotient greater than 1. Since the SSL is based on an anticipated residential use, the Tier 1 remediation objectives for the industrial/commercial and construction workers had to be calculated from the SSL equations using the different exposure assumptions appropriate for these populations.

The Tier 1 objectives for the soil component of the groundwater ingestion route were derived from two different sources. For organic chemical contaminants, the Tier 1 objectives were derived using SSL equations with separate objectives established for Class I and Class II groundwaters. For inorganics, the proposed Tier 1 objectives establish two alternative approaches to setting soil objectives for the soil component of the groundwater ingestion route. The first alternative is based on the Toxicity Characteristic Leaching Procedure (TCLP) or the Synthetic Precipitation Leaching Procedure (SPLP) test. The second alternative is to allow pH-specific remediation objectives.

Section 742.510(a)(4) allows the use of the SPLP or the TCLP to evaluate the soil component of the groundwater ingestion exposure route. The TCLP method is used in the Underground Storage Tank program; the SPLP method is new under the TACO rules. The

---

accuracy for a specific laboratory analytical method during routine laboratory operating conditions. See Section 742.200 for these definitions in their entirety.

latter has been adopted by the USEPA for determining compliance with remediation objectives. It is designed to mimic the pH of rainwater that percolates through a contaminated site. (January 15, 1997 Transcript and January 16, 1997 Transcript at pages 237-250; Exh. 18 at 7.) Under either method, the remediation objective is the same as the Part 620 groundwater quality standard for the chemical of concern for the applicable groundwater classification. The applicant must perform a TCLP or SPLP analysis on a soil sample from the site and compare the concentration of the inorganic chemical of concern in the TCLP or the SPLP extract to the applicable groundwater standard. Additionally, an applicant still may evaluate the soil component on the basis of the total amount of contaminant in the soil sample result. This alternative is provided at subsection (a)(5) of Section 742.510.

The second alternative for establishing soil objectives for the soil component of the groundwater ingestion appears at Section 742.510(a)(5) and allows the applicant to establish pH-specific remediation objectives appropriate for the conditions at the site. (Exh. 5 at 20.) In addition to inorganics, this alternative may be used to establish remediation objectives for certain ionizable organics. The pH-specific objectives for Class I and Class II groundwaters are set forth in Appendix B, Tables C and D, respectively. These alternative remediation objectives allow the applicant to elect to evaluate the soil component of the groundwater ingestion exposure route based on the total amount of contaminant in a soil sample result, rather than the TCLP or SPLP analysis. In order to use this alternative approach, the applicant must determine the soil pH at the site and then select the appropriate soil remediation objectives based on Class I and Class II groundwaters. This method cannot be used if the soil pH is less than 4.5 or greater than 8.0.

Separate soil remediation objectives based on pH for identified ionizable organics or inorganics are adopted because the solubility of metals in water is highly dependent on pH of the solution. (See Section 742.505(a)(3)(C).) Generally, at lower pH, all metals are more soluble than at higher pH. To account for this phenomenon, the proposed standards have tables (Appendix B, Tables C and D) that list inorganic and ionizable organic compounds for pH values ranging from 4.5 to 8.0 in nine intervals of 0.25 increments. There are 15 metals and eight organic compounds listed in each table. There are separate tables for Class I and Class II groundwater.

Finally, pursuant to Section 742.510(a)(6), the applicant must review the soil remediation objectives determined for each remaining exposure routes, and select the most stringent of those remediation objectives, and then compare that one to the soil concentrations measured at the site. When using Appendix B, Table B for evaluating industrial/commercial properties, the remediation objectives for the ingestion and inhalation exposure routes shall be the more stringent of the industrial/commercial populations and the construction worker populations. If the soil remediation objective for a chemical is less than the ADL, the ADL shall serve as the remediation objective. Based upon this analysis, the applicant will be able to identify the applicable soil remediation objective under a Tier 1 evaluation.

Tier 1 Groundwater Remediation Objectives. Identifying the applicable groundwater remediation objective is more straightforward. Appendix B, Table E contains the groundwater remediation objectives for the groundwater component of the groundwater ingestion exposure route. The table contains separate values for Class I and Class II groundwaters and therefore the applicant must determine the Part 620 classification for groundwater at the site. The applicant must then compare the concentrations of groundwater contaminants at the site to the applicable Tier 1 groundwater remediation objectives set forth in Table E. Because this exposure route is based on direct ingestion of groundwater, an exposure route which is not impacted by the land use at the site, no distinction is made between residential and industrial/commercial use.

On an interim basis, the Board also adopts a requirement that the effect of similar-acting chemicals be evaluated when determining Tier 1 groundwater remediation objectives for noncarcinogenic chemicals. When more than one contaminant of concern which affects the same target organ is detected at the site, the mixture rule is applicable. That rule, which is essentially the same required under Tier 2, is found at Section 742.505(b)(3). The groundwater remediation objectives listed at Appendix B, Table E must be corrected to take into account the cumulative effect of mixtures of noncarcinogenic chemicals. There are two optional procedures for doing so. The first option at Section 742.505(b)(3) requires the calculation of weighted average using the formula set out therein. The remediation objectives are met if the weighted average is less than or equal to 1. If the weighted average is greater than 1, more remediation must be carried out until the weighted average is less than or equal to 1.

Alternatively, Section 742.505(b)(3) provides that the individual remediation objective be divided by the number of chemicals detected in the groundwater that affect specific target organs or organ system.

Finally, for any contaminants of concern not listed in Appendix B, Tables A through E, the applicant may request site-specific remediation objectives from the Agency, or propose site-specific objectives in accordance with 35 Ill. Adm. Code 620, Tier 3 at Subpart I of TACO, or both.

Subpart F: Tier 2 General Evaluation

Under Tier 2, the applicant can develop soil and groundwater remediation objectives applying site-specific information to pre-established modeling equations. The Tier 2 equations are set forth in Appendix C, Tables A and C. These equations are from the SSL and the RBCA approaches. (See Appendix C, Tables A and C, respectively for the equations to be used. The values to be used in the calculations, and the appropriate units are found at Appendix C, Tables B and D.) Table B contains the values for use in the SSL equations, and Table D contains the values for the RBCA equations. These tables also contain the acceptable exposure factors for the residential, industrial/commercial and construction worker populations when the present and post-remediation land uses are evaluated. As in Tier 1, the remediation objectives in Tier 2 cannot exceed an excess cancer risk of 1 in 1,000,000 for carcinogens, or a hazard quotient of 1 for noncarcinogens. Additionally, as in Tier 1, there is a mixture rule requiring that the cumulative effect of similar-acting noncarcinogenic chemicals be evaluated in developing remediation objectives. However, unlike Tier 1, this rule is applicable to soil as well as groundwater remediation objectives.

The rules generally applicable to a Tier 2 evaluation are set forth at Section 742.600. Similar to a Tier 1 analysis, the soil saturation and soil attenuation capacity restrictions found at Subpart B: Sections 742.215 and 742.220 apply. In other words, free product must be removed. (See subsections (d)(1) and (3) of Section 742.600.) Section 742.600 also instructs the applicant about how to choose the correct remediation objective if there is more than one exposure route requiring a remediation objective. This selection process is described at Section 742.600(e), (f), and (g), and presumes that the applicant has chosen to forgo the Tier 1 fixed numerical remediation objectives.

At the December 3, 1996 hearing, John Sherrill, an Agency witness, gave examples demonstrating when to use a Tier 1 or 2 remediation objective. For the purposes of illustration, the remediation objectives for benzene are used. The groundwater is Class I, none of the exposure routes are excluded and the numbers from Appendix B, Table A are used. The remediation objective for benzene for the ingestion route is 22 mg/kg, for the inhalation route is 0.8 mg/kg, and the migration to groundwater route is 0.03 mg/kg. The migration to groundwater remediation objective applies as the Tier 1 soil remediation objective because it is the most stringent out of the three values. If the calculated Tier 2 soil remediation objective for an exposure route is more stringent than the Tier 1 remediation objective for the same exposure route, then the Tier 1 remediation objective applies. In the hypothetical, within Tier 2, the applicant calculates a soil remediation objective of 0.02 mg/kg. The remediation objective would then be 0.03 mg/kg because it is less stringent than the calculated Tier 2 soil remediation objective.

If the calculated Tier 2 soil remediation objective for an exposure route is more stringent than one or more of the Tier 1 soil remediation objective(s) for a different exposure route, then the calculated Tier 2 soil remediation objective applies and the Tier 2 remediation objective for other exposure routes do not need to be calculated. For example, within Tier 2, the applicant calculates a migration to groundwater remediation objective of 0.1 mg/kg. The remediation objective would then be 0.1 because it is more stringent than the Tier 1 ingestion and inhalation remediation objectives (22 mg/kg and 0.8 mg/kg respectively).

If the calculated Tier 2 soil remediation objective is less stringent than one or more of the Tier 1 soil remediation objectives for the remaining exposure routes, then the other Tier 2 remediation objectives are calculated and the most stringent calculated Tier 2 value applies. Within Tier 2, the applicant calculates a migration to groundwater remediation objective of 1.2 mg/kg. This is less stringent than the inhalation soil remediation objective in Tier 1. So the applicant then calculates the Tier 2 ingestion and inhalation remediation objectives. The applicant calculates an ingestion remediation objective of 30 mg/kg and an inhalation remediation objective of 11 mg/kg. Since the Tier 2 migration to groundwater remediation objective is the most stringent (1.2 is less than 30 and 11), 1.2 mg/kg is the remediation objective.

As in Tier 1, the proposed land use for the site is considered in establishing Tier 2 soil remediation objectives. In a Tier 2 evaluation, the proposed land use for the site will determine the appropriate exposure factors contained in the applicable equation. The appropriate exposure factors for residential, industrial/commercial, and construction worker populations are set forth in Appendix C, Tables B and D. The established exposure factors can only be varied in a Tier 3 analysis. If a Tier 2 evaluation is based on an industrial/commercial property use, the construction worker scenario must also be evaluated. Additionally, the applicant must obtain an institutional control in accordance with the requirements of Subpart J.

Finally, the mixture rule originally proposed for Tier 2 is adopted at Section 742.720 which addresses soil remediation objectives, and at Section 742.805 which addresses groundwater remediation objectives. These procedures are the same as those adopted under Tier 1 for noncarginogenic chemicals detected in groundwater. The distinction is that the mixture rule is applicable to soil remediation objectives, as well as groundwater remediation objectives. Again, these mixture rules are subject to further consideration under Docket B.

Subpart G: Tier 2 Soil Evaluation

Tier 2 provides the applicant with two options for establishing soil remediation objectives: reliance on the SSL equations or on the RBCA equations. Because the RBCA equations combine the soil ingestion, inhalation of vapors and particulates, and dermal contact exposure routes, while the SSL equations treat the soil ingestion, inhalation of volatiles, and inhalation of fugitive dust exposure routes separately, the applicant must choose only one of these approaches, and the two approaches cannot be combined. However, both methods treat the soil component of the groundwater exposure route separately, so the applicant can choose to use either method to calculate the remediation objectives for this exposure route, no matter which approach was used to establish the other soil objectives.

SSL Equations. The SSL equations are set forth in Appendix C, Table A. The parameters for these equations are set forth in Appendix C, Table B. The equations are divided into separate categories by exposure route: ingestion, inhalation of volatiles, inhalation of fugitive dust, and migration to groundwater.

Within each exposure route's set of equations, there are separate sets of equations for noncarcinogens and carcinogens. The equations for carcinogens reflect an expected excess cancer risk of 1 in 1,000,000, while the equations for noncarcinogens reflect a hazard quotient of 1. Within these categories, separate equations are set forth for residential, industrial/commercial, and construction worker populations. The different equations for each type of land use reflects the expected differences in the exposure factor, exposure duration, averaging time, and ingestion rate. Default values and parameters for these equations are listed in Appendix C, Table B.

RBCA Equations. Appendix C, Table C contains the RBCA equations used in Tier 2. The RBCA equations for establishing soil remediation objectives are separated into three categories. The first category examines the combined exposures of soil ingestion, inhalation of vapors and particulates, or dermal contact with soil. The second category examines the ambient vapor inhalation (outdoor) route from subsurface soils. The third category is for the migration to groundwater pathway. Within each of these categories, separate equations are set forth for noncarcinogens and carcinogens. Since RBCA offers two different ways to evaluate the soil remediation objectives, the smaller, or more stringent, of the two values will be the remediation objective (either equation R1 or R7 for carcinogens and equation R2 or R8 for noncarcinogens).

Unlike the SSL approach, which sets forth separate equations for each exposure route for each type of land use, RBCA does not have separate equations for each type of land use. Instead, RBCA allows for differences in the type of land use to be reflected in the values of certain parameters. Because RBCA groups ingestion and inhalation into the same equations, one model is to be used in Tier 2 for the inhalation and ingestion exposure routes. Either the RBCA or SSL models can be used for the soil component of the groundwater exposure route.

The requirements concerning the cumulative effect of noncarcinogenic chemicals for soil remediation are at Section 742.720. Appendix A, Table E sets forth groups of chemicals from Appendix B, Tables A and B, that have remediation objectives based on noncarcinogenic toxicity, and that affect the same target organ. If more than one chemical detected at the site affects the same target organ, the applicant must correct the initially calculated remediation value for each chemical in the group.

Subpart H: Tier 2 Groundwater Evaluation

Subpart H contains the procedures for developing Tier 2 groundwater remediation objectives. If the contaminants of concern exceed the applicable Tier 1 remediation objectives, an applicant has several choices. As preliminary to a Tier 1 analysis, the applicant can as a preliminary matter, demonstrate that the pathway is excluded, that the contamination is at or below the area background concentration in accordance with Subpart D, or conduct a Tier 3 analysis. There are also two alternatives distinctive to groundwater remediation objectives available. An applicant can seek from the Board reclassification of the contaminated groundwater pursuant to 35 Ill. Adm. Code 620.260, or an adjusted standard pursuant to Section 28.1 of the Act. However, should an applicant choose to develop Tier 2 groundwater remediation objectives, the applicant must use RBCA Equation R26. Using this equation, the applicant can develop remediation objectives which exceed the applicable Part 620 groundwater standards at the site, but which will meet the applicable groundwater standards at the point of human exposure.

Pursuant to Section 742.805, before developing a Tier 2 groundwater remediation objective, the applicant must first identify the horizontal and vertical extent of the contamination, and, to the extent practicable, take remedial action to remove any free product. The applicant can then use RBCA Equation R26 to demonstrate that the applicable groundwater standards will be achieved at the point of human exposure. The basis and application of Equation R26 to predict impacts from remaining groundwater contamination are explained in Section 742.810.

Equation R26 predicts the concentration of a contaminant along the centerline of a plume, taking into account the three dimensional dispersion and biodegradation. Using Equation R26, the applicant can demonstrate that, although the concentration of a contaminant exceeds the applicable Tier 1 objective at the source, the concentration at the point of human exposure will meet either the applicable Tier 1 groundwater remediation objective, or if no Tier 1 objective exists, the applicable Health Advisory concentration as determined in accordance with the procedures set forth in 35 Ill. Adm. Code 620, Subpart F. If the applicant determines that the applicable Tier 1 objective will be exceeded at the point of human exposure, the applicant can back-calculate the concentration that must be achieved at the source in order for the compliance to be achieved.

In addition to demonstrating that the applicable Tier 1 objective will be achieved at the point of human exposure, in order to demonstrate compliance pursuant to Tier 2, the applicant must demonstrate that five additional requirements are satisfied. First, using Equation R26, the applicant must demonstrate that the concentration of any contaminant in groundwater within the minimum or designated maximum setback zone of an existing potable water supply well will meet the applicable Tier 1 groundwater objective or the Health Advisory concentration[4]. Second, the applicant must demonstrate that the source of the release is not located within the minimum or designated maximum setback zone of a potable water supply well. Third, the applicant must use Equation R26 to demonstrate that the concentration of any contaminant in groundwater discharging into surface water will meet the applicable water quality standard pursuant to 35 Ill. Adm. Code 302. Fourth, the applicant must demonstrate that any groundwater remediation objective established pursuant to this procedure does not exceed the water solubility for that contaminant. Finally, if the remediation relies on an engineered barrier, the applicant must demonstrate that an institutional control is in place requiring that the barrier remain in place. These requirements are set forth at Section 742.805 and illustrated at Appendix C: Illustration B.

During the public comment period, the Board received a public comment questioning whether the R26 equation in the RBCA guidelines has been properly adapted for use in establishing risk based remediation objectives under Tier 2. The commentator urged the Board to change the last *erf* (error function) term in the denominator of Equation R26 from 4

---

[4] Health Advisory concentrations are established in accordance with 35 Ill. Adm. Code 620, Subpart F for contaminants that do not have a groundwater quality standard under Part 620, Subpart D.

to 2. (PC 6.) The Agency supports the change. (PC 10 at 12-13; PC 22.) Since the supporting documents indicate that the use of the incorrect number (4) for the vertical dispersion results in under-prediction of concentrations of contaminants along the centerline of a plume, the Board adopts Equation R26 by changing the constant value for the *erf* relating to vertical dispersion from 4 to 2. Equation R15, essentially the same equation, is also changed where it appears in the rules.

Section 742.810 contains another provision regarding the distance between an existing potable water supply well and the source of contamination. Minimum setback zones are established pursuant to Section 14.2(g) of the Act and maximum setback zones are established on a site-specific basis. Either the minimum or maximum setback zone can be used, whichever is closer. However, the minimum setback zone is to be used unless the maximum setback zone is established.

Section 742.810(b)(1) provides the procedure for demonstrating that no existing potable water supply is adversely impacted by a remediation site. Essentially, the revisions to subsection (b)(1) require an applicant to calculate the distance "X" from the downgradient edge of the source to the point where the contaminant concentration is equal to the Tier 1 groundwater remediation objective or Health Advisory concentration. If there are any potable water supply wells located within the distance X downgradient of the source, then the applicable groundwater remediation objectives must be met at the edge of the minimum or designated maximum setback zone. If no potable water supply wells exist within the calculated distance X, then it can be determined that no potable water supply wells are adversely impacted.

Subpart I: Tier 3 Evaluation

Tier 3 allows the applicant to develop remediation objectives using alternative parameters not found in Tier 1 or Tier 2. It allows the applicant great flexibility in developing remediation objectives appropriate for a particular site based upon site-specific information, rather than relying on general categories of information. The options available under Tier 3 include: use of modified parameters in the Tier 2 equations, use of alternative models, conducting a site-specific risk assessment, use of probabilistic analysis and sophisticated fate and transport models, assessment of impractical remediation, and variation of the target risk level. If any contaminants of concern are found to exceed the remediation objectives developed using the Tier 3 analysis, the applicant would be required to remediate the contamination until the objectives are achieved. The applicant must provide appropriate justification for the use and application of any alternative parameters, models, or analysis relied upon in a Tier 3 evaluation.

Tier 1 remediation objectives and Tier 2 equations are based upon a one-in-a-million individual excess cancer risk for carcinogens and a hazard quotient of one for noncarcinogens. (Exh. 4 at 12 & 21.) Changes in the target risk levels are allowed under Tier 3. Section 742.900(d) clearly sets forth that requests for changes in target risk levels at the point of human exposure level must be supported with a formal risk assessment conducted in accordance with Section 742.915. Section 742.915(h) contains four factors critical in such a risk assessment. Those factors are: 1) the presence of sensitive populations; 2) the number of receptors potentially impacted; 3) the duration of risk at the differing target levels; and 4) the characteristics of the contaminants of concern.

Pursuant to Section 58.5(d)(4) of the Act, an applicant can seek site-specific remediation objectives which exceed the Tier 1 remediation objectives, *i.e.*, the Board's groundwater quality standards at Part 620, under Tier 3. To obtain such an exception, the Act requires two demonstrations. First, the applicant must demonstrate that the exceedence of the groundwater quality standard has been minimized and the beneficial use of the groundwater has been returned, and that any threat to human health has also been minimized. These two statutory requirements are incorporated verbatim into the rules at Section 742.900(c)(9). Since the right to exceed groundwater quality standards, adopted by the Board as the level at which human health is protected, is available only due to this statutory exception, it is most important that the demonstration to obtain such a right comply with the these statutory requirements.

Subpart J: Institutional Controls

Institutional controls are defined under the proposed rules as "a legal mechanism for imposing a restriction on land use." The Agency testified that institutional controls are a fundamental part of the proposal, and are the key to assuring long-term protection of human health, while providing flexibility in developing practical, risk-based remediation objectives. (Exh. 3 at 1.) The applicant must obtain an institutional control whenever the applicant seeks to take any of the following measures, or any combination thereof: 1) restrict a property to industrial/commercial use; 2) establish remediation objectives based on a target cancer risk greater than 1 in 1,000,000; 3) establish a target hazard quotient greater than 1 for a noncarcinogen under a Tier 3 analysis; 4) rely on an engineered barrier; 5) set the point of human exposure at a location other than at the source; or 6) exclude exposure pathways under Subpart C. An institutional control is transferred with the property.

Pursuant to Subpart J, the following types of institutional controls are recognized under these rules: 1) No Further Remediation Letters; 2) restrictive covenants, deed restrictions, and negative easements; 3) ordinances adopted and administered by a unit of local government; and 4) agreements between a property owner and a highway authority with respect to any contamination remaining under highways. The requirements for each of these categories are set forth in a separate section.

The requirements for a No Further Remediation Letter are set forth at Section 742.1005. Subsection (b) therein provides that "a request for approval of a No Further Remediation Letter as an institutional control shall follow the requirements applicable to the remediation program under which the remediation is performed." The TACO rules are intended to establish a method for deriving corrective action objectives for remedial programs.

Specific provisions concerning the effectiveness and limitations on No Further Remediation Letters and other instruments memorializing a no further remediation determination by the Agency are more appropriately set forth in the specific programs pursuant to which such a determination is made.

The requirements for restrictive covenants, deed restrictions, and negative easements are set forth at Section 742.1010. These measures are to be used only in situations where a No Further Remediation Letter is not available, and it is not necessary to obtain restrictive covenants, deed restrictions, or negative easements which duplicate conditions set forth in a No Further Remediation Letter that is appropriately recorded. Restrictive covenants, deed restrictions, and negative easements approved by the Agency in accordance with this Section must be appropriately recorded, together with the instrument memorializing the Agency's no further remediation determination, *e.g.*, a No Further Remediation Letter.

Section 742.1015 sets forth the requirements for ordinances used as an institutional control. The use of an ordinance as an institutional control is specifically limited to ordinances that effectively prohibit the installation of potable water supply wells in order to meet the requirements of Section 742.320(d) or 742.805(a)(3). Unless the Agency and the unit of local government have entered into a Memorandum of Understanding (MOU), this section places the burden on the owner or successor in interest for monitoring the local governments activities with respect to the ordinance. If the ordinance is modified, or if a variance or other site-specific request is granted that allows use of the groundwater at the site as a potable water supply, or if the terms of another institutionally control at the site are violated, the use of the ordinance as an institutional control can be voided.

Given the potential human health risk and the cost of groundwater remediation and installation of potable water supply wells, it makes sense to be forewarned about potential problems concerning a groundwater source. Moreover, most of the existing ordinances were not enacted by considering environmental concerns. In this regard, the regulation provides an opportunity to a unit of local government to adopt an ordinance based upon the consideration of environmental concerns. The requirements concerning the MOU at Section 742.1015(i) require the unit of local government to make a commitment to: maintain a registry of all sites within its boundaries which have received a No Further Remediation Letter and review the registry of sites prior to siting potable water supply well; consider if groundwater contamination from the sites on the registry may be present at potential well sites; and take appropriate protective measures if wells are sited in the vicinity of such locations. By requiring the local government to enter into an MOU and abiding by the commitments contained therein will forewarn communities of the existence of contamination plumes and may prevent costly mistakes in siting, construction, and use of public potable water supply wells. This provision is cross referenced at Sections 742.320(d) and 742.805(a)(3).

Finally, Section 742.1020 sets forth the requirements applicable to highway authority agreements used as an institutional control. When contamination level of the groundwater exceed the Tier 1 residential levels, the highway authority must agree to prohibit the use of groundwater under the highway right-of-way as a potable water supply. When the contamination of the soil under the highway right of way exceeds the Tier 1 residential level, the highway authority must agree to limit access to soil contamination, and in the event access is allowed, human health and the environment must be protected.

Subpart K: Engineered Barriers

An engineered barrier is defined in Section 742.200 as a barrier designed or verified using engineering practices that limits exposure to or controls migration of the contaminants of concern. The Agency testified that, in addition to including man-made structures designed using engineering practices, engineered barriers could include native or *in-situ* materials if their effectiveness is verified using engineering practices. (Exh. 3 at 7) The use and maintenance of an engineered barrier must be accompanied by an institutional control in accordance with Subpart J. Furthermore, any no further remediation determination by the Agency based on the use of the engineered barrier must be conditioned upon maintenance of the engineered barrier, and the institutional control must address provisions for temporary breaches of the engineered barrier. Failure to maintain an engineered barrier in accordance with the terms of the no further remediation determination constitutes grounds for voidance of that determination.

Section 742.1105 sets forth the requirements for engineered barriers and limitations on their use in achieving remediation objectives. It makes clear that natural attenuation, access controls and point of use treatment do not fall within the definition of engineered barriers, and that engineered barriers cannot be relied upon in determining compliance with Tier 1 remediation objectives. Subsection (c) of this Section sets forth a list of engineered barriers accepted for each exposure route. For the soil component of the groundwater ingestion exposure route, these include caps constructed of clay, asphalt, or concrete, and permanent structures, such as buildings or highways. For the soil ingestion and inhalation exposure routes, the acceptable engineered barriers include clean soil at least three feet in depth, as well as caps and permanent structures. Finally, for the groundwater component, the acceptable engineered barriers include slurry walls and hydraulic control of groundwater. Subsection (d) of this Section makes it clear that the list of accepted measures is not intended to be exhaustive and that other methods will be accepted by the Agency if they are shown to be as effective as the listed options.

SUMMARY

The Board hereby adopts as final the new Part 742, Tiered Approach to Corrective Action Objectives. The Board has examined the substantive issues concerning the proposal initially filed by the Agency and the record and modification to the Agency's proposal developed during this rulemaking. The Board has reserved one issue for a separate Docket B. That issue is to what extent the mixture rule adopted today should be modified to insure that risk based remediation objectives determined using TACO are protective of human health. The Board concludes that the TAOC rules set forth in the attached order provide a tiered approach for assessing risk to human health when determining remediation objectives and the methodologies acceptable for doing so on a site-specific basis. The Board further finds that these rules are economically reasonable and technically feasible. The Board anticipates that these rules will be used successfully in conjunction with the remediation programs contained in other Board rules to remediate contamination at properties throughout Illinois based on risk of human exposure and that future uses of such properties will not be modified without consideration of the adequacy of the risk-based remediation. Finally, the Board looks forward to a proposal from the Agency addressing the risk-based methodology necessary to protect the environment.

ORDER

The Board hereby directs that the final notice of the following adopted rules be submitted to the Secretary of State for publication in the Illinois Register.

TITLE 35: ENVIRONMENTAL PROTECTION
SUBTITLE G: WASTE DISPOSAL
CHAPTER I: POLLUTION CONTROL BOARD
SUBCHAPTER f: RISK BASED CLEANUP OBJECTIVES

PART 742
TIERED APPROACH TO CORRECTIVE ACTION OBJECTIVES

SUBPART A: INTRODUCTION

SUBPART B: GENERAL

SUBPART C: EXPOSURE ROUTE EVALUATIONS

AUTHORITY: Implementing Sections 22.4, 22.12, Title XVI, and Title XVII and authorized by Sections 27, 57.14, and 58.5 of the Environmental Protection Act [415 ILCS 5/22.4, 22.12, Title XVI and Title VII] (see P.A. 88-496, effective September 13, 1993 and P.A. 89-0431, effective December 15, 1995).

SOURCE: Adopted at 21 Ill. Reg. _____ , effective July 1, 1997.

NOTE: Capitalization indicates statutory language.

SUBPART A: INTRODUCTION

Section 742.100   Intent and Purpose

a)   This Part sets forth procedures for evaluating the risk to human health posed by environmental conditions and developing remediation objectives that achieve acceptable risk levels.

b)   The purpose of these procedures is to provide for the adequate protection of human health and the environment based on the risks to human health posed by environmental conditions while incorporating site related information.

Section 742.105   Applicability

a)   Any person, including a person required to perform an investigation pursuant to the Illinois Environmental Protection Act (415 ILCS 5/1 et seq.) (Act), may elect to proceed under this Part to the extent allowed by State or federal law and regulations and the provisions of this Part. A person proceeding under this Part may do so to the extent such actions are consistent with the requirements of the program under which site remediation is being addressed.

b)   This Part is to be used in conjunction with the procedures and requirements applicable to the following programs:

1)   Leaking Underground Storage Tanks (35 Ill. Adm. Code 731 and 732);

2)   Site Remediation Program (35 Ill. Adm. Code 740); and

3)   RCRA Part B Permits and Closure Plans (35 Ill. Adm. Code 724 and 725).

c)   The procedures in this Part may not be used if their use would delay response action to address imminent and substantial threats to human health and the environment. This Part may only be used after actions to address such threats have been completed.

d)   This Part may be used to develop remediation objectives to protect surface waters, sediments or ecological concerns, when consistent with the regulations of other programs, and as approved by the Agency.

e) A no further remediation determination issued by the Agency prior to July 1, 1997 pursuant to Section 4(y) of the Act or one of the programs listed in subsection (b) of this Section that approves completion of remedial action relative to a release shall remain in effect in accordance with the terms of that determination.

f) Site specific groundwater remediation objectives determined under this Part for contaminants of concern may exceed the groundwater quality standards established pursuant to the rules promulgated under the Illinois Groundwater Protection Act (415 ILCS 55) as long as done in accordance with Sections 742.805(a) and 742.900(c)(9). [See 415 ILCS 5/58.5(d)(4).]

g) Where contaminants of concern include polychlorinated biphenyls (PCBs), a person may need to evaluate the applicability of regulations adopted under the Toxic Substances Control Act. (15 U.S.C. 2601)

Section 742.110 Overview of Tiered Approach

a) This Part presents an approach for developing remediation objectives (see Appendix A, Illustrations A and B) that include an option for exclusion of pathways from further consideration, use of area background concentrations as remediation objectives and three tiers for selecting applicable remediation objectives. An understanding of human exposure routes is necessary to properly conduct an evaluation under this approach. In some cases, applicable human exposure route(s) can be excluded from further consideration prior to any tier evaluation. Selecting which tier or combination of tiers to be used to develop remediation objectives is dependent on the site-specific conditions and remediation goals. Tier 1 evaluations and Tier 2 evaluations are not prerequisites to conducting Tier 3 evaluations.

b) A Tier 1 evaluation compares the concentration of contaminants detected at a site to the corresponding remediation objectives for residential and industrial/commercial properties contained in Appendix B, Tables A, B, C, D and E. To complete a Tier 1 evaluation, the extent and concentrations of the contaminants of concern, the groundwater class, the land use classification, human exposure routes at the site, and, if appropriate, soil pH, must be known. If remediation objectives are developed based on industrial/commercial property use, then institutional controls under Subpart J are required.

c) A Tier 2 evaluation uses the risk based equations from the Soil Screening Level (SSL) and Risk Based Corrective Action (RBCA) listed in Appendix C, Tables A and C, respectively. In addition to the information that is required for a Tier 1 evaluation, site-specific information is used to calculate Tier 2 remediation objectives. As in Tier 1, Tier 2 evaluates residential and industrial/commercial properties only. If remediation objectives are developed based on industrial/commercial property use, then institutional controls under Subpart J are required.

d)   A Tier 3 evaluation allows alternative parameters and factors, not available under a Tier 1 or Tier 2 evaluation, to be considered when developing remediation objectives. Remediation objectives developed for conservation and agricultural properties can only be developed under Tier 3.

e)   Remediation objectives may be developed using area background concentrations or any of the three tiers if the evaluation is conducted in accordance with applicable requirements in Subparts D through I. When contaminant concentrations do not exceed remediation objectives developed under one of the tiers or area background procedures under Subpart D, further evaluation under any of the other tiers is not required.

Section 742.115   Key Elements

To develop remediation objectives under this Part, the following key elements shall be addressed.

a)   Exposure Routes

1)   This Part identifies the following as potential exposure routes to be addressed:

A)   Inhalation;

B)   Soil ingestion;

C)   Groundwater ingestion; and

D)   Dermal contact with soil.

2)   The evaluation of exposure routes under subsections (a)(1)(A),(a)(1)(B) and (a)(1)(C) of this Section is required for all sites when developing remediation objectives or excluding exposure pathways. Evaluation of the dermal contact exposure route is required for use of RBCA equations in Appendix C, Table C or use of formal risk assessment under Section 742.915.

3)   The groundwater ingestion exposure route is comprised of two components:

        A)   Migration from soil to groundwater (soil component); and

        B)   Direct ingestion of groundwater (groundwater component).

b)   Contaminants of Concern

The contaminants of concern to be remediated depend on the following:

1)   The materials and wastes managed at the site;

2)   The extent of the no further remediation determination being requested from the Agency pursuant to a specific program; and

3)   The requirements applicable to the specific program, as listed at Section 742.105(b) under which the remediation is being performed.

c)   Land Use

The present and post-remediation uses of the site where exposures may occur shall be evaluated. The land use of a site, or portion thereof, shall be classified as one of the following:

1)   Residential property;

2)   Conservation property;

3)   Agricultural property; or

4)   Industrial/commercial property.

Section 742.120   Site Characterization

Characterization of the extent and concentrations of contamination at a site shall be performed before beginning development of remediation objectives. The actual steps and methods taken to characterize a site are determined by the requirements applicable to the specific program under which site remediation is being addressed.

SUBPART B: GENERAL

Section 742.200   Definitions

Except as stated in this Section, or unless a different meaning of a word or term is clear from the context, the definition of words or terms in this Part shall be the same as that applied to the same words or terms in the Act.

"Act" means the Illinois Environmental Protection Act (415 ILCS 5/1 et seq.).

"ADL" means Acceptable Detection Limit, which is the detectable concentration of a substance which is equal to the lowest appropriate Practical Quantitation Limit (PQL) as defined in this Section.

"Agency" means the Illinois Environmental Protection Agency.

"Agricultural Property" means any real property for which its present or post-remediation use is for growing agricultural crops for food or feed either as harvested crops, cover crops or as pasture. This definition includes, but is not limited to, properties used for confinement or grazing of livestock or poultry and for silviculture operations. Excluded from this definition are farm residences, farm outbuildings and agrichemical facilities.

"Area Background" means CONCENTRATIONS OF REGULATED SUBSTANCES THAT ARE CONSISTENTLY PRESENT IN THE ENVIRONMENT IN THE VICINITY OF A SITE THAT ARE THE RESULT OF NATURAL CONDITIONS OR HUMAN ACTIVITIES, AND NOT THE RESULT SOLELY OF RELEASES AT THE SITE. (Section 58.2 of the Act)

"ASTM" means the American Society for Testing and Materials.

"Board" means the Illinois Pollution Control Board.

"Cancer Risk" means a unitless probability of an individual developing cancer from a defined exposure rate and frequency.

"Cap" means a barrier designed to prevent the infiltration of precipitation or other surface water, or impede the ingestion or inhalation of contaminants.

"Carcinogen" means A CONTAMINANT THAT IS CLASSIFIED AS A CATEGORY A1 OR A2 CARCINOGEN BY THE AMERICAN CONFERENCE OF GOVERNMENTAL INDUSTRIAL HYGIENISTS; A CATEGORY 1 OR 2A/2B CARCINOGEN BY THE WORLD HEALTH

ORGANIZATION'S INTERNATIONAL AGENCY FOR RESEARCH ON
CANCER; A "HUMAN CARCINOGEN" OR "ANTICIPATED HUMAN
CARCINOGEN" BY THE UNITED STATES DEPARTMENT OF HEALTH
AND HUMAN SERVICE NATIONAL TOXICOLOGICAL PROGRAM; OR
A CATEGORY A OR B1/B2 CARCINOGEN BY THE UNITED STATES
ENVIRONMENTAL PROTECTION AGENCY IN the INTEGRATED RISK
INFORMATION SYSTEM OR A FINAL RULE ISSUED IN A FEDERAL
REGISTER NOTICE BY THE USEPA. (Section 58.2 of the Act)

"Class I Groundwater" means groundwater that meets the Class I: Potable
Resource Groundwater criteria set forth in 35 Illinois Administrative Code 620.

"Class II Groundwater" means groundwater that meets the Class II: General
Resource Groundwater criteria set forth in 35 Illinois Administrative Code 620.

"Conservation Property" means any real property for which present or post-
remediation use is primarily for wildlife habitat.

"Construction Worker" means a person engaged on a temporary basis to
perform work involving invasive construction activities including, but not
limited to, personnel performing demolition, earth-moving, building, and
routine and emergency utility installation or repair activities.

"Contaminant of Concern" or "Regulated Substance of Concern" means ANY
CONTAMINANT THAT IS EXPECTED TO BE PRESENT AT THE SITE
BASED UPON PAST AND CURRENT LAND USES AND ASSOCIATED
RELEASES THAT ARE KNOWN TO THE person conducting a remediation
BASED UPON REASONABLE INQUIRY. (Section 58.2 of the Act)

"Engineered Barrier" means a barrier designed or verified using engineering
practices that limits exposure to or controls migration of the contaminants of
concern.

"Exposure Route" means the transport mechanism by which a contaminant of
concern reaches a receptor.

"Free Product" means a contaminant that is present as a non-aqueous phase
liquid for chemicals whose melting point is less than 30°C (e.g., liquid not
dissolved in water).

"GROUNDWATER" MEANS UNDERGROUND WATER WHICH OCCURS
WITHIN THE SATURATED ZONE AND GEOLOGIC MATERIALS
WHERE THE FLUID PRESSURE IN THE PORE SPACE IS EQUAL TO OR
GREATER THAN ATMOSPHERIC PRESSURE. (Section 3.64 of the Act)

"Groundwater Quality Standards" means the standards for groundwater as set forth in 35 Illinois Administrative Code 620.

"Hazard Quotient" means the ratio of a single substance exposure level during a specified time period to a reference dose for that substance derived from a similar exposure period.

"Highway" means ANY PUBLIC WAY FOR VEHICULAR TRAVEL WHICH HAS BEEN LAID OUT IN PURSUANCE OF ANY LAW OF THIS STATE, OR OF THE TERRITORY OF ILLINOIS, OR WHICH HAS BEEN ESTABLISHED BY DEDICATION, OR USED BY THE PUBLIC AS A HIGHWAY FOR 15 YEARS, OR WHICH HAS BEEN OR MAY BE LAID OUT AND CONNECT A SUBDIVISION OR PLATTED LAND WITH A PUBLIC HIGHWAY AND WHICH HAS BEEN DEDICATED FOR THE USE OF THE OWNERS OF THE LAND INCLUDED IN THE SUBDIVISION OR PLATTED LAND WHERE THERE HAS BEEN AN ACCEPTANCE AND USE UNDER SUCH DEDICATION BY SUCH OWNERS, AND WHICH HAS NOT BEEN VACATED IN PURSUANCE OF LAW. THE TERM "HIGHWAY" INCLUDES RIGHTS OF WAY, BRIDGES, DRAINAGE STRUCTURES, SIGNS, GUARD RAILS, PROTECTIVE STRUCTURES AND ALL OTHER STRUCTURES AND APPURTENANCES NECESSARY OR CONVENIENT FOR VEHICULAR TRAFFIC. A HIGHWAY IN A RURAL AREA MAY BE CALLED A "ROAD", WHILE A HIGHWAY IN A MUNICIPAL AREA MAY BE CALLED A "STREET". (Illinois Highway Code [605 ILCS 5/2-202])

"Highway Authority" means THE DEPARTMENT of Transportation WITH RESPECT TO A STATE HIGHWAY; THE COUNTY BOARD WITH RESPECT TO A COUNTY HIGHWAY OR A COUNTY UNIT DISTRICT ROAD IF A DISCRETIONARY FUNCTION IS INVOLVED AND THE COUNTY SUPERINTENDENT OF HIGHWAYS IF A MINISTERIAL FUNCTION IS INVOLVED; THE HIGHWAY COMMISSIONER WITH RESPECT TO A TOWNSHIP OR DISTRICT ROAD NOT IN A COUNTY UNIT ROAD DISTRICT; OR THE CORPORATE AUTHORITIES OF A MUNICIPALITY WITH RESPECT TO A MUNICIPAL STREET. (Illinois Highway Code [605 ILCS 5/2-213])

"Human Exposure Pathway" means a physical condition which may allow for a risk to human health based on the presence of all of the following: contaminants of concern; an exposure route; and a receptor activity at the point of exposure that could result in contaminant of concern intake.

"Industrial/Commercial Property" means any real property that does not meet the definition of residential property, conservation property or agricultural property.

"Infiltration" means the amount of water entering into the ground as a result of precipitation.

"Institutional Control" means a legal mechanism for imposing a restriction on land use, as described in Subpart J.

"Man-Made Pathways" means CONSTRUCTED physical conditions THAT MAY ALLOW FOR THE TRANSPORT OF REGULATED SUBSTANCES INCLUDING, BUT NOT LIMITED TO, SEWERS, UTILITY LINES, UTILITY VAULTS, BUILDING FOUNDATIONS, BASEMENTS, CRAWL SPACES, DRAINAGE DITCHES, OR PREVIOUSLY EXCAVATED AND FILLED AREAS. (Section 58.2 of the Act)

"Natural Pathways" means NATURAL physical conditions that may allow FOR THE TRANSPORT OF REGULATED SUBSTANCES INCLUDING, BUT NOT LIMITED TO, SOIL, GROUNDWATER, SAND SEAMS AND LENSES, AND GRAVEL SEAMS AND LENSES. (Section 58.2 of the Act)

"Negative Easement" means a right of the owner of the dominant or benefitted estate or property to restrict the property rights of the owner of the servient or burdened estate or property.

"Person" means an INDIVIDUAL, TRUST, FIRM, JOINT STOCK COMPANY, JOINT VENTURE, CONSORTIUM, COMMERCIAL ENTITY, CORPORATION (INCLUDING A GOVERNMENT CORPORATION), PARTNERSHIP, ASSOCIATION, STATE, MUNICIPALITY, COMMISSION, POLITICAL SUBDIVISION OF A STATE, OR ANY INTERSTATE BODY INCLUDING THE UNITED STATES GOVERNMENT AND EACH DEPARTMENT, AGENCY, AND INSTRUMENTALITY OF THE UNITED STATES. (Section 58.2 of the Act)

"Point of Human Exposure" means the point(s) at which human exposure to a contaminant of concern may reasonably be expected to occur. The point of human exposure is at the source, unless an institutional control limiting human exposure for the applicable exposure route has been or will be in place, in which case the point of human exposure will be the boundary of the institutional control. Point of human exposure may be at a different location than the point of compliance.

"PQL" means Practical Quantitation Limit or estimated quantitation limit, which is the lowest concentration that can be reliably measured within specified limits of precision and accuracy for a specific laboratory analytical method during routine laboratory operating conditions in accordance with "Test Methods for Evaluating Solid Wastes, Physical/Chemical Methods", EPA Publication No. SW-846, incorporated by reference in Section 742.210. When applied to filtered water samples, PQL includes the method detection limit or estimated detection limit in accordance with the applicable method revision in: "Methods for the Determination of Organic Compounds in Drinking Water", Supplement II", EPA Publication No. EPA/600/4-88/039; "Methods for the Determination of Organic Compounds in Drinking Water, Supplement III", EPA Publication No. EPA/600/R-95/131, all of which are incorporated by reference in Section 742.210.

"RBCA" means Risk Based Corrective Action as defined in ASTM E-1739-95, as incorporated by reference in Section 742.210.

"RCRA" means the Resource Conservation and Recovery Act of 1976. (42 U.S.C. 6921)

"Reference Concentration (RfC)" means an estimate of a daily exposure, in units of milligrams of chemical per cubic meter of air (mg/m^3), to the human population (including sensitive subgroups) that is likely to be without appreciable risk of deleterious effects during a portion of a lifetime (up to approximately seven years, subchronic) or for a lifetime (chronic).

"Reference Dose (RfD)" means an estimate of a daily exposure, in units of milligrams of chemical per kilogram of body weight per day (mg/kg/d), to the human population (including sensitive subgroups) that is likely to be without appreciable risk of deleterious effects during a portion of a lifetime (up to approximately seven years, subchronic) or for a lifetime (chronic).

"Regulated Substance" means ANY HAZARDOUS SUBSTANCE AS DEFINED UNDER SECTION 101(14) OF THE COMPREHENSIVE ENVIRONMENTAL RESPONSE, COMPENSATION, AND LIABILITY ACT OF 1980 (P.L. 96-510) AND PETROLEUM PRODUCTS INCLUDING CRUDE OIL OR ANY FRACTION THEREOF, NATURAL GAS, NATURAL GAS LIQUIDS, LIQUEFIED NATURAL GAS, OR SYNTHETIC GAS USABLE FOR FUEL (OR MIXTURES OF NATURAL GAS AND SUCH SYNTHETIC GAS). (Section 58.2 of the Act)

"Residential Property" MEANS ANY REAL PROPERTY THAT IS USED FOR HABITATION BY INDIVIDUALS, OR where children have the opportunity for exposure to contaminants through soil ingestion or inhalation at educational facilities, health care facilities, child care facilities or outdoor recreational areas.

"Restrictive Covenant or Deed Restriction" means a provision placed in a deed limiting the use of the property and prohibiting certain uses. (Black's Law Dictionary, 5th Edition)

"Right of Way" means THE LAND, OR INTEREST THEREIN, ACQUIRED FOR OR DEVOTED TO A HIGHWAY. (Illinois Highway Code [605 ILCS 5/2-217])

"Site" means ANY SINGLE LOCATION, PLACE, TRACT OF LAND OR PARCEL OF PROPERTY, OR PORTION THEREOF, INCLUDING CONTIGUOUS PROPERTY SEPARATED BY A PUBLIC RIGHT-OF-WAY. (Section 58.2 of the Act)

"Slurry Wall" means a man-made barrier made of geologic material which is constructed to prevent or impede the movement of contamination into a certain area.

"Soil Saturation Limit (C_{sat})" means the contaminant concentration at which soil pore air and pore water are saturated with the chemical and the adsorptive limits of the soil particles have been reached.

"Solubility" means a chemical specific maximum amount of solute that can dissolve in a specific amount of solvent (groundwater) at a specific temperature.

"SPLP" means Synthetic Precipitation Leaching Procedure (Method 1312) as published in "Test Methods for Evaluating Solid Waste, Physical/Chemical Methods", USEPA Publication No. SW-846, as incorporated by reference in Section 742.210.

"SSL" means Soil Screening Levels as defined in USEPA's Soil Screening Guidance: User's Guide and Technical Background Document, as incorporated by reference in Section 742.210.

"Stratigraphic Unit" means a site-specific geologic unit of native deposited material and/or bedrock of varying thickness (e.g., sand, gravel, silt, clay, bedrock, etc.). A change in stratigraphic unit is recognized by a clearly distinct contrast in geologic material or a change in physical features within a zone of gradation. For the purposes of this Part, a change in stratigraphic unit is identified by one or a combination of differences in physical features such as texture, cementation, fabric, composition, density, and/or permeability of the native material and/or bedrock.

"TCLP" means Toxicity Characteristic Leaching Procedure (Method 1311) as published in "Test Methods for Evaluating Solid Waste, Physical/Chemical Methods," USEPA Publication No. SW-846, as incorporated by reference in Section 742.210.

"Total Petroleum Hydrocarbon (TPH)" means the additive total of all petroleum hydrocarbons found in an analytical sample.

"Volatile Organic Compounds (VOCs)" means organic chemical analytes identified as volatiles as published in "Test Methods for Evaluating Solid Waste, Physical/Chemical Methods,"USEPA Publication No. SW-846 (incorporated by reference in Section 742.210), method numbers 8010, 8011, 8015, 8020, 8021, 8030, 8031, 8240, 8260, 8315, and 8316. For analytes not listed in any category in those methods, those analytes which have a boiling point less than 200^0C and a vapor pressure greater than 0.1 Torr (mm Hg) at 20^0C.

Section 742.205   Severability

If any provision of this Part or its application to any person or under any circumstances is adjudged invalid, such adjudication shall not affect the validity of this Part as a whole or any portion not adjudged invalid.

Section 742.210   Incorporations by Reference

a)   The Board incorporates the following material by reference:

ASTM. American Society for Testing and Materials, 1916 Race Street, Philadelphia, PA 19103 (215) 299-5400

ASTM D 2974-87, Standard Test Methods for Moisture, Ash and Organic Matter of Peat and Other Organic Soils, approved May 29, 1987 (reapproved 1995).

ASTM D 2488-93, Standard Practice for Description and Identification of Soils (Visual-Manual Procedure), approved September 15, 1993.

ASTM D 1556-90, Standard Test Method for Density and Unit Weight of Soil in Place by the Sand-Cone Method, approved June 29, 1990.

ASTM D 2167-94, Standard Test Method for Density and Unit Weight of Soil in Place by the Rubber Balloon Method, approved March 15, 1994.

ASTM D 2922-91, Standard Test Methods for Density of Soil and Soil-Aggregate in Place by Nuclear Methods (Shallow Depth), approved December 23, 1991.

ASTM D 2937-94, Standard Test Method for Density of Soil in Place by the Drive-Cylinder Method, approved June 15, 1994.

ASTM D 854-92, Standard Test Method for Specific Gravity of Soils, approved November 15, 1992.

ASTM D 2216-92, Standard Method for Laboratory Determination of Water (Moisture) Content of Soil and Rock, approved June 15, 1992.

ASTM D 4959-89, Standard Test Method for Determination of Water (Moisture) Content of Soil by Direct Heating Method, approved June 30, 1989 (reapproved 1994).

ASTM D 4643-93, Standard Test Method for Determination of Water (Moisture) Content of Soil by the Microwave Oven Method, approved July 15, 1993.

ASTM D 5084-90, Standard Test Method for Measurement of Hydraulic Conductivity of Saturated Porous Materials Using a Flexible Wall Permeameter, approved June 29, 1990.

ASTM D 422-63, Standard Test Method for Particle-Size Analysis of Soils, approved November 21, 1963 (reapproved 1990).

ASTM D 1140-92, Standard Test Method for Amount of Material in Soils Finer than the No. 200 (75 µm) Sieve, approved November 15, 1992.

ASTM D 3017-88, Standard Test Method for Water Content of Soil and Rock in Place by Nuclear Methods (Shallow Depth), approved May 27, 1988.

ASTM D 4525-90, Standard Test Method for Permeability of Rocks by Flowing Air, approved May 25, 1990.

ASTM D 2487-93, Standard Test Method for Classification of Soils for Engineering Purposes, approved September 15, 1993.

ASTM E 1527-93, Standard Practice for Environmental Site Assessments: Phase I Environmental Site Assessment Process, approved March 15, 1993. Vol. 11.04.

ASTM E 1739-95, Standard Guide for Risk-Based Corrective Action Applied at Petroleum Release Sites, approved September 10, 1995.

Barnes, Donald G. and Dourson, Michael. (1988). Reference Dose (RfD): Description and Use in Health Risk Assessments. *Regulatory Toxicology and Pharmacology*. 8, 471-486.

GPO. Superintendent of Documents, U.S. Government Printing Office, Washington, DC 20401, (202) 783-3238.

USEPA Guidelines for Carcinogenic Risk Assessment, 51 Fed. Reg. 33992-34003 (September 24, 1986).

"Test Methods for Evaluating Solid Waste, Physical/Chemical Methods," USEPA Publication number SW-846 (Third Edition, November 1986), as amended by Updates I and IIA (Document No. 955-001-00000-1)(contact USEPA, Office of Solid Waste, for Update IIA).

"Methods for the Determination of Organic Compounds in Drinking Water", EPA Publication No. EPA/600/4-88/039 (December 1988 (Revised July 1991)).

"Methods for the Determination of Organic Compounds in Drinking Water, Supplement II", EPA Publication No. EPA/600/R-92/129 (August 1992).

"Methods for the Determination of Organic Compounds in Drinking Water, Supplement III", EPA Publication No. EPA/600/R-95/131 (August 1995).

IRIS. Integrated Risk Information System, National Center for Environmental Assessment, U.S. Environmental Protection Agency, 26 West Martin Luther King Drive, MS-190, Cincinnati, OH 45268. (513) 569-7254.

"Reference Dose (RfD): Description and Use in Health Risk Assessments", Background Document 1A (March 15, 1993).

"EPA Approach for Assessing the Risks Associated with Chronic Exposures to Carcinogens", Background Document 2 (January 17, 1992).

Nelson, D.W., and L.E. Sommers. 1982. Total carbon, organic carbon, and organic matter. In: A.L. Page (ed.), *Methods of Soil Analysis. Part 2. Chemical and Microbiological Properties. 2nd Edition*, pp. 539-579, American Society of Agronomy. Madison, WI.

NTIS. National Technical Information Service, 5285 Port Royal Road, Springfield, VA 22161, (703) 487-4600.

"Dermal Exposure Assessment: Principles and Applications", EPA Publication No. EPA/600/8-91/011B (January 1992).

"Exposure Factors Handbook", EPA Publication No. EPA/600/8-89/043 (July 1989).

"Risk Assessment Guidance for Superfund, Vol. I; Human Health Evaluation Manual, Supplemental Guidance: Standard Default Exposure Factors", OSWER Directive 9285.6-03 (March 1991).

"Rapid Assessment of Exposure to Particulate Emissions from Surface Contamination Sites," EPA Publication No. EPA/600/8-85/002 (February 1985), PB 85-192219.

"Risk Assessment Guidance for Superfund, Volume I; Human Health Evaluation Manual (Part A)", Interim Final, EPA Publication No. EPA/540/1-89/002 (December 1989).

"Risk Assessment Guidance for Superfund, Volume I; Human Health Evaluation Manual, Supplemental Guidance, Dermal Risk Assessment Interim Guidance", Draft (August 18, 1992).

"Soil Screening Guidance: Technical Background Document", EPA Publication No. EPA/540/R-95/128, PB96-963502 (May 1996).

"Soil Screening Guidance: User's Guide", EPA Publication No. EPA/540/R-96/018, PB96-963505 (April 1996).

"Superfund Exposure Assessment Manual", EPA Publication No. EPA/540/1-88/001 (April 1988).

RCRA Facility Investigation Guidance, Interim Final, developed by USEPA (EPA 530/SW-89-031), 4 volumes (May 1989).

b)   CFR (Code of Federal Regulations). Available from the Superintendent of Documents, U.S. Government Printing Office, Washington, D.C. 20402 (202) 783-3238:

   40 CFR 761.120 (1993).

c)   This Section incorporates no later editions or amendments.

Section 742.215   Determination of Soil Attenuation Capacity

a)   The concentrations of organic contaminants of concern remaining in the soil shall not exceed the attenuation capacity of the soil, as determined under subsection (b) of this Section.

b)   The soil attenuation capacity is not exceeded if:

   1)   The sum of the organic contaminant residual concentrations analyzed for the purposes of the remediation program for which the analysis is performed, at each discrete sampling point, is less than the natural organic carbon fraction of the soil. If the information relative to the concentration of other organic contaminants is available, such information shall be included in the sum. The natural organic carbon fraction (f_{oc}) shall be either:

      A)   A default value of 6000 mg/kg for soils within the top meter and 2000 mg/kg for soils below one meter of the surface; or

      B)   A site-specific value as measured by ASTM D2974-87, Nelson and Sommers, or by SW-846 Method 9060: Total Organic Carbon, as incorporated by reference in Section 742.210;

   2)   The total petroleum hydrocarbon concentration is less than the natural organic carbon fraction of the soil as demonstrated using a method approved by the Agency. The method selected shall be appropriate for the contaminants of concern to be addressed; or

   3)   Another method, approved by the Agency, shows that the soil attenuation capacity is not exceeded.

Section 742.220   Determination of Soil Saturation Limit

a)   For any organic contaminant that has a melting point below 30^{0}C, the remediation objective for the inhalation exposure route developed under Tier 2 or Tier 3 shall not exceed the soil saturation limit, as determined under subsection (c) of this Section.

b)   For any organic contaminant, the remediation objective under Tier 2 or Tier 3 for the soil component of the groundwater ingestion exposure route shall not exceed the soil saturation limit, as determined under subsection (c) of this Section.

c)   The soil saturation limit shall be:

  1)   The value listed in Appendix A, Table A for that specific contaminant;

  2)   A value derived from Equation S29 in Appendix C, Table A; or

  3)   A value derived from another method approved by the Agency.

Section 742.225   Demonstration of Compliance with Remediation Objectives

Compliance is achieved if each sample result does not exceed that respective remediation objective unless a person elects to proceed under subsections (c), (d) and (e) of this Section.

a)   Compliance with groundwater remediation objectives developed under Subparts D through F and H through I shall be demonstrated by comparing the contaminant concentrations of discrete samples at each sample point to the applicable groundwater remediation objective. Sample points shall be determined by the program under which remediation is performed.

b)   Unless the person elects to composite samples or average sampling results as provided in subsections (c) and (d) of this Section, compliance with soil remediation objectives developed under Subparts D through G and I shall be demonstrated by comparing the contaminant concentrations of discrete samples to the applicable soil remediation objective.

  1)   Except as provided in subsections (c) and (d) of this Section, compositing of samples is not allowed.

  2)   Except as provided in subsections (c) and (d) of this Section, averaging of sample results is not allowed.

  3)   Notwithstanding subsections (c) and (d) of this Section, compositing of samples and averaging of sample results is not allowed for the construction worker population.

4)   The number of sampling points required to demonstrate compliance is determined by the requirements applicable to the program under which remediation is performed.

c)   If a person chooses to composite soil samples or average soil sample results to demonstrate compliance relative to the soil component of the groundwater ingestion exposure route, the following requirements apply:

     1)   A minimum of two sampling locations for every 0.5 acre of contaminated area is required, with discrete samples at each sample location obtained at every two feet of depth, beginning at six inches below the ground surface and continuing through the zone of contamination. Alternatively, a sampling method may be approved by the Agency based on an appropriately designed site-specific evaluation. Samples obtained at or below the water table shall not be used in compositing or averaging.

     2)   For contaminants of concern other than volatile organic contaminants:

          A)   Discrete samples from the same boring may be composited.

          B)   Discrete sample results from the same boring may be averaged.

     3)   For volatile organic contaminants:

          A)   Compositing of samples is not allowed.

          B)   Discrete sample results from the same boring may be averaged.

d)   If a person chooses to composite soil samples or average soil sample results to demonstrate compliance relative to the inhalation exposure route or ingestion exposure route, the following requirements apply:

     1)   A person shall submit a sampling plan for Agency approval, based upon a site-specific evaluation;

     2)   For volatile organic compounds, compositing of samples is not allowed; and

     3)   All samples shall be collected within the contaminated area.

e)   When averaging under this Section, if no more than 50% of sample results are reported as "non-detect", "no contamination", "below detection limits", or similar terms, such results shall be included in the averaging calculation as one-half of the reported analytical detection limit for the contaminant. If more than 50% of sample results are "non-detect", another statistically valid procedure approved by the Agency may be used to determine an average.

Section 742.230   Agency Review and Approval

a)   Documents and requests filed with the Agency under this Part shall be submitted in accordance with the procedures applicable to the specific program under which remediation is performed.

b)   Agency review and approval of documents and requests under this Part shall be performed in accordance with the procedures applicable to the specific program under which the remediation is performed.

SUBPART C: EXPOSURE ROUTE EVALUATIONS

Section 742.300   Exclusion of Exposure Route

a)   This Subpart sets forth requirements to demonstrate that an actual or potential impact to a receptor or potential receptor from a contaminant of concern can be excluded from consideration from one or more exposure routes. If an evaluation under this Part demonstrates the applicable requirements for excluding an exposure route are met, then the exposure route is excluded from consideration and no remediation objective(s) need be developed for that exposure route.

b)   No exposure route may be excluded from consideration until characterization of the extent and concentrations of contaminants of concern at a site has been performed. The actual steps and methods taken to characterize a site shall be determined by the specific program requirements under which the site remediation is being addressed.

c)   As an alternative to the use of the requirements in this Part, a person may use the procedures for evaluation of exposure routes under Tier 3 as set forth in Section 742.925.

Section 742.305   Contaminant Source and Free Product Determination

No exposure route shall be excluded from consideration relative to a contaminant of concern unless the following requirements are met:

a)   The sum of the concentrations of all organic contaminants of concern shall not exceed the attenuation capacity of the soil as determined under Section 742.215;

b)   The concentrations of any organic contaminants of concern remaining in the soil shall not exceed the soil saturation limit as determined under Section 742.220;

c)   Any soil which contains contaminants of concern shall not exhibit any of the characteristics of reactivity for hazardous waste as determined under 35 Ill. Adm. Code 721.123;

d)   Any soil which contains contaminants of concern shall not exhibit a pH less than or equal to 2.0 or greater than or equal to 12.5, as determined by SW-846 Method 9040B:pH Electrometric for soils with 20 % or greater aqueous (moisture) content or by SW-846 Method 9045C:Soil pH for soils with less than 20% aqueous (moisture) content as incorporated by reference in Section 742.210; and

e)   Any soil which contains contaminants of concern in the following list of inorganic chemicals or their salts shall not exhibit any of the characteristics of toxicity for hazardous waste as determined by 35 Ill. Adm. Code 721.124, or an alternative method approved by the Agency: arsenic, barium, cadmium, chromium, lead, mercury, selenium or silver.

Section 742.310   Inhalation Exposure Route

The inhalation exposure route may be excluded from consideration if:

a)   The requirements of Sections 742.300 and 742.305 are met; and

b)   An institutional control, in accordance with Subpart J, is in place that meets the following requirements:

   1)   Either:

      A)   The concentration of any contaminant of concern the land surface or any man-made pathway shall not exceed the Tier 1 remediation objective under Subpart E for the inhalation exposure route; or

      B)   An engineered barrier, as set forth in Subpart K and approved by the Agency, is in place; and

   2)   Requires safety precautions for the construction worker if the Tier 1 construction worker remediation objectives are exceeded.

Section 742.315   Soil Ingestion Exposure Route

The soil ingestion exposure route may be excluded from consideration if:

a)   The requirements of Sections 742.300 and 742.305 are met; and

b)   An institutional control, in accordance with Subpart J, is in place that meets the following requirements:

    1)   Either:

        A)   The concentration of any contaminant of concern within three feet of the land surface shall not exceed the Tier 1 remediation objective under Subpart E for the ingestion of soil exposure route; or

        B)   An engineered barrier, as set forth in Subpart K and approved by the Agency, is in place; and

    2)   Requires safety precautions for the construction worker if the Tier 1 construction worker remediation objectives are exceeded.

Section 742.320   Groundwater Ingestion Exposure Route

The groundwater ingestion exposure route may be excluded from consideration if:

a)   The requirements of Sections 742.300 and 742.305 are met;

b)   The corrective action measures have been completed to remove any free product to the maximum extent practicable;

c)   The source of the release is not located within the minimum or designated maximum setback zone or within a regulated recharge area of a potable water supply well;

d)   As demonstrated in accordance with Section 742.1015, for any area within 2500 feet from the source of the release, an ordinance adopted by a unit of local government is in place that effectively prohibits the installation of potable water supply wells (and the use of such wells);

e)   As demonstrated using Equation R26, in Appendix C, Table C, in accordance with Section 742.810, the concentration of any contaminant of concern in groundwater within the minimum or designated maximum setback zone of an existing potable water supply well will meet the applicable Tier 1 groundwater remediation objective; and

f)   As demonstrated using Equation R26, in Appendix C, Table C, in accordance with Section 742.810, the concentration of any contaminant of concern in groundwater discharging into a surface water will meet the applicable surface water quality standard under 35 Ill. Adm. Code 302.

SUBPART D: DETERMINING AREA BACKGROUND

Section 742.400   Area Background

This Subpart provides procedures for determining area background concentrations for contaminants of concern. Except as described in Section 742.415(c) and (d) of this Subpart, area background concentrations may be used as remediation objectives for contaminants of concern at a site.

Section 742.405   Determination of Area Background for Soil

a)   Soil sampling results shall be obtained for purposes of determining area background levels in accordance with the following procedures:

1)   For volatile organic contaminants, sample results shall be based on discrete samples;

2)   Unless an alternative method is approved by the Agency, for contaminants other than volatile organic contaminants, sample results shall be based on discrete samples or composite samples. If a person elects to use composite samples, each 0.5 acre of the area to be sampled shall be divided into quadrants and 5 aliquots of equal volume per quadrant shall be composited into 1 sample;

3)   Samples shall be collected from similar depths and soil types, which shall be consistent with the depths and soil types in which maximum levels of contaminants are found in the areas of known or suspected releases; and

4)   Samples shall be collected from areas of the site or adjacent to the site that are unaffected by known or suspected releases at or from the site. If the sample results show an impact from releases at or from the site, then the sample results shall not be included in determining area background levels under this Part.

b)   Area background shall be determined according to one of the following
        approaches:

   1)   Statewide Area Background Approach:

      A)   The concentrations of inorganic chemicals in background soils
              listed in Appendix A, Table G may be used as the upper limit of
              the area background concentration for the site. The first column
              to the right of the chemical name presents inorganic chemicals in
              background soils for counties within Metropolitan Statistical
              Areas. Counties within Metropolitan Statistical Areas are
              identified in Appendix A, Table G, Footnote a. Sites located in
              counties outside Metropolitan Statistical Areas shall use the
              concentrations of inorganic chemicals in background soils shown
              in the second column to the right of the chemical name.

      B)   Soil area background concentrations determined according to this
              statewide area background approach shall be used as provided in
              Section 742.415(b) of this Part. For each parameter whose
              sampling results demonstrate concentrations above those in
              Appendix A, Table G, the person shall develop appropriate soil
              remediation objectives in accordance with this Part, or may
              determine area background in accordance with subsection (b)(2)
              of this Section.

   2)   A statistically valid approach for determining area background
           concentrations appropriate for the characteristics of the data set, and
           approved by the Agency.

Section 742.410   Determination of Area Background for Groundwater

   a)   Groundwater sampling results shall be obtained for purposes of determining
           area background in accordance with the following procedures:

   1)   Samples shall be collected from areas of the site or adjacent to the site
           that are unaffected by releases at the site;

   2)   The background monitoring wells shall be sufficient in number to
           account for the spatial and temporal variability, size, and number of
           known or suspected off-site releases of contaminants of concern, and the
           hydrogeological setting of the site;

3)   The samples shall be collected in consecutive quarters for a minimum of one year for each well unless another sample schedule is approved by the Agency;

4)   The samples shall be collected from the same stratigraphic unit(s) as the groundwater contamination at the site; and

5)   The background monitoring wells shall be located hydraulically upgradient from the release(s) of contaminants of concern, unless a person demonstrates to the Agency that the upgradient location is undefinable or infeasible.

b)   Area background shall be determined according to one of the following approaches:

1)   Prescriptive Approach:

A)   If more than 15% of the groundwater sampling results for a chemical obtained in accordance with subsection (a) of this Section are less than the appropriate detection limit for that chemical, the Prescriptive Approach may not be used for that chemical. If 15% or less of the sampling results are less than the appropriate detection limit, a concentration equal to one-half the detection limit shall be used for that chemical in the calculations contained in this Prescriptive Approach.

B)   The groundwater sampling results obtained in accordance with subsection (a) of this Section shall be used to determine if the sample set is normally distributed. The Shapiro-Wilk Test of Normality shall be used to determine whether the sample set is normally distributed, if the sample set for the background well(s) contains 50 or fewer samples. Values necessary for the Shapiro-Wilk Test of Normality shall be determined using Appendix A, Tables C and D. If the computed value of W is greater than the 5% Critical Value in Appendix A, Table D, the sample set shall be assumed to be normally distributed, and the Prescriptive Approach is allowed. If the computed value of W is less than 5% Critical Value in Appendix A, Table D, the sample set shall be assumed to not be normally distributed, and the Prescriptive Approach shall not be used.

C)   If the sample set contains at least ten sample results, the Upper Tolerance Limit (UTL) of a normally distributed sample set may be calculated using the mean (x) and standard deviation(s), from:

$$UTL = x + (K \bullet s),$$

where K = the one-sided normal tolerance factor for estimating the 95% upper confidence limit of the 95th percentile of a normal distribution. Values for K shall be determined using Appendix A, Table B.

D)   If the sample set contains at least ten sample results, the UTL shall be the upper limit of the area background concentration for the site. If the sample set contains fewer than ten sample results, the maximum value of the sample set shall be the upper limit of the area background concentration for the site.

E)   This Prescriptive Approach shall not be used for determining area background for the parameter pH.

2)   Another statistically valid approach for determining area background concentrations appropriate for the characteristics of the data set, and approved by the Agency.

Section 742.415   Use of Area Background Concentrations

a)   A person may request that area background concentrations determined pursuant to Sections 742.405 and 742.410 be used according to the provisions of subsection (b) of this Section. Such request shall address the following:

1)   The natural or man-made pathways of any suspected off-site contamination reaching the site;

2)   Physical and chemical properties of suspected off-site contaminants of concern reaching the site; and

3)   The location and justification of all background sampling points.

b)   Except as specified in subsections (c) and (d) of this Section, an area background concentration may be used as follows:

1)   To support a request to exclude a chemical as a contaminant of concern from further consideration for remediation at a site due to its presence as a result of background conditions; or

2)   As a remediation objective for a contaminant of concern at a site in lieu of an objective developed pursuant to the other procedures of this Part.

c) An area background concentration shall not be used IN THE EVENT THAT THE AGENCY HAS DETERMINED IN WRITING THAT THE BACKGROUND LEVEL FOR A REGULATED SUBSTANCE POSES AN ACUTE THREAT TO HUMAN HEALTH OR THE ENVIRONMENT AT THE SITE WHEN CONSIDERING THE POST-REMEDIAL ACTION LAND USE. (Section 58.5(b)(3) of the Act)

d) IN THE EVENT THAT THE CONCENTRATION OF A REGULATED SUBSTANCE OF CONCERN ON THE SITE EXCEEDS A REMEDIATION OBJECTIVE ADOPTED BY THE BOARD FOR RESIDENTIAL LAND USE, THE PROPERTY MAY NOT BE CONVERTED TO RESIDENTIAL USE UNLESS SUCH REMEDIATION OBJECTIVE OR AN ALTERNATIVE RISK-BASED REMEDIATION OBJECTIVE FOR THAT REGULATED SUBSTANCE OF CONCERN IS FIRST ACHIEVED. If the land use is restricted, there shall be an institutional control in place in accordance with Subpart J. (Section 58.5(b)(2) of the Act)

SUBPART E: TIER 1 EVALUATION

Section 742.500   Tier 1 Evaluation Overview

a) A Tier 1 evaluation compares the concentration of each contaminant of concern detected at a site to the baseline remediation objectives provided in Appendix B, Tables A, B, C, D and E. Use of Tier 1 remediation objectives requires only limited site-specific information: concentrations of contaminants of concern, groundwater classification, land use classification, and, if appropriate, soil pH. (See Appendix B, Illustration A.)

b) Although Tier 1 allows for differentiation between residential and industrial/commercial property use of a site, an institutional control under Subpart J is required where remediation objectives are based on an industrial/commercial property use.

c) Any given exposure route is not a concern if the concentration of each contaminant of concern detected at the site is below the Tier 1 value of that given route. In such a case, no further evaluation of that route is necessary.

Section 742.505   Tier 1 Soil and Groundwater Remediation Objectives

a) Soil

1) Inhalation Exposure Route

A)   The Tier 1 soil remediation objectives for this exposure route based upon residential property use are listed in Appendix B, Table A.

B)   The Tier 1 soil remediation objectives for this exposure route based upon industrial/commercial property use are listed in Appendix B, Table B. Soil remediation objective determinations relying on this table require use of institutional controls in accordance with Subpart J.

2)   Ingestion Exposure Route

A)   The Tier 1 soil remediation objectives for this exposure route based upon residential property use are listed in Appendix B, Table A.

B)   The Tier 1 soil remediation objectives for this exposure route based upon industrial/commercial property use are listed in Appendix B, Table B. Soil remediation objective determinations relying on this table require use of institutional controls in accordance with Subpart J.

3)   Soil Component of the Groundwater Ingestion Route

A)   The Tier 1 soil remediation objectives for this exposure route based upon residential property use are listed in Appendix B, Table A.

B)   The Tier 1 soil remediation objectives for this exposure route based upon industrial/commercial property use are listed in Appendix B, Table B.

C)   The pH-dependent Tier 1 soil remediation objectives for identified ionizable organics or inorganics for the soil component of the groundwater ingestion exposure route (based on the total amount of contaminants present in the soil sample results and groundwater classification) are provided in Appendix B, Tables C and D.

D)   Values used to calculate the Tier 1 soil remediation objectives for this exposure route are listed in Appendix B, Table F.

4)   Evaluation of the dermal contact with soil exposure route is not required under Tier 1.

b)   Groundwater

1)   The Tier 1 groundwater remediation objectives for the groundwater component of the groundwater ingestion route are listed in Appendix B, Table E.

2)   The Tier 1 groundwater remediation objectives for this exposure route are given for Class I and Class II groundwaters, respectively.

3)   The Class I groundwater remediation objectives set forth in Appendix B, Table E shall be corrected for cumulative effect of mixtures of similar-acting noncarcinogenic chemicals in accordance with the methodologies set forth in either subsection (b)(3)(A) or (B), if more than one chemical listed in Appendix A, Table E is detected at a site and if such chemicals affect the same target organ (*i.e.*, has the same critical effect as defined by the RfD):

A)   Calculate the weighted average using the following equations:

$$W_{ave} = \frac{x_1}{CUO_{x_1}} + \frac{x_2}{CUO_{x_2}} + \frac{x_3}{CUO_{x_3}} + ... + \frac{x_a}{CUO_{x_a}}$$

where:

W_{ave} = Weighted Average

x_1 through x_a =   Concentration of each individual contaminant at the location of concern. Note that, depending on the target organ/mode of action, the actual number of contaminants will range from 2 to 14.

CUO_{x_a} =   A Tier 1 remediation objective each x_a from Appendix B, Table E.

ii)   If the value of the weighted average calculated in accordance with the equations above is less than or equal to 1.0, then the remediation objectives are met for those chemicals.

ii)   If the value of the weighted average calculated in accordance with the equations above is greater than 1.0, then additional remediation must be carried out until the level of contaminants remaining in the remediated area

have a weighted average calculated in accordance with the
equation above less than or equal to one;

B)   Divide each individual chemical's remediation objective by the
number of chemicals in that specific target organ group that were
detected at the site. Each of the contaminant concentrations at
the site is then compared to the remediation objectives that have
been adjusted to account for this potential additivity.

Section 742.510   Tier 1 Remediation Objectives Tables

a)   Soil remediation objectives are listed in Appendix B, Tables A, B, C and D.

1)   Appendix B, Table A is based upon residential property use.

A)   The first column to the right of the chemical name lists soil
remediation objectives for the soil ingestion exposure route.

B)   The second column lists the soil remediation objectives for the
inhalation exposure route.

C)   The third and fourth columns list soil remediation objectives for
the soil component of the groundwater ingestion exposure route
for the respective classes of groundwater:

i)   Class I groundwater; and

ii)   Class II groundwater.

D)   The final column lists the Acceptable Detection Limit (ADL),
only where applicable.

2)   Appendix B, Table B is based upon industrial/commercial property use.

A)   The first and third columns to the right of the chemical name list
the soil remediation objectives for the soil ingestion exposure
route based on two receptor populations:

i)   Industrial/commercial; and

ii)   Construction worker.

B)   The second and fourth columns to the right of the chemical name list the soil remediation objectives for the inhalation exposure route based on two receptor populations:

    i)   Industrial/commercial; and

    ii)   Construction worker.

C)   The fifth and sixth columns to the right of the chemical name list the soil remediation objectives for the soil component of the groundwater ingestion exposure route for two classes of groundwater:

    i)   Class I groundwater; and

    ii)   Class II groundwater.

3)   Appendix B, Tables C and D set forth pH specific soil remediation objectives for inorganic and ionizing organic chemicals for the soil component of the groundwater ingestion route.

A)   Table C sets forth remediation objectives based on Class I groundwater and Table D sets forth remediation objectives based on Class II groundwater.

B)   The first column in Tables C and D lists the chemical names.

C)   The second through ninth columns to the right of the chemical names list the pH based soil remediation objectives.

4)   For the inorganic chemicals listed in Appendix B, Tables A and B, the soil component of the groundwater ingestion exposure route shall be evaluated using TCLP (SW-846 Method 1311) or SPLP (SW-846 Method 1312), incorporated by reference at Section 742.210 unless a person chooses to evaluate the soil component on the basis of the total amount of contaminant in a soil sample result in accordance with subsection (a)(5) of this Section.

5)   For those inorganic and ionizing organic chemicals listed in Appendix B, Tables C and D, if a person elects to evaluate the soil component of the groundwater ingestion exposure route based on the total amount of contaminant in a soil sample result (rather than TCLP or SPLP analysis), the person shall determine the soil pH at the site and then select the appropriate soil remediation objectives based on Class I and Class II

groundwaters from Tables C and D, respectively. If the soil pH is less than 4.5 or greater than 8.0, then Tables C and D cannot be used.

6) Unless one or more exposure routes are excluded from consideration under Subpart C, the most stringent soil remediation objective of the exposure routes (*i.e.*, soil ingestion exposure route, inhalation exposure route, and soil component of the groundwater ingestion exposure route) shall be compared to the concentrations of soil contaminants of concern measured at the site. When using Appendix B, Table B to select soil remediation objectives for the ingestion exposure route and inhalation exposure route, the remediation objective shall be the more stringent soil remediation objective of the industrial/commercial populations and construction worker populations.

7) Confirmation sample results may be averaged or soil samples may be composited in accordance with Section 742.225.

8) If a soil remediation objective for a chemical is less than the ADL, the ADL shall serve as the soil remediation objective.

b) Groundwater remediation objectives for the groundwater component of the groundwater ingestion exposure route are listed in Appendix B, Table E. However, Appendix B, Table E must be corrected for cumulative effect of mixtures of similar-acting noncarcinogenic chemicals as set forth in Section 742.505(b)(3).

1) The first column to the right of the chemical name lists groundwater remediation objectives for Class I groundwater, and the second column lists the groundwater remediation objectives for Class II groundwater.

2) To use Appendix B, Table E of this Part, the 35 Ill. Adm. Code 620 classification for groundwater at the site shall be determined. The concentrations of groundwater contaminants of concern at the site are compared to the applicable Tier 1 groundwater remediation objectives for the groundwater component of the groundwater ingestion exposure route in Appendix B, Table E.

c) For contaminants of concern not listed in Appendix B, Tables A, B and E, a person may request site-specific remediation objectives from the Agency or propose site-specific remediation objectives in accordance with 35 Ill. Adm. Code 620, Subpart I of this Part, or both.

SUBPART F: TIER 2 GENERAL EVALUATION

Section 742.600   Tier 2 Evaluation Overview

a)   Tier 2 remediation objectives are developed through the use of equations which allow site-specific data to be used. (See Appendix C, Illustrations A and B.) The equations, identified in Appendix C, Tables A and C may be used to develop Tier 2 remediation objectives.

b)   Tier 2 evaluation is only required for contaminants of concern and corresponding exposure routes (except where excluded from further consideration under Subpart C) exceeding the Tier 1 remediation objectives. When conducting Tier 2 evaluations, the values used in the calculations must have the appropriate units of measure as identified in Appendix C, Tables B and D.

c)   Any development of remediation objectives using site-specific information or equations outside the Tier 2 framework shall be evaluated under Tier 3.

d)   Any development of a remediation objective under Tier 2 shall not use a target hazard quotient greater than one at the point of human exposure or a target cancer risk greater than 1 in 1,000,000 at the point of human exposure.

e)   In conducting a Tier 2 evaluation, the following conditions shall be met:

   1)   For each discrete sample, the total soil contaminant concentration of either a single contaminant or multiple contaminants of concern shall not exceed the attenuation capacity of the soil as provided in Section 742.215.

   2)   Remediation objectives for noncarcinogenic compounds which affect the same target organ, organ system or similar mode of action shall meet the requirements of Section 742.720.

   3)   The soil remediation objectives based on the inhalation and the soil component of the groundwater ingestion exposure routes shall not exceed the soil saturation limit as provided in Section 742.220.

f)   If the calculated Tier 2 soil remediation objective for an applicable exposure route is more stringent than the corresponding Tier 1 remediation objective, then the Tier 1 remediation objective applies.

g) If the calculated Tier 2 soil remediation objective for an exposure route is more stringent than the Tier 1 soil remediation objective(s) for the other exposure routes, then the Tier 2 calculated soil remediation objective applies and Tier 2 soil remediation objectives for the other exposure routes are not required.

h) If the calculated Tier 2 soil remediation objective is less stringent than one or more of the soil remediation objectives for the remaining exposure routes, then the Tier 2 values are calculated for the remaining exposure route(s) and the most stringent Tier 2 calculated value applies.

Section 742.605   Land Use

a) Present and post-remediation land use is evaluated in a Tier 2 evaluation. Acceptable exposure factors for the Tier 2 evaluation for residential, industrial/commercial, and construction worker populations are provided in the far right column of both Appendix C, Tables B and D. Use of exposure factors different from those in Appendix C, Tables B and D must be approved by the Agency as part of a Tier 3 evaluation.

b) If a Tier 2 evaluation is based on an industrial/commercial property use, then:

1) Construction worker populations shall also be evaluated; and

2) Institutional controls are required in accordance with Subpart J.

Section 742.610   Chemical and Site Properties

a) Physical and Chemical Properties of Contaminants

Tier 2 evaluations require information on the physical and chemical properties of the contaminants of concern. The physical and chemical properties used in a Tier 2 evaluation are contained in Appendix C, Table E. If the site has contaminants not included in this table, a person may request the Agency to provide the applicable physical and chemical input values or may propose input values under Subpart I. If a person proposes to apply values other than those in Appendix C, Table E, or those provided by the Agency, the evaluation shall be considered under Tier 3.

b)   Soil and Groundwater Parameters

1)   A Tier 2 evaluation requires examination of soil and groundwater parameters. The parameters that may be varied, and the conditions under which these parameters are determined as part of Tier 2, are summarized in Appendix C, Tables B and D. If a person proposes to vary site-specific parameters outside of the framework of these tables, the evaluation shall be considered under Tier 3.

2)   To determine site-specific physical soil parameters, a minimum of one boring per 0.5 acre of contamination shall be collected. This boring must be deep enough to allow the collection of the required field measurements. The site-specific physical soil parameters must be determined from the portion of the boring representing the stratigraphic unit(s) being evaluated. For example, if evaluating the soil component of the groundwater ingestion exposure route, two samples from the boring will be required:

A)   A sample of the predominant soil type for the vadose zone; and

B)   A sample of the predominant soil type for the saturated zone.

3)   A site-specific SSL dilution factor (used in developing soil remediation objectives based upon the protection of groundwater) may be determined by substituting site information in Equation S22 in Appendix C, Table A. To make this demonstration, a minimum of three monitoring wells shall be used to determine the hydraulic gradient. As an alternative, the default dilution factor value listed in Appendix C, Table B may be used. If monitoring wells are used to determine the hydraulic gradient, the soil taken from the borings shall be visually inspected to ensure there are no significant differences in the stratigraphy. If there are similar soil types in the field, one boring shall be used to determine the site-specific physical soil parameters. If there are significant differences, all of the borings shall be evaluated before determining the site-specific physical soil parameters for the site.

4)   Not all of the parameters identified in Appendix C, Tables B and D need to be determined on a site-specific basis. A person may choose to collect partial site-specific information and use default values as listed in Appendix C, Tables B and D for the rest of the parameters.

SUBPART G: TIER 2 SOIL EVALUATION

Section 742.700   Tier 2 Soil Evaluation Overview

a)   Tier 2 remediation objectives are developed through the use of models which
       allow site-specific data to be considered. Appendix C, Tables A and C list
       equations that shall be used under a Tier 2 evaluation to calculate soil
       remediation objectives prescribed by SSL and RBCA models, respectively.
       (See also Appendix C, Illustration A.)

b)   Appendix C, Table A lists equations that are used under the SSL model. (See
       also Appendix C, Illustration A.) The SSL model has equations to evaluate the
       following human exposure routes:

       1)   Soil ingestion exposure route;

       2)   Inhalation exposure route for:

              A)   Volatiles;

              B)   Fugitive dust; and

       3)   Soil component of the groundwater ingestion exposure route.

c)   Evaluation of the dermal exposure route is not required under the SSL model.

d)   Appendix C, Table C lists equations that are used under the RBCA model. (See
       also Appendix C, Illustration A.) The RBCA model has equations to evaluate
       human exposure based on the following:

       1)   The combined exposure routes of inhalation of vapors and particulates,
              soil ingestion and dermal contact with soil;

       2)   The ambient vapor inhalation (outdoor) route from subsurface soils;

       3)   Soil component of the groundwater ingestion route; and

       4)   Groundwater ingestion exposure route.

e)   The equations in either Appendix C, Table A or C may be used to calculate
       remediation objectives for each contaminant of concern under Tier 2, if the
       following requirements are met:

1)   The Tier 2 soil remediation objectives for the ingestion and inhalation exposure routes shall use the applicable equations from the same approach (i.e., SSL equations in Appendix C, Table C).

2)   The equations used to calculate soil remediation objectives for the soil component of the groundwater ingestion exposure route are not dependent on the approach utilized to calculate soil remediation objectives for the other exposure routes. For example, it is acceptable to use the SSL equations for calculating Tier 2 soil remediation objectives for the ingestion and inhalation exposure routes, and the RBCA equations for calculating Tier 2 soil remediation objectives for the soil component of the groundwater ingestion exposure route.

3)   Combining equations from Appendix C, Tables A and C to form a new model is not allowed. In addition, Appendix C, Tables A and C must use their own applicable parameters identified in Appendix C, Tables B and D, respectively.

f)   In calculating soil remediation objectives for industrial/commercial property use, applicable calculations shall be performed twice: once using industrial/commercial population default values and once using construction worker population default values. The more stringent soil remediation objectives derived from these calculations must be used for further Tier 2 evaluations.

g)   Tier 2 data sheets provided by the Agency shall be used to present calculated Tier 2 remediation objectives, if required by the particular program for which remediation is being performed.

h)   The RBCA equations which rely on the parameter Soil Water Sorption Coefficient (k_s) can only be used for ionizing organics and inorganics by substituting values for k_s from Appendix C, Tables I and J, respectively. This will also require the determination of a site-specific value for soil pH.

Section 742.705   Parameters for Soil Remediation Objective Equations

a)   Appendix C, Tables B and D list the input parameters for the SSL and RBCA equations, respectively. The first column lists each symbol as it is presented in the equation. The next column defines the parameters. The third column shows the units for the parameters. The fourth column identifies where information on the parameters can be obtained (i.e., field measurement, applicable equation(s), reference source, or default value). The last column identifies how the parameters can be generated.

b)   Default Values

Default values are numerical values specified for use in the Tier 2 equations.
The fourth column of Appendix C, Tables B and D denotés if the default values
are from the SSL model, RBCA model, or some other source. The last column
of Appendix C, Tables B and D lists the numerical values for the default values
used in the SSL and RBCA equations, respectively.

c)   Site-specific Information

Site-specific information is a parameter measured, obtained, or determined from
the site to calculate Tier 2 remediation objectives. The fourth column of
Appendix C, Tables B and D identifies those site-specific parameters that may
require direct field measurement. For some parameters, numerical default
inputs have been provided in the last column of Appendix C, Tables B and D to
substitute for site-specific information. In some cases, information on the
receptor or soil type is required to select the applicable numerical default inputs.
Site-specific information includes:

1)   Physical soil parameters identified in Appendix C, Table F. The second
column identifies the location where the sample is to be collected.
Acceptable methods for measuring or calculating these soil parameters
are identified in the last column of Appendix C, Table F;

2)   Institutional controls or engineered barriers, pursuant to Subparts J and
K, describe applicable institutional controls and engineered barriers
under a Tier 2 evaluation; and

3)   Land use classification

d)   Toxicological-specific Information

1)   Toxicological-specific information is used to calculate Tier 2 remediation
objectives for the following parameters, if applicable:

A)   Oral Chronic Reference Dose (RfD_o, expressed in mg/kg-d);

B)   Oral Subchronic Reference Dose (RfD_s, expressed in mg/kg-d,
shall be used for construction worker remediation objective
calculations);

C)   Oral Slope Factor (SF_o, expressed in $(mg/kg-d)^{-1}$);

D)   Inhalation Unit Risk Factor (URF expressed in $(\mu g/m^3)^{-1}$);

E) Inhalation Chronic Reference Concentration (RfC, expressed in mg/m^3);

F) Inhalation Subchronic Reference Concentration (RfC$_s$, expressed in mg/m^3, shall be used for construction worker remediation objective calculations);

G) Inhalation Chronic Reference Dose (RfD$_i$, expressed in mg/kg-d);

H) Inhalation Subchronic Reference Dose (RfD$_{is}$, expressed in mg/kg-d, shall be used for construction worker remediation objective calculations); and

I) Inhalation Slope Factor (SF$_i$, expressed in (mg/kg-d)$^{-1}$);

2) Toxicological information can be obtained from IRIS, as incorporated by reference in Section 742.210, or the program under which the remediation is being performed.

e) Chemical-specific Information

Chemical-specific information used to calculate Tier 2 remediation objectives is listed in Appendix C, Table E.

f) Calculations

Calculating numerical values for some parameters requires the use of equations listed in Appendix C, Table A or C. The parameters that are calculated are listed in Appendix C, Tables B and D.

Section 742.710 SSL Soil Equations

a) This Section sets forth the equations and parameters used to develop Tier 2 soil remediation objectives for the three exposure routes using the SSL approach.

b)   Soil Ingestion Exposure Route

    1)   Equations S1 through S3 form the basis for calculating Tier 2
        remediation objectives for the soil ingestion exposure route using the
        SSL approach. Equation S1 is used to calculate soil remediation
        objectives for noncarcinogenic contaminants. Equations S2 and S3 are
        used to calculate soil remediation objectives for carcinogenic
        contaminants for residential populations and industrial/commercial and
        construction worker populations, respectively.

    2)   For Equations S1 through S3, the SSL default values cannot be modified
        with site-specific information.

c)   Inhalation Exposure Route

    1)   Equations S4 through S16, S26 and S27 are used to calculate Tier 2 soil
        remediation objectives for the inhalation exposure route using the SSL
        approach. To address this exposure route, volatiles must be evaluated
        separately from fugitive dust using their own equations set forth in
        subsections (c)(2) and (c)(3) of this Section, respectively.

    2)   Volatiles

        A)   Equations S4 through S10 are used to calculate Tier 2 soil
            remediation objectives for volatile contaminants based on the
            inhalation exposure route. Equation S4 is used to calculate soil
            remediation objectives for noncarcinogenic volatile contaminants
            in soil for residential and industrial/commercial populations.
            Equation S5 is used to calculate soil remediation objectives for
            noncarcinogenic volatile contaminants in soil for construction
            worker populations. Equation S6 is used to calculate soil
            remediation objectives for carcinogenic volatile contaminants in
            soil for residential and industrial/commercial populations.
            Equation S7 is used to calculate soil remediation objectives for
            carcinogenic volatile contaminants in soil for construction worker
            populations. Equations S8 through S10, S27 and S28 are used
            for calculating numerical values for some of the parameters in
            Equations S4 through S7.

        B)   For Equation S4, a numerical value for the Volatilization Factor
            (VF) can be calculated in accordance with subsection (c)(2)(F) of
            this Section. The remaining parameters in Equation S4 have
            either SSL default values listed in Appendix C, Table B or

toxicological-specific information (i.e., RfC), which can be obtained from IRIS or requested from the program under which the remediation is being performed.

C)   For Equation S5, a numerical value for the Volatilization Factor adjusted for Agitation (VF') can be calculated in accordance with subsection (c)(2)(G) of this Section. The remaining parameters in Equation S5 have either SSL default values listed in Appendix C, Table B or toxicological-specific information (i.e., RfC), which can be obtained from IRIS or requested from the program under which the remediation is being performed.

D)   For Equation S6, a numerical value for VF can be calculated in accordance with subsection (c)(2)(F) of this Section. The remaining parameters in Equation S6 have either default values listed in Appendix C, Table B or toxicological-specific information (i.e., URF), which can be obtained from IRIS or requested from the program under which the remediation is being performed.

E)   For Equation S7, a numerical value for VF' can be calculated in accordance with subsection (c)(2)(G) of this Section. The remaining parameters in Equation S7 have either default values listed in Appendix C, Table B or toxicological-specific information (i.e., URF), which can be obtained from IRIS or requested from the program under which the remediation is being performed.

F)   The VF can be calculated for residential and industrial/commercial populations using one of the following equations based on the information known about the contaminant source and receptor population:

   i)   Equation S8, in conjunction with Equation S10, is used to calculate VF assuming an infinite source of contamination; or

   ii)   If the area and depth of the contaminant source are known or can be estimated reliably, mass limit considerations may be used to calculate VF using Equation S26.

G)   The VF' can be calculated for the construction worker populations using one of the following equations based on the information known about the contaminant source:

i)   Equation S9 is used to calculate VF' assuming an infinite
        source of contamination; or

ii)   If the area and depth of the contaminant source are known
        or can be estimated reliably, mass limit considerations
        may be used to calculate VF' using Equation S27.

3)   Fugitive Dust

A)   Equations S11 through S16 are used to calculate Tier 2 soil
        remediation objectives using the SSL fugitive dust model for the
        inhalation exposure route. Equation S11 is used to calculate soil
        remediation objectives for noncarcinogenic contaminants in
        fugitive dust for residential and industrial/commercial
        populations. Equation S12 is used to calculate soil remediation
        objectives for noncarcinogenic contaminants in fugitive dust for
        construction worker populations. Equation S13 is used to
        calculate soil remediation objectives for carcinogenic
        contaminants in fugitive dust for residential and
        industrial/commercial populations. Equation S14 is used to
        calculate soil remediation objectives for carcinogenic
        contaminants in fugitive dust for construction worker
        populations. Equations S15 and S16 are used for calculating
        numerical quantities for some of the parameters in Equations S11
        through S14.

B)   For Equation S11, a numerical value can be calculated for the
        Particulate Emission Factor (PEF) using Equation S15. This
        equation relies on various input parameters from a variety of
        sources. The remaining parameters in Equation S11 have either
        SSL default values listed in Appendix C, Table B or
        toxicological-specific information (i.e., RfC), which can be
        obtained from IRIS or requested from the program under which
        the remediation is being performed.

C)   For Equation S12, a numerical value for the Particulate Emission
        Factor for Construction Worker (PEF') can be calculated using
        Equation S16. The remaining parameters in Equation S12 have
        either SSL default values listed in Appendix C, Table B or
        toxicological-specific information (i.e., RfC), which can be
        obtained from IRIS or requested from the program under which
        the remediation is being performed.

D)   For Equation S13, a numerical value for PEF can be calculated using Equation S15. The remaining parameters in Equation S13 have either default values listed in Appendix C, Table B or toxicological-specific information (i.e., URF), which can be obtained from IRIS or requested from the program under which the remediation is being performed.

E)   For Equation S14, a numerical value for PEF' can be calculated using Equation S16. The remaining parameters in Equation S14 have either default values listed in Appendix C, Table B or toxicological-specific information (i.e., URF), which can be obtained from IRIS or requested from the program under which the remediation is being performed.

d)   Soil Component of the Groundwater Ingestion Exposure Route

The Tier 2 remediation objective for the soil component of the groundwater ingestion exposure route can be calculated using one of the following equations based on the information known about the contaminant source and receptor population:

1)   Equation S17 is used to calculate the remediation objective assuming an infinite source of contamination.

A)   The numerical quantities for four parameters in Equation S17, the Target Soil Leachate Concentration (C_w), Soil-Water Partition Coefficient (K_d) for non-ionizing organics, Water-Filled Soil Porosity (θ_w) and Air-Filled Soil Porosity (θ_a), are calculated using Equations S18, S19, S20 and S21, respectively. Equations S22, S23, S24 and S25 are also needed to calculate numerical values for Equations S18 and S21. The pH-dependent K_d values for ionizing organics can be calculated using Equation S19 and the pH-dependent K_{oc} values in Appendix C, Table I.

B)   The remaining parameters in Equation S17 are Henry's Law Constant (H'), a chemical specific value listed in Appendix C, Table E and Dry Soil Bulk Density (ρ_b), a site-specific based value listed in Appendix C, Table B.

C)   The default value for GW_{obj} is the Tier 1 groundwater objective. For chemicals for which there is no Tier 1 groundwater remediation objective, the value for GW_{obj} shall be the Health Advisory concentration determined according to the procedures specified in 35 Ill. Adm. Code 620, Subpart F. As an alternative to using Tier 1 groundwater remediation objectives or Health Advisory concentrations, GW_{obj} may be developed using Equations R25 and R26, if approved institutional controls are in place as required in Subpart J

2)   If the area and depth of the contaminant source are known or can be estimated reliably, mass limit considerations may be used to calculate the remediation objective for this exposure route using Equation S28. The parameters in Equation S28 have default values listed in Appendix C, Table B.

Section 742.715   RBCA Soil Equations

a)   This Section presents the RBCA model and describes the equations and parameters used to develop Tier 2 soil remediation objectives.

b)   Ingestion, Inhalation, and Dermal Contact

1)   The two sets of equations in subsections (b)(2) and (b)(3) of this Section shall be used to generate Tier 2 soil remediation objectives for the combined ingestion, inhalation, and dermal contact with soil exposure routes.

2)   Combined Exposure Routes of Soil Ingestion, Inhalation of Vapors and Particulates, and Dermal Contact with Soil

A)   Equations R1 and R2 form the basis for deriving Tier 2 remediation objectives for the set of equations that evaluates the combined exposure routes of soil ingestion, inhalation of vapors and particulates, and dermal contact with soil using the RBCA approach. Equation R1 is used to calculate soil remediation objectives for carcinogenic contaminants. Equation R2 is used to calculate soil remediation objectives for noncarcinogenic contaminants. Soil remediation objectives for the ambient vapor inhalation (outdoor) route from subsurface soils must also be calculated in accordance with the procedures outlined in subsection (b)(3) of this Section and compared to the values generated from Equations R1 or R2. The smaller value (i.e., R1 and R2 compared to R7 and R8, respectively) from these calculations is the Tier 2 soil remediation objective for the combined exposure routes of soil ingestion, inhalation, and dermal contact with soil.

B)   In Equation R1, numerical values are calculated for two parameters:

   i)   The volatilization factor for surficial soils (VF_{ss}) using Equations R3 and R4; and

   ii)   The volatilization factor for subsurface soils regarding particulates (VF_p) using Equation R5.

C) VF_{ss} uses Equations R3 and R4 to derive a numerical value. Equation R3 requires the use of Equation R6. Both equations must be used to calculate the VF_{ss}. The lowest calculated value from these equations must be substituted into Equation R1.

D)   The remaining parameters in Equation R1 have either default values listed in Appendix C, Table D or toxicological-specific information (i.e., SF_o, SF_i), which can be obtained from IRIS or requested from the program under which the remediation is being performed.

E)   For Equation R2, the parameters VF_{ss} and VF_p are calculated. The remaining parameters in Equation R2 have either default values listed in Appendix C, Table D or toxicological-specific information (i.e., RfD_o, RfD_i), which can be obtained from IRIS or requested from the program under which the remediation is being performed.

F)   For chemicals other than inorganics which do not have default values for the dermal absorption factor (RAF_d) in Appendix C, Table D, a dermal absorption factor of 0.5 shall be used for Equations R1 and R2. For inorganics, dermal absorption may be disregarded (i.e., $RAF_d = 0$).

3)   Ambient Vapor Inhalation (outdoor) route from Subsurface Soils (soil below one meter)

A)   Equations R7 and R8 form the basis for deriving Tier 2 remediation objectives for the ambient vapor inhalation (outdoor) route from subsurface soils using the RBCA approach. Equation R7 is used to calculate soil remediation objectives for carcinogenic contaminants. Equation R8 is used to calculate soil remediation objectives for noncarcinogenic contaminants.

B)   For Equation R7, the carcinogenic risk-based screening level for air (RBSL$_{air}$) and the volatilization factor for soils below one meter to ambient air (VF$_{samb}$) have numerical values that are calculated using Equations R9 and R11, respectively. Both equations rely on input parameters from a variety of sources.

C)   The noncarcinogenic risk-based screening level for air (RBSL$_{air}$) and the volatilization factor for soils below one meter to ambient air (VF$_{samb}$) in Equation R8 have numerical values that can be calculated using Equations R10 and R11, respectively.

c)   Soil Component of the Groundwater Ingestion Exposure Route

1)   Equation R12 forms the basis for deriving Tier 2 remediation objectives for the soil component of the groundwater ingestion exposure route using the RBCA approach. The parameters, groundwater at the source (GW$_{source}$), and Leaching Factor (LF$_{sw}$), have numerical values that are calculated using Equations R13 and R14, respectively.

2)   Equation R13 requires numerical values that are calculated using Equation R15.

3)   Equation R14 requires numerical values that are calculated using Equations R21, R22, and R24. For non-ionizing organics, the Soil Water Sorption Coefficient (k$_s$) shall be calculated using Equation R20. For ionizing organics and inorganics, the values for k$_s$ are listed in Appendix C, Tables I and J, respectively. The pH-dependent k$_s$ values for ionizing organics can be calculated using Equation R20 and the pH-dependent K$_{oc}$ values in Appendix C, Table I. The remaining parameters in Equation R14 are field measurements or default values listed in Appendix C, Table D.

d)   The default value for GW$_{comp}$ is the Tier 1 groundwater remediation objective. For chemicals for which there is no Tier 1 groundwater remediation objective, the value for GW$_{comp}$ shall be the Health Advisory concentration determined according to the procedures specified in 35 Ill. Adm. Code 620, Subpart F. As an alternative to using the Tier 1 groundwater remediation objectives or Health Advisory concentrations, GW$_{comp}$ may be developed using Equations R25 and R26, if approved institutional controls are in place as may be required in Subpart J.

Section 742.720   Chemicals with Cumulative Noncarcinogenic Effects

Appendix A, Table E lists the groups of chemicals from Appendix B, Tables A and B that have remediation objectives based on noncarcinogenic toxicity and that affect the same target organ. If more than one chemical detected at a site affects the same target organ (i.e., has the same critical effect as defined by the RfD), the initially calculated remediation value for each chemical in the group shall be corrected for cumulative effects by one of the following two methods:

a)   Calculate the weighted average using the following equations:

$$W_{ave} = \frac{x_1}{CUO_{x_1}} + \frac{x_2}{CUO_{x_2}} + \frac{x_3}{CUO_{x_3}} + \ldots + \frac{x_a}{CUO_{x_a}}$$

where:

W_{ave} = Weighted Average

x_1 through x_a =   Concentration of each individual contaminant at the location of concern. Note that, depending on the target organ/mode of action, the actual number of contaminants will range from 2 to 14.

CUO_{x_a} =   A Tier 2 remediation objective must be developed for each x_a.

If the value of the weighted average calculated in accordance with the equations above is less than or equal to 1.0, then the remediation objectives are met for those chemicals.

If the value of the weighted average calculated in accordance with the equations above is greater than 1.0, then additional remediation must be carried out until the level of contaminants remaining in the remediated area has a weighted average calculated in accordance with the equation above less than or equal to one.

b)   Divide each individual chemical's remediation objective by the number of chemicals in that specific target organ group that were detected at the site. Each of the contaminant concentrations at the site is then compared to the remediation objectives that have been adjusted to account for this potential additivity. For the noncarcinogenic contaminants listed in Appendix A, Table E, a respective soil remediation objective need be no lower than the respective value listed in Appendix B, Table A or B.

SUBPART H: TIER 2 GROUNDWATER EVALUATION

Section 742.800   Tier 2 Groundwater Evaluation Overview

If the contaminant concentrations in the groundwater exceed the applicable Tier 1 remediation objectives, a person has the following options:

a)   Demonstrate that the groundwater ingestion exposure route is excluded from consideration pursuant to Subpart C;

b)   Demonstrate that the groundwater contamination is at or below area background concentrations in accordance with Subpart D and, if necessary, an institutional control restricting usage of the groundwater is in place in accordance with Subpart J;

c)   Remediate to Tier 1 remediation objectives;

d)   Propose and obtain approval of Tier 2 groundwater remediation objectives in accordance with Section 742.805 and remediate to that level, if necessary;

e)   Conduct a Tier 3 evaluation in accordance with Subpart I; or

f)   Obtain approval from the Board to:

   1)   Reclassify the groundwater pursuant to 35 Ill. Adm. Code 620.260; or

   2)   Use an adjusted standard pursuant to Section 28.1 of the Act.   [415 ILCS 5/28.1].

Section 742.805   Tier 2 Groundwater Remediation Objectives

a)   To develop a groundwater remediation objective under this Section that exceeds the applicable Tier 1 groundwater remediation objective, a person may request approval from the Agency if the person has performed the following:

   1)   Identified the horizontal and vertical extent of groundwater for which the Tier 2 groundwater remediation objective is sought;

   2)   Taken corrective action, to the maximum extent practicable to remove any free product;

   3)   Using Equation R26 in accordance with Section 742.810, demonstrated that the concentration of any contaminant of concern in groundwater will meet:

A)   The applicable Tier 1 groundwater remediation objective at the point of human exposure; or

B)   For any contaminant of concern for which there is no Tier 1 groundwater remediation objective, the Health Advisory concentration determined according to the procedures specified in 35 Ill. Adm. Code 620, Subpart F at the point of human exposure. A person may request the Agency to provide these concentrations or may propose these concentrations under Subpart I;.

4)   Using Equation R26 in accordance with Section 742.810, demonstrated that the concentration of any contaminant of concern in groundwater within the minimum or designated maximum setback zone of an existing potable water supply well will meet the applicable Tier 1 groundwater remediation objective or if there is no Tier 1 groundwater remediation objective, the Health Advisory concentration;

5)   Using Equation R26 in accordance with Section 742.810, demonstrated that the concentration of any contaminant of concern in groundwater discharging into a surface water will meet the applicable water quality standard under 35 Ill. Adm. Code 302;

6)   Demonstrated that the source of the release is not located within the minimum or designated maximum setback zone or within a regulated recharge area of an existing potable water supply well; and

7)   If the selected corrective action includes an engineered barrier as set forth in Subpart K to minimize migration of contaminant of concern from the soil to the groundwater, demonstrated that the engineered barrier will remain in place for post-remediation land use through an institutional control as set forth in Subpart J.

b)   A groundwater remediation objective that exceeds the water solubility of that chemical (refer to Appendix C, Table E for solubility values) is not allowed.

c)   Groundwater remediation objectives for chemicals which affect the same target organ, organ system or similar mode of action shall meet the requirements of Section 742.505(b)(3). Contaminants of concern for which a Tier 1 remediation objective has been developed shall be included in any mixture of similar-acting substances under consideration in Tier 2.

Section 742.810   Calculations to Predict Impacts from Remaining Groundwater
Contamination

a)   Equation R26 predicts the contaminant concentration along the centerline of a
        plume emanating from a vertical planar source in the aquifer (dimensions S_w
        wide and S_d deep). This model accounts for both three-dimensional dispersion
        (x is the direction of groundwater flow, y is the other horizontal direction, and z
        is the vertical direction) and biodegradation.

   1)   The parameters in this equation are:

            $X =$ distance from the planar source to the location of
                           concern, along the centerline of the plume (i.e., $y=0$,
                           $z=0$)

            $C_x =$ the concentration of the contaminant at a distance X
                           from the source, along the centerline of the plume

            $C_{source} =$ the greatest potential concentration of the contaminant
                           of concern in the groundwater at the source of the
                           contamination, based on the concentrations of
                           contaminants in groundwater due to the release and
                           the projected concentration of the contaminant
                           migrating from the soil to the groundwater. As
                           indicated above, the model assumes a planar source
                           discharging groundwater at a concentration equal to
                           C_{source}

            $\alpha_x =$ dispersivity in the x direction (i.e., Equation R16)

            $\alpha_y =$ dispersivity in the y direction (i.e., Equation R17)

            $\alpha_z =$ dispersivity in the z direction (i.e., Equation R18)

            $U =$ specific discharge (i.e., actual groundwater flow
                           velocity through a porous medium; takes into account
                           the fact that the groundwater actually flows only
                           through the pores of the subsurface materials) where
                           the aquifer hydraulic conductivity (K), the hydraulic
                           gradient (I) and the total soil porosity (θ_T) must be
                           known (i.e., Equation R19)

$\lambda =$ first order degradation constant obtained from Appendix C, Table E or from measured groundwater data

$S_w =$ width of planar source in the y direction

$S_d =$ depth of planar source in the z direction

2)   The following parameters are determined through field measurements: U, K, I, θ_T, S_w, S_d.

    A)   The determination of values for U, K, I and θ_T can be obtained through the appropriate laboratory and field techniques;

    B)   From the immediate down-gradient edge of the source of the groundwater contamination values for S_w and S_d shall be determined. S_w is defined as the width of groundwater at the source which exceeds the Tier 1 groundwater remediation objective. S_d is defined as the depth of groundwater at the source which exceeds the Tier 1 groundwater remediation objective; and

    C)   Total soil porosity can also be calculated using Equation R23.

b)   Once values are obtained for all the input parameters identified in subsection (a) of this Section, the contaminant concentration along the centerline of the plume at a distance X from the source shall be calculated such that that distance from the down-gradient edge of the source of the contamination at the site to the point where the contaminant concentration is equal to the Tier 1 groundwater remediation objective or Health Advisory concentration.

1)   If there are any potable water supply wells located within the calculated distance X, then the Tier 1 groundwater remediation objective or Health Advisory concentration shall be met at the edge of the minimum or designated maximum setback zone of the nearest potable water supply well down-gradient of the source. If no potable water supply wells exist within the calculated distance X, then it can be determined that no existing potable water supply wells are adversely impacted.

2)   To demonstrate that no surface water is adversely impacted, X shall be the distance from the down-gradient edge of the source of the contamination at the site to the nearest surface water body. This calculation must show that the contaminant in the groundwater at this location (C_x) does not exceed the applicable water quality standard.

SUBPART I: TIER 3 EVALUATION

Section 742.900   Tier 3 Evaluation Overview

a)   Tier 3 sets forth a flexible framework to develop remediation objectives outside
     of the requirements of Tiers 1 and 2. Although Tier 1 and Tier 2 evaluations
     are not prerequisites to conduct Tier 3 evaluations, data from Tier 1 and Tier 2
     can assist in developing remediation objectives under a Tier 3 evaluation.

b)   The level of detail required to adequately characterize a site depends on the
     particular use of Tier 3. Tier 3 can require additional investigative efforts
     beyond those described in Tier 2 to characterize the physical setting of the site.
     However, in situations where remedial efforts have simply reached a physical
     obstruction additional investigation may not be necessary for a Tier 3 submitta[1]

c)   Situations that can be considered for a Tier 3 evaluation include, but are not
     limited to:

     1)   Modification of parameters not allowed under Tier 2;

     2)   Use of models different from those used in Tier 2;

     3)   Use of additional site data to improve or confirm predictions of exposed
          receptors to contaminants of concern;

     4)   Analysis of site-specific risks using formal risk assessment, probabilistic
          data analysis, and sophisticated fate and transport models (e.g.,
          requesting a target hazard quotient greater than 1 or a target cancer risk
          greater than 1 in 1,000,000);

     5)   Requests for site-specific remediation objectives because an assessment
          indicates further remediation is not practical;

     6)   Incomplete human exposure pathway(s) not excluded under Subpart C;

     7)   Use of toxicological-specific information not available from the sources
          listed in Tier 2;

     8)   Land uses which are substantially different from the assumed residential
          or industrial/commercial property uses of a site (e.g., a s site will be
          used for recreation in the future and cannot be evaluated in Tiers 1 or 2);
          and

    9)   Requests for site-specific remediation objectives which exceed Tier 1 groundwater remediation objectives so long as the following is demonstrated:

        A) TO THE EXTENT PRACTICAL, THE EXCEEDANCE OF THE GROUNDWATER QUALITY STANDARD HAS BEEN MINIMIZED AND BENEFICIAL USE APPROPRIATE TO THE GROUNDWATER THAT WAS IMPACTED HAS BEEN RETURNED; AND

        B) ANY THREAT TO HUMAN HEALTH OR THE ENVIRONMENT HAS BEEN MINIMIZED. (Section 58.5(D)(4)(A) of the Act)

d)   For requests of a target cancer risk ranging between 1 in 1,000,000 and 1 in 10,000 at the point of human exposure or a target hazard quotient greater than 1 at the point of human exposure, the requirements of Section 742.915 shall be followed. Requests for a target cancer risk exceeding 1 in 10,000 at the point of human exposure are not allowed.

e)   Requests for approval of a Tier 3 evaluation must be submitted to the Agency for review under the specific program under which remediation is performed. When reviewing a submittal under Tier 3, the Agency shall consider WHETHER THE INTERPRETATIONS AND CONCLUSIONS REACHED ARE SUPPORTED BY THE INFORMATION GATHERED. (Section 58.7(e)(1) of the Act) The Agency shall approve a Tier 3 evaluation if the person submits the information required under this Part and establishes through such information that public health is protected and that specified risks to human health and the environment have been minimized.

Section 742.905   Modifications of Parameters

Any proposed changes to Tier 2 parameters which are not provided for in Tier 2 shall be submitted to the Agency for review and approval. A submittal under this Section shall include the following information:

a)   The justification for the modification; and

b)   The technical and mathematical basis for the modification.

Section 742.910 Alternative Models

Any proposals for the use of models other than those specified in Tier 2 shall be submitted to the Agency for review and approval. A submittal under this Section shall include the following information:

a)   Physical and chemical properties of contaminants of concern;

b)   Contaminant movement properties;

c)   Contaminant availability to receptors;

d)   Receptor exposure to the contaminants of concern;

e)   Mathematical and technical justification for the model proposed;

f)   A licensed copy of the model, if the Agency does not have a licensed copy of the model currently available for use; and

g)   Demonstration that the models were correctly applied.

Section 742.915   Formal Risk Assessments

A comprehensive site-specific risk assessment shall demonstrate that contaminants of concern at a site do not pose a significant risk to any human receptor. All site-specific risk assessments shall be submitted to the Agency for review and approval. A submittal under this Section shall address the following factors:

a)   Whether the risk assessment procedure used is nationally recognized and accepted including, but not limited to, those procedures incorporated by reference in Section 742.210;

b)   Whether the site-specific data reflects actual site conditions;

c)   The adequacy of the investigation of present and post-remediation exposure routes and risks to receptors identified at the site;

d)   The appropriateness of the sampling and analysis;

e)   The adequacy and appropriateness of toxicity information;

f)   The extent of contamination;

g)   Whether the calculations were accurately performed; and

h)   Proposals seeking to modify the target risk consistent with Section 742.900(d) shall address the following factors:

1)   the presence of sensitive populations;

2)   the number of receptors potentially impacted;

3)   the duration of risk at the differing target levels; and

4)   the characteristics of the chemical of concern.

Section 742.920   Impractical Remediation

Any request for site-specific remediation objectives due to impracticality of remediation shall be submitted to the Agency for review and approval. A submittal under this Section shall include the following information:

a)   The reason(s) why the remediation is impractical;

b)   The extent of contamination;

c)   Geology, including soil types;

d)   The potential impact to groundwater;

e)   Results and locations of sampling events;

f)   Map of the area, including all utilities and structures; and

g)   Present and post-remediation uses of the area of contamination, including human receptors at risk.

Section 742.925   Exposure Routes

Technical information may demonstrate that there is no actual or potential impact of contaminants of concern to receptors from a particular exposure route. In these instances, a demonstration excluding an exposure route shall be submitted to the Agency for review and approval. A submittal under this Section shall include the following information:

a)   A description of the route evaluated;

b)   Technical support including a discussion of the natural or man-made barriers to exposure through that route, calculations, and modeling results;

c)   Physical and chemical properties of contaminants of concern;

d)   Contaminant migration properties;

e) Description of the site and physical site characteristics; and

f) Discussion of the result and possibility of the route becoming active in the future.

Section 742.930   Derivation of Toxicological Data

If toxicological-specific information is not available for one or more contaminants of concern from the sources incorporated by reference in Section 742.210, the derivations of toxicological-specific information shall be submitted for Agency review and approval.

SUBPART J: INSTITUTIONAL CONTROLS

Section 742.1000   Institutional Controls

a) Institutional controls in accordance with this Subpart must be placed on the property when remediation objectives are based on any of the following assumptions:

   1) Industrial/Commercial property use;

   2) Target cancer risk greater than 1 in 1,000,000;

   3) Target hazard quotient greater than 1;

   4) Engineered barrier(s);

   5) The point of human exposure is located at a place other than at the source;

   6) Exclusion of exposure routes under Subpart C; or

   7) Any combination of the above.

b) The Agency shall not approve any remediation objective under this Part that is based on the use of institutional controls unless the person has proposed institutional controls meeting the requirements of this Subpart and the requirements of the specific program under which the institutional control is proposed. A proposal for approval of institutional controls shall provide identification of the selected institutional controls from among the types recognized in this Subpart.

c)   The following instruments may be institutional controls, subject to the requirements of this Subpart J and the requirements of the specific program under which the institutional control is proposed:

   1)   No Further Remediation Letters;

   2)   Restrictive covenants and deed restrictions;

   3)   Negative easements;

   4)   Ordinances adopted and administered by a unit of local government; and

   5)   Agreements between a property owner and a highway authority with respect to any contamination remaining under highways.

d)   An institutional control is transferred with the property.

Section 742.1005 No Further Remediation Letters

a)   A No Further Remediation Letter issued by the Agency under 35 Ill. Adm. Code 732 or 742 may be used as an institutional control under this Part if the requirements of subsection (b) of this Section are met.

b)   A request for approval of a No Further Remediation Letter as an institutional control shall meet the requirements applicable to the specific program under which the remediation is performed.

Section 742.1010 Restrictive Covenants, Deed Restrictions and Negative Easements

a)   A restrictive covenant, deed restriction or negative easement may be used as an institutional control under this Part if the requirements of this Section are met and the Agency has determined that no further remediation is required as to the property(ies) to which the institutional control is to apply.

b)   A request for approval of a restrictive covenant, deed restriction or negative easement as an acceptable institutional control shall provide the following:

   1)   A copy of the restrictive covenant, deed restriction, or negative easement in the form it will be recorded with the Office of the Recorder or Registrar of Titles in the county where the site is located;

   2)   A scaled map showing the horizontal extent of contamination above the applicable remediation objectives;

3)   Information showing the concentration of contaminants of concern in which the applicable remediation objectives are exceeded;

4)   A scaled map showing the legal boundaries of all properties under which contamination is located that exceeds the applicable remediation objectives and which are subject to the restrictive covenant, deed restriction, or negative;

5)   Information identifying the current owner(s) of each property identified in subsection (b)(4) of this Section; and

6)   Authorization by the current owner(s), or person authorized by law to act on behalf of the owner, of each property identified in subsection (b)(5) of this Section to record the restrictive covenant or deed restriction.

c)   Any restrictive covenant, deed restriction, or negative easement approved by the Agency pursuant to this Section shall be recorded in the Office of the Recorder or Registrar of Titles of the county in which the site is located together with the instrument memorializing the Agency's no further remediation determination pursuant to the specific program within 45 days after receipt of the Agency's no further remediation determination.

d)   An institutional control approved under this Section shall not become effective until officially recorded in accordance with subsection (c) of this Section. The person receiving the approval shall obtain and submit to the Agency within 30 days after recording a copy of the institutional control demonstrating that it has been recorded.

e)   At no time shall any site for which land use has been restricted under an institutional control approved under this Section be used in a manner inconsistent with such land use limitation unless further investigation or remedial action has been conducted that documents the attainment of remediation objectives appropriate for such land use and a new institutional control, if necessary, is approved and recorded in accordance with subsection (c) of this Section.

f)   Violation of the terms of an institutional control approved under this Section shall be grounds for voidance of the institutional control and the instrument memorializing the Agency's no further remediation determination.

Section 742.1015   Ordinances

a)   An ordinance adopted by a unit of local government that effectively prohibits the installation of potable water supply wells (and the use of such wells) may be used as an institutional control to meet the requirements of Section 742.320(d) or 742.805(a)(3) if the requirements of this Section are met. Ordinances prohibiting the installation of potable water supply wells (and the use of such wells) that do not expressly prohibit the installation of potable water supply wells (and the use of such wells) by units of local government may be acceptable as institutional controls if the requirements of this Section are met and a Memorandum of Understanding (MOU) is entered into under subsection (i) of this Section.

b)   A request for approval of a local ordinance as an institutional control shall provide the following:

   1)   A copy of the ordinance restricting groundwater use certified by an official of the unit of local government in which the site is located that it is the latest, most current copy of the ordinance, unless the Agency and the unit of local government have entered an agreement under subsection (i) of this Section, in which case the request may alternatively reference the MOU. The ordinance must demonstrate that potable use of groundwater from potable water supply wells is prohibited;

   2)   A scaled map(s) delineating the areal extent of groundwater contamination (measured or modeled) above the applicable remediation objectives;

   3)   Information showing the concentration of contaminants of concern in which the applicable remediation objectives are exceeded;

   4)   A scaled map delineating the boundaries of all properties under which groundwater is located which exceeds the applicable groundwater remediation objectives;

   5)   Information identifying the current owner(s) of each property identified in subsection (b)(4) of this Section; and

   6)   A copy of the proposed submission of the information to the current owners identified in subsection (b)(5) of this Section of the information required in subsections (b)(1) through (b)(5) of this Section and proof that the notification required in subsection (c) of this Section has been submitted.

c)   Each of the property owners identified in subsection (b)(5) of this Section and
      the unit of local government must receive written notification from the party
      desiring to use the institutional control that groundwater remediation objectives
      have been approved by the Agency. Written proof of this notification shall be
      submitted to the Agency within 45 days from the date of the instrument
      memorializing the Agency's no further remediation determination. The
      notification shall include:

      1)   The name and address of the unit of local government;

      2)   The citation to the ordinance;

      3)   A description of the property being sent notice by adequate legal
            description or by reference to a plat showing the boundaries;

      4)   A statement that the ordinance restricting groundwater use has been used
            by the Agency in reviewing a request for a groundwater remediation
            objective;

      5)   A statement as to the nature of the release and response action with the
            site name, address, and Agency site number or Illinois inventory
            identification number; and

      6)   A statement as to where more information may be obtained regarding the
            ordinance.

d)   Unless the Agency and the unit of local government have entered into a MOU
      under subsection (i) of this Section, the current owner or successors in interest
      of a site who have received approval of use of an ordinance as an institutional
      control under this Section shall:

      1)   Monitor activities of the unit of local government relative to variance
            requests or changes in the ordinance relative to the use of potable
            groundwater at properties identified in subsection (b)(4) of this Section;
            and

      2)   Notify the Agency of any approved variance requests or ordinance
            changes within 30 days after the date such action has been approved.

e)   The information required in subsections (b)(1) through (b)(6) of this Section and
      the Agency letter approving the groundwater remediation objective shall be
      submitted to the unit of local government. Proof that the information has been
      filed with the unit of local government shall be provided to the Agency.

f)   Any ordinance or MOU used as an institutional control pursuant to this Section shall be recorded in the Office of the Recorder or Registrar of Titles of the county in which the site is located together with the instrument memorializing the Agency's no further remediation determination pursuant to the specific program within 45 days after receipt of the Agency's no further remediation.

g)   An institutional control approved under this Section shall not become effective until officially recorded in accordance with subsection (f) of this Section. The person receiving the approval shall obtain and submit to the Agency within 30 days after recording a copy of the institutional control demonstrating that it has been recorded.

h)   The following shall be grounds for voidance of the ordinance as an institutional control and the instrument memorializing the Agency's no further remediation determination:

   1)   Modification of the ordinance by the unit of local government to allow potable use of groundwater;

   2)   Approval of a site-specific request, such as a variance, to allow potable use of groundwater at a site identified in subsection (b)(4) of this Section; or

   3)   Violation of the terms of an institutional control recorded under Section 742.1005 or Section 742.1010.

i)   The Agency and a unit of local government may enter into a MOU under this Section if the unit of local government has adopted an ordinance satisfying subsection (a) of this Section and if the requirements of this subsection are met. The MOU shall include the following:

   1)   Identification of the authority of the unit of local government to enter the MOU;

   2)   Identification of the legal boundaries, or equivalent, under which the ordinance is applicable;

   3)   A certified copy of the ordinance;

   4)   A commitment by the unit of local government to notify the Agency of any variance requests or proposed ordinance changes at least 30 days prior to the date the local government is scheduled to take action on the request or proposed change;

5)   A commitment by the unit of local government to maintain a registry of all sites within the unit of local government that have received no further remediation determinations pursuant to specific programs and

6)   If the ordinance does not expressly prohibit the installation of potable water supply wells (and the use of such wells) by units of local government, a commitment by the unit of local government:

   A)   To review the registry of sites established under subsection (i)(5) of this Section prior to siting potable water supply wells within the area covered by the ordinance;

   B)   To determine whether the potential source of potable water may be or has been affected by contamination left in place at those sites; and

   C)   To take whatever steps are necessary to ensure that the potential source of potable water is protected from the contamination or treated before it is used as a potable water supply.

Section 742.1020 Highway Authority Agreements

a)   An agreement with a highway authority may be used as an institutional control where the requirements of this Section are met and the Agency has determined that no further remediation is required as to the property(ies) to which the agreement is to apply.

b)   As part of the agreement the highway authority shall agree to:

   1)   Prohibit the use of groundwater under the highway right of way that is contaminated above residential Tier 1 remediation objectives from the release as a potable supply of water.

   2)   Limit access to soil contamination under the highway right of way that is contaminated above residential Tier 1 remediation objectives from the release. Access to soil contamination may be allowed if, during and after any access, public health and the environment are protected.

c)   A request for approval of an agreement as an institutional control shall provide the following:

   1)   A copy of the agreement executed by the highway authority and the owner of the property from which the release occurred;

2) A scaled map delineating the areal extent of soil and groundwater contamination above the applicable Tier 1 remediation objectives;

3) Information showing the concentration of contaminants of concern within the zone in which the applicable Tier 1 remediation objectives are exceeded;

4) A stipulation of the information required by subsection (b) of this Section in the agreement if it is not practical to obtain the information by sampling the highway right-of-way; and

5) Information identifying the current fee owner of the highway right-of-way and highway authority having jurisdiction.

d) Violation of the terms of an Agreement approved by the Agency as an institutional control under this Section shall be grounds for voidance of the Agreement as an institutional control and the instrument memorializing the Agency's no further remediation determination.

SUBPART K: ENGINEERED BARRIERS

Section 742.1100 Engineered Barriers

a) Any person who develops remediation objectives under this Part based on engineered barriers shall meet the requirements of this Subpart and the requirements of Subpart J relative to institutional controls.

b) The Agency shall not approve any remediation objective under this Part that is based on the use of engineered barriers unless the person has proposed engineered barriers meeting the requirements of this Subpart.

c) The use of engineered barriers can be recognized in calculating remediation objectives only if the engineered barriers are intended for use as part of the final corrective action.

d) Any no further remediation determination based upon the use of engineered barriers shall require effective maintenance of the engineered barrier. The maintenance requirements shall be included in an institutional control under Subpart J. This institutional control shall address provisions for temporary breaches of the barrier by requiring the following if intrusive construction work is to be performed in which the engineered barrier is to be temporarily breached:

1)   The construction workers shall be notified by the site owner/operator in advance of intrusive activities. Such notification shall enumerate the contaminant of concern known to be present; and

2)   The site owner/operator shall require construction workers to implement protective measures consistent with good industrial hygiene practice.

e)   Failure to maintain an engineered barrier in accordance with the no further remediation determination shall be grounds for voidance of that determination and the instrument memorializing the Agency's no further remediation determination.

Section 742.1105 Engineered Barrier Requirements

a)   Natural attenuation, access controls, and point of use treatment shall not be considered engineered barriers. Engineered barriers may not be used to prevent direct human exposure to groundwater without the use of institutional controls.

b)   For purposes of determining remediation objectives under Tier 1, engineered barriers are not recognized.

c)   The following engineered barriers are recognized for purposes of calculating remediation objectives that exceed residential remediation objectives:

1)   For the soil component of the groundwater ingestion exposure route, the following engineered barriers are recognized:

A)   Caps, covering the contaminated media, constructed of compacted clay, asphalt, concrete or other material approved by the Agency; and

B)   Permanent structures such as buildings and highways.

2)   For the soil ingestion exposure route, the following engineered barriers are recognized:

A)   Caps, covering the contaminated media, constructed of compacted clay, asphalt, concrete, or other material approved by the Agency;

B)   Permanent structures such as buildings and highways; and

C)   Clean soil, covering the contaminated media, that is a minimum of 3 feet in depth.

3)   For the inhalation exposure route, the following engineered barriers are recognized:

A)   Caps, covering the contaminated media, constructed of compacted clay, asphalt, concrete, or other material approved by the Agency;

B)   Permanent structures such as buildings and highways; and

C)   Clean soil covering the contaminated media, that is a minimum of 10 feet in depth and not within 10 feet of any manmade pathway.

4)   For the ingestion of groundwater exposure route, the following engineered barriers are recognized:

A)   Slurry walls; and

B)   Hydraulic control of groundwater.

d) Unless otherwise prohibited under Section 742.1100, any other type of engineered barrier may be proposed if it will be as effective as the options listed in subsection (c) of this Section.

Section 742.APPENDIX A: General

Section 742.Illustration A: Developing Soil Remediation Objectives Under the Tiered Approach

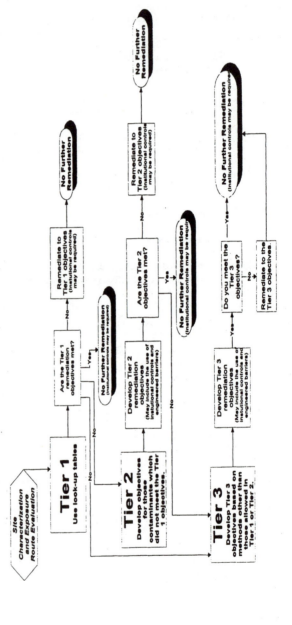

Section 742.APPENDIX A: General

Section 742.Illustration B: Developing Groundwater Remediation Objectives Under the Tiered Approach

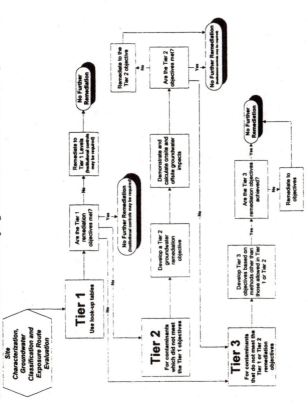

Section 742.APPENDIX A: General

Section 742.TABLE A: Soil Saturation Limits (C_{sat})for Chemicals Whose Melting Point is Less than 30° C

CAS No.	Chemical Name	C_{sat} (mg/kg)
67-64-1	Acetone	100,000
71-43-2	Benzene	870
111-44-4	Bis(2-chloroethyl)ether	3,300
117-81-7	Bis(2-ethylhexyl)phthalate	31,000
75-27-4	Bromodichloromethane (Dichlorobromomethane)	3,000
75-25-2	Bromoform	1,900
71-36-3	Butanol	10,000
85-68-7	Butyl benzyl phthalate	930
75-15-0	Carbon disulfide	720
56-23-5	Carbon tetrachloride	1,100
108-90-7	Chlorobenzene (Monochlorobenzene)	680
124-48-1	Chlorodibromomethane (Dibromochloromethane)	1,300
67-66-3	Chloroform	2,900
96-12-8	1,2-Dibromo-3-chloropropane	1,400
106-93-4	1,2-Dibromoethane (Ethylene dibromide)	2,800
84-74-2	Di-*n*-butyl phthalate	2,300
95-50-1	1,2-Dichlorobenzene (o-Dichlorobenzene)	560
75-34-3	1,1-Dichloroethane	1,700
107-06-2	1,2-Dichloroethane (Ethylene dichloride)	1,800
75-35-4	1,1-Dichloroethylene	1,500
156-59-2	*cis*-1,2-Dichloroethylene	1,200
156-60-5	*trans*-1,2-Dichloroethylene	3,100
78-87-5	1,2-Dichloropropane	1,100
542-75-6	1,3-Dichloropropene (1,3-Dichloropropylene, *cis* + *trans*)	1,400

CAS No.	Chemical Name	C_{sat} (mg/kg)
84-66-2	Diethyl phthalate	2,000
117-84-0	Di-*n*-octyl phthalate	10,000
100-41-4	Ethylbenzene	400
77-47-4	Hexachlorocyclopentadiene	2,200
78-59-1	Isophorone	4,600
74-83-9	Methyl bromide (Bromomethane)	3,200
75-09-2	Methylene chloride (Dichloromethane)	2,400
98-95-3	Nitrobenzene	1,000
100-42-5	Styrene	1,500
127-18-4	Tetrachloroethylene (Perchloroethylene)	240
108-88-3	Toluene	650
120-82-1	1,2,4-Trichlorobenzene	3,200
71-55-6	1,1,1-Trichloroethane	1,200
79-00-5	1,1,2-Trichloroethane	1,800
79-01-6	Trichloroethylene	1,300
108-05-4	Vinyl acetate	2,700
75-01-4	Vinyl chloride	1,200
108-38-3	m-Xylene	420
95-47-6	o-Xylene	410
106-42-3	p-Xylene	460
1330-20-7	Xylenes (total)	410
	Ionizable Organics	
95-57-8	2-Chlorophenol	53,000

Section 742. APPENDIX A: General

Section 742. TABLE B: Tolerance Factor (K)

Tolerance factors (K) for one-sided normal tolerance intervals with probability level (confidence factor) Y = 0.95 and coverage P = 95%. n = number of samples collected.

n	K		n	K
3	7.655		175	1.850
4	5.145		200	1.836
5	4.202		225	1.824
6	3.707		250	1.814
7	3.399		275	1.806
8	3.188		300	1.799
9	3.031		325	1.792
10	2.911		350	1.787
11	2.815		375	1.782
12	2.736		400	1.777
13	2.670		425	1.773
14	2.614		450	1.769
15	2.566		475	1.766
16	2.523		500	1.763
17	2.486		525	1.760
18	2.543		550	1.757
19	2.423		575	1.754
20	2.396		600	1.752
21	2.371		625	1.750
22	2.350		650	1.748
23	2.329		675	1.746
24	2.309		700	1.744
25	2.292		725	1.742
30	2.220		750	1.740
35	2.166		n	K
40	2.126			
45	2.092		775	1.739
50	2.065		800	1.737
55	2.036		825	1.736
60	2.017		850	1.734
65	2.000		875	1.733
70	1.986		900	1.732
75	1.972		925	1.731
100	1.924		950	1.729
125	1.891		975	1.728
150	1.868		1000	1.727

Section 742. APPENDIX A: General

Section 742. TABLE C: Coefficients $\{A_{N-I+1}\}$ for W Test of Normality, for $N=2(1)50$

i/n	2	3	4	5	6	7	8	9	10
1	0.7071	0.7071	0.6872	0.6646	0.6431	0.6233	0.6052	0.5888	0.5739
2	---	.0000	.1677	.2413	.2806	.3031	.3164	.3244	.3291
3	---	---	---	.0000	.0875	.1401	.1743	.1976	.2141
4	---	---	---	---	---	.0000	.0561	.0947	.1224
5	---	---	---	---	---	---	---	.0000	.0399

i/n	11	12	13	14	15	16	17	18	19	20
1	0.5601	0.5475	0.5359	0.5251	0.5150	0.5056	0.4968	0.4886	0.4808	0.4734
2	.3315	.3325	.3325	.3318	.3306	.3290	.3273	.3253	.3232	.3211
3	.2260	.2347	.2412	.2460	.2495	.2521	.2540	.2553	.2561	.2565
4	.1429	.1586	.1707	.1802	.1878	.1939	.1988	.2027	.2059	.2085
5	.0695	.0922	.1099	.1240	.1353	.1447	.1524	.1587	.1641	.1686
6	0.0000	.0303	.0539	.0727	.0880	.1005	.1109	.1197	.1271	.1334
7	---	---	.0000	.0240	.0433	.0593	.0725	.0837	.0932	.1013
8	---	---	---	---	.0000	.0196	.0359	.0496	.0612	.0711
9	---	---	---	---	---	---	.0000	.0163	.0303	.0422
10	---	---	---	---	---	---	---	---	.0000	.0140

i/n	21	22	23	24	25	26	27	28	29	30
1	0.4643	0.4590	0.4542	0.4493	0.4450	0.4407	0.4366	0.4328	0.4291	0.4254
2	.3185	.3156	.3126	.3098	.3069	.3043	.3018	.2992	.2968	.2944
3	.2578	.2571	.2563	.2554	.2543	.2533	.2522	.2510	.2499	.2487
4	.2119	.2131	.2139	.2145	.2148	.2151	.2152	.2151	.2150	.2148
5	.1736	.1764	.1787	.1807	.1822	.1836	.1848	.1857	.1864	.1870
6	0.1399	0.1443	0.1480	0.1512	0.1539	0.1563	0.1584	0.1601	0.1616	0.1630
7	.1092	.1150	.1201	.1245	.1283	.1316	.1346	.1372	.1395	.1415
8	.0804	.0878	.0941	.0997	.1046	.1089	.1128	.1162	.1192	.1219
9	.0530	.0618	.0696	.0764	.0823	.0876	.0923	.0965	.1002	.1036
10	.0263	.0368	.0459	.0539	.0610	.0672	.0728	.0778	.0822	.0862
11	0.0000	0.0122	0.0228	0.0321	0.0403	0.0476	0.0540	0.0598	0.0650	0.0697
12	----	----	0.0000	.0107	.0200	.0284	.0358	.0424	.0483	.0537
13	----	----	----	----	.0000	.0094	.0178	.0253	.0320	.0381
14	----	----	----	----	----	----	.0000	.0084	.0159	.0227
15	----	----	----	----	----	----	----	----	.0000	.0076

i/n	31	32	33	34	35	36	37	38	39	40
1	0.4220	0.4188	0.4156	0.4127	0.4096	0.4068	0.4040	0.4015	0.3989	0.3964
2	.2921	.2898	.2876	.2854	.2834	.2813	.2794	.2774	.2755	.2737
3	.2475	.2463	.2451	.2439	.2427	.2415	.2403	.2391	.2380	.2368
4	.2145	.2141	.2137	.2132	.2127	.2121	.2116	.2110	.2104	.2098
5	.1874	.1878	.1880	.1882	.1883	.1883	.1883	.1881	.1880	.1878

i/n	31	32	33	34	35	36	37	38	39	40
6	0.1641	0.1651	0.1660	0.1667	0.1673	0.1678	0.1683	0.1686	0.1689	0.1691
7	.1433	.1449	.1463	.1475	.1487	.1496	.1503	.1513	.1520	.1526
8	.1243	.1265	.1284	.1301	.1317	.1331	.1344	.1356	.1366	.1376
9	.1066	.1093	.1118	.1140	.1160	.1179	.1196	.1211	.1225	.1237
10	.0899	.0931	.0961	.0988	.1013	.1036	.1056	.1075	.1092	.1108
11	0.0739	0.0777	0.0812	0.0844	0.0873	0.0900	0.0924	0.0947	0.0967	0.0986
12	.0585	.0629	.0669	.0706	.0739	.0770	.0798	.0824	.0848	.0870
13	.0435	.0485	.0530	.0572	.0610	.0645	.0677	.0706	.0733	.0759
14	.0289	.0344	.0395	.0441	.0484	.0523	.0559	.0592	.0622	.0651
15	.0144	.0206	.0262	.0314	.0361	.0404	.0444	.0481	.0515	.0546

i	41	42	43	44	45	46	47	48	49	50
16	0.0000	0.0068	0.0131	0.0187	0.0239	0.0287	0.0331	0.0372	0.0409	0.0444
17	----	----	0000	0062	0119	0172	0220	0264	0305	0343
18	----	----	----	----	0000	0057	0110	0158	0203	0244
19	----	----	----	----	----	----	0000	0053	0101	0146
20	----	----	----	----	----	----	----	----	0000	0049

i/n	41	42	43	44	45	46	47	48	49	50
1	0.3940	0.3917	0.3894	0.3872	0.3850	0.3830	0.3808	0.3789	0.3770	0.3751
2	2719	2701	2684	2667	2651	2635	2620	2604	2589	2574
3	2357	2345	2334	2323	2313	2302	2291	2281	2271	2260
4	2091	2085	2078	2072	2065	2058	2052	2045	2038	2032
5	1876	1874	1871	1868	1865	1862	1859	1855	1851	1847

i/n	41	42	43	44	45	46	47	48	49	50
6	0.1693	0.1694	0.1695	0.1695	0.1695	0.1695	0.1695	0.1693	0.1692	0.1691
7	1531	1535	1539	1542	1545	1548	1550	1551	1553	1554
8	1384	1392	1398	1405	1410	1415	1420	1423	1427	1430
9	1249	1259	1269	1278	1286	1293	1300	1306	1312	1317
10	1123	1136	1149	1160	1170	1180	1189	1197	1205	1212

	0.1004	0.1020	0.1035	0.1049	0.1062	0.1073	0.1085	0.1095	0.1105	0.1113
11	0.1004	0.1020	0.1035	0.1049	0.1062	0.1073	0.1085	0.1095	0.1105	0.1113
12	.0891	.0909	.0927	.0943	.0959	.0972	.0986	.0998	.1010	.1020
13	.0782	.0804	.0824	.0842	.0860	.0876	.0892	.0906	.0919	.0932
14	.0677	.0701	.0724	.0745	.0775	.0785	.0801	.0817	.0832	.0846
15	.0575	.0602	.0628	.0651	.0673	.0694	.0713	.0731	.0748	.0764
16	0.0476	0.0506	0.0534	0.0560	0.0584	0.0607	0.0628	0.0648	0.0667	0.0685
17	.0379	.0411	.0442	.0471	.0497	.0522	.0546	.0568	.0588	.0608
18	.0283	.0318	.0352	.0383	.0412	.0439	.0465	.0489	.0511	.0532
19	.0188	.0227	.0263	.0296	.0328	.0357	.0385	.0411	.0436	.0459
20	.0094	.0136	.0175	.0211	.0245	.0277	.0307	.0335	.0361	.0386
21	0.0000	0.0045	0.0087	0.0126	0.0163	0.0197	0.0229	0.0259	0.0288	0.0314
22	----	----	.0000	.0042	.0081	.0118	.0153	.0185	.0215	.0244
23	----	----	----	----	.0000	.0039	.0076	.0111	.0143	.0174
24	----	----	----	----	----	----	.0000	.0037	.0071	.0104
25	----	----	----	----	----	----	----	----	.0000	.0035

Section 742.APPENDIX A: General

Section 742.TABLE D: Percentage Points of the W Test for N=3(1)50

n	0.01	0.05
3	0.753	0.767
4	0.687	0.748
5	0.686	0.762
6	0.713	0.788
7	0.730	0.803
8	0.749	0.818
9	0.764	0.829
10	0.781	0.842
11	0.792	0.850
12	0.805	0.859
13	0.814	0.866
14	0.825	0.874
15	0.835	0.881
16	0.844	0.887
17	0.851	0.892
18	0.858	0.897
19	0.863	0.901
20	0.868	0.905
21	0.873	0.908
22	0.878	0.911
23	0.881	0.914
24	0.884	0.916
25	0.888	0.918
26	0.891	0.920
27	0.894	0.923
28	0.896	0.924
29	0.898	0.926
30	0.900	0.927
31	0.902	0.929
32	0.904	0.930
33	0.906	0.931
34	0.908	0.933
35	0.910	0.934

Section 742.APPENDIX A: General

Section 742.TABLE E: Chemicals with Noncarcinogenic Toxic Effects on Specific Target Organs/Organ Systems or Similar Modes of Action

Kidney
Acetone
Cadmium (Ingestion only)
Chlorobenzene
Dalapon
1,1-Dichloroethane
Di-n-octyl phthalate
Endosulfan
Ethylbenzene
Fluoranthene
Nitrobenzene
Pyrene
Toluene
2,4,5-Trichlorophenol
Vinyl acetate

Liver
Acenaphthene
Acetone
Butylbenzyl phthalate
1,1-Dichloroethylene
Chlorobenzene
Di-n-octyl phthalate
Endrin
Ethylbenzene
Fluoranthene
Nitrobenzene
Picloram
Styrene
2,4,5-TP (Silvex)
Toluene
2,4,5-Trichlorophenol

Central Nervous System
Butanol
Cyanide (amenable)
2,4-Dimethylphenol
Endrin
Manganese
2-Methylphenol
Mercury
Styrene
Xylenes

Circulatory System
Antimony
Barium
2,4-D
cis-1,2-Dichloroethylene
Nitrobenzene
trans-1,2-Dichloroethylene
2,4-Dimethylphenol
Fluoranthene
Fluorene
Styrene
Zinc

Gastrointestinal System
Endothall
Hexachlorocyclopentadiene
Methyl bromide

Reproductive System
Barium
Boron
Carbon disulfide
2-Chlorophenol
1,2 Dibromo-3-Chloropropane (Inhalation only)
Dinoseb
Methoxychlor
Phenol

Cholinesterase Inhibition
Aldicarb
Carbofuran

Decreased Body Weight Gains
and Circulatory System Effects
Atrazine
Simazine

Adrenal Gland
Nitrobenzene
1,2,4-Trichlorobenzene

Respiratory System
1,2-Dichloropropane
Hexachlorocyclopentadiene
Methyl bromide
Vinyl acetate

Immune System
2,4-Dichlorophenol
p-Chloroaniline

Section 742.APPENDIX A: General

Section 742.TABLE F: Chemicals With Carcinogenic Toxic Effects on Specific Target Organs/Organ Systems or Similar Modes of Action

Kidney
Bromodichloromethane
Chloroform
1,2-Dibromo-3-chloropropane
2,4-Dinitrotoluene
2,6-Dinitrotoluene
Hexachlorobenzene

Liver
Aldrin
Bis(2-chloroethyl)ether
Bis(2-ethylhexyl)phthalate
Carbazole
Carbon tetrachloride
Chlordane
Chloroform
DDD
DDE
DDT
1,2-Dibromo-3-chloropropane
1,2-Dibromoethane
3,3'-Dichlorobenzidine
1,2-Dichloroethane
1,3-Dichloropropane (Ingestion only)
1,3-Dichloropropylene
Dieldrin
2,4-Dinitrotoluene
2,6-Dinitrotoluene
Heptachlor
Heptachlor epoxide
Hexachlorobenzene
alpha-HCH
gamma-HCH (Lindane)
Methylene chloride
N-Nitrosodiphenylamine
N-Nitrosodi-n-propylamine
Pentachlorophenol
Tetrachloroethylene
Trichloroethylene
2,4,6-Trichlorophenol

Toxaphene
Vinyl chloride

Circulatory System
Benzene
2,4,6-Trichlorophenol

Gastrointestinal System
Benzo(a)anthracene
Benzo(b)fluoranthene
Benzo(k)fluoranthene
Benzo(a)pyrene
Chrysene
Dibenzo(a,h)anthracene
Indeno(1,2,3-c,d)pyrene
Bromodichloromethane
Bromoform
1,2-Dibromo-3-chloropropane
1,2-Dibromoethane
1,3-Dichloropropylene

Lung
Arsenic
Beryllium (Inhalation only)
Cadmium (Inhalation only)
Chromium, hexavalent (Inhalation only)
1,3-Dichloropropylene
Methylene chloride
N-Nitrosodi-n-propylamine
Vinyl chloride

Nasal Cavity
1,2-Dibromo-3-chloropropane
    (Inhalation only)
1,2-Dibromoethane (Inhalation only)
N-Nitrosodi-n-propylamine

Bladder
3,3'-Dichlorobenzidine
1,3-Dichloropropylene
N-Nitrosodiphenylamine

Section 742.APPENDIX A: General

Section 742.TABLE G: Concentrations of Inorganic Chemicals in Background Soils

Chemical Name	Counties Within Metropolitan Statistical Areas[a] (mg/kg)	Counties Outside Metropolitan Statistical Areas (mg/kg)
Aluminum	9,500	9,200
Antimony	4.0	3.3
Arsenic	7.2	5.2
Barium	110	122
Beryllium	0.59	0.56
Cadmium	0.6	0.50
Calcium	9,300	5,525
Chromium	16.2	13.0
Cobalt	8.9	8.9
Copper	19.6	12.0
Cyanide	0.51	0.50
Iron	15,900	15,000
Lead	36.0	20.9
Magnesium	4,820	2,700
Manganese	636	630
Mercury	0.06	0.05

---

[a]Counties within Metropolitan Statistical Areas: Boone, Champaign, Clinton, Cook, DuPage, Grundy, Henry, Jersey, Kane, Kankakee, Kendall, Lake, Macon, Madison, McHenry, McLean, Menard, Monroe, Peoria, Rock Island, Sangamon, St. Clair, Tazewell, Will, Winnebago and Woodford.

Chemical Name	Counties Within Metropolitan Statistical Areas[a] (mg/kg)	Counties Outside Metropolitan Statistical Areas (mg/kg)
Nickel	18.0	13.0
Potassium	1,268	1,100
Selenium	0.48	0.37
Silver	0.55	0.50
Sodium	130	130.0
Sulfate	85.5	110
Sulfide	3.1	2.9
Thallium	0.32	0.42
Vanadium	25.2	25.0
Zinc	95.0	60.2

Section 742.APPENDIX B: Tier 1 Tables and Illustrations

Section 742.Illustration A: Tier 1 Evaluation

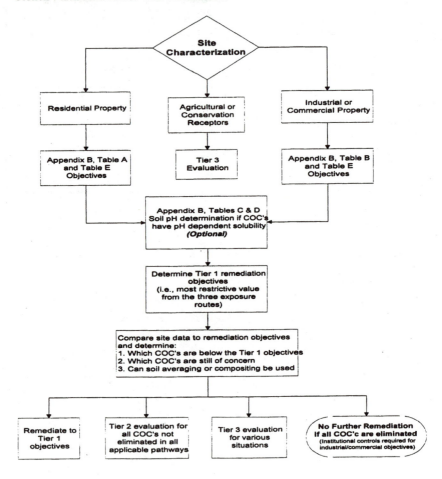

Section 742.TABLE A: Tier 1 Soil Remediation Objectives[a] for Residential Properties

CAS No.	Chemical Name	Exposure Route-Specific Values for Soils		Soil Component of the Groundwater Ingestion Exposure Route Values		ADL (mg/kg)
		Ingestion (mg/kg)	Inhalation (mg/kg)	Class I (mg/kg)	Class II (mg/kg)	
83-32-9	Acenaphthene	4,700[b]	---[c]	570[b]	2,900	*
67-64-1	Acetone	7,800[b]	100,000[d]	16[b]	16	*
15972-60-8	Alachlor[o]	8[e]	---[c]	0.04	0.2	NA
116-06-3	Aldicarb[o]	78[b]	---[c]	0.013	0.07	NA
309-00-2	Aldrin	0.04[e]	3[e]	0.5[e]	2.5	*
120-12-7	Anthracene	23,000[b]	---[c]	12,000[b]	59,000	*
1912-24-9	Atrazine[o]	2700[b]	---[c]	0.066	0.33	NA
71-43-2	Benzene	22[e]	0.8[e]	0.03	0.17	*
56-55-3	Benzo(a)anthracene	0.9[e]	---[c]	2	8	*
205-99-2	Benzo(b)fluoranthene	0.9[e]	---[c]	5	25	*

		Exposure Route-Specific Values for Soils		Soil Component of the Groundwater Ingestion Exposure Route Values		
CAS No.	Chemical Name	Ingestion (mg/kg)	Inhalation (mg/kg)	Class I (mg/kg)	Class II (mg/kg)	ADL (mg/kg)
207-08-9	Benzo(k)fluoranthene	9[e]	---[c]	49	250	*
50-32-8	Benzo(a)pyrene	0.09[e,f]	---[c]	8	82	*
111-44-4	Bis(2-chloroethyl)ether	0.6[e]	0.2[e,f]	0.0004[e,f]	0.0004	0.66
117-81-7	Bis(2-ethylhexyl)phthalate	46[e]	31,000[d]	3,600	31,000[d]	*
75-27-4	Bromodichloromethane (Dichlorobromomomethane)	10[e]	3,000[d]	0.6	0.6	*
75-25-2	Bromoform	81[e]	53[e]	0.8	0.8	*
71-36-3	Butanol	7,800[b]	10,000[d]	17[b]	17	NA
85-68-7	Butyl benzyl phthalate	16,000[b]	930[d]	930[d]	930[d]	*
86-74-8	Carbazole	32[e]	---[c]	0.6[e]	2.8	NA
1563-66-2	Carbofuran[o]	390[b]	---[c]	0.22	1.1	NA
75-15-0	Carbon disulfide	7,800[b]	720[d]	32[b]	160	*

452

| CAS No. | Chemical Name | Exposure Route-Specific Values for Soils | | Soil Component of the Groundwater Ingestion Exposure Route Values | | ADL (mg/kg) |
		Ingestion (mg/kg)	Inhalation (mg/kg)	Class I (mg/kg)	Class II (mg/kg)	
56-23-5	Carbon tetrachloride	5[e]	0.3[e]	0.07	0.33	*
57-74-9	Chlordane	0.5[e]	20[e]	10	48	*
106-47-8	4-Chloroaniline (p-Chloroaniline)	310[b]	---[c]	0.7[b]	0.7	1.3
108-90-7	Chlorobenzene (Monochlorobenzene)	1,600[b]	130[b]	1	6.5	*
124-48-1	Chlorodibromomethane (Dibromochloromethane)	1,600[b]	1,300[d]	0.4	0.4	*
67-66-3	Chloroform	100[e]	0.3[e]	0.6	2.9	*
218-01-9	Chrysene	88[e]	---[c]	160	800	*
94-75-7	2,4-D	780[b]	---[c]	1.5	7.7	*
75-99-0	Dalapon	2,300[b]	---[c]	0.85	8.5	1.2
72-54-8	DDD	3[e]	---[c]	16[e]	80	*
72-55-9	DDE	2[e]	---[c]	54[e]	270	*

		Exposure Route-Specific Values for Soils		Soil Component of the Groundwater Ingestion Exposure Route Values		
CAS No.	Chemical Name	Ingestion (mg/kg)	Inhalation (mg/kg)	Class I (mg/kg)	Class II (mg/kg)	ADL (mg/kg)
50-29-3	DDT	2[e]	—[g]	32[e]	160	*
53-70-3	Dibenzo(a,h)anthracene	0.09[e,f]	—[c]	2	7.6	*
96-12-8	1,2-Dibromo-3-chloropropane	0.46[e]	11[b]	0.002	0.002	*
106-93-4	1,2-Dibromoethane (Ethylene dibromide)	0.0075[e]	0.17[e]	0.0004	0.004	0.005
84-74-2	Di-n-butyl phthalate	7,800[b]	2,300[d]	2,300[d]	2,300[d]	*
95-50-1	1,2-Dichlorobenzene (o - Dichlorobenzene)	7,000[b]	560[d]	17	43	*
106-46-7	1,4-Dichlorobenzene (p - Dichlorobenzene)	—[c]	—[g]	2	11	*
91-94-1	3,3'-Dichlorobenzidine	1[e]	—[c]	0.007[e,f]	0.033	1.3
75-34-3	1,1-Dichloroethane	7,800[b]	1,300[b]	23[b]	110	*

CAS No.	Chemical Name	Exposure Route-Specific Values for Soils		Soil Component of the Groundwater Ingestion Exposure Route Values		
		Ingestion (mg/kg)	Inhalation (mg/kg)	Class I (mg/kg)	Class II (mg/kg)	ADL (mg/kg)
107-06-2	1,2-Dichloroethane (Ethylene dichloride)	7e	0.4e	0.02	0.1	*
75-35-4	1,1-Dichloroethylene	700b	1,500d	0.06	0.3	*
156-59-2	cis-1,2-Dichloroethylene	780b	1,200d	0.4	1.1	*
156-60-5	trans-1,2-Dichloroethylene	1,600b	3,100d	0.7	3.4	*
78-87-5	1,2-Dichloropropane	9e	15b	0.03	0.15	*
542-75-6	1,3-Dichloropropene (1,3-Dichloropropylene, cis + trans)	4e	0.1e	0.004e	0.02	0.005
60-57-1	Dieldrin[n]	0.04e	1e	0.004e	0.02	*
84-66-2	Diethyl phthalate	63,000b	2,000d	470b	470	*
105-67-9	2,4-Dimethylphenol	1,600b	---c	9b	9	*
121-14-2	2,4-Dinitrotoluene	0.9e	---c	0.0008e,f	0.0008	0.013

| CAS No. | Chemical Name | Exposure Route-Specific Values for Soils | | Soil Component of the Groundwater Ingestion Exposure Route Values | | ADL (mg/kg) |
		Ingestion (mg/kg)	Inhalation (mg/kg)	Class I (mg/kg)	Class II (mg/kg)	
606-20-2	2,6-Dinitrotoluene	0.9[e]	---[c]	0.0007[e,f]	0.0007	0.0067
117-84-0	Di-n-octyl phthalate	1,600[b]	10,000[d]	10,000[d]	10,000[d]	*
115-29-7	Endosulfan	470[b]	---[c]	18[b]	90	*
145-73-3	Endothall[o]	1,600[b]	---[c]	0.4	0.4	NA
72-20-8	Endrin	23[b]	---[c]	1	5	*
100-41-4	Ethylbenzene	7,800[b]	400[d]	13	19	*
206-44-0	Fluoranthene	3,100[b]	---[c]	4,300[b]	21,000	*
86-73-7	Fluorene	3,100[b]	---[c]	560[b]	2,800	*
76-44-8	Heptachlor	0.1[e]	0.1[e]	23	110	*
1024-57-3	Heptachlor epoxide	0.07[e]	5[e]	0.7	3.3	*
118-74-1	Hexachlorobenzene	0.4[e]	1[e]	2	11	*
319-84-6	alpha-HCH (alpha-BHC)	0.1[e]	0.8[e]	0.0005[e,f]	0.003	0.002

CAS No.	Chemical Name	Exposure Route-Specific Values for Soils		Soil Component of the Groundwater Ingestion Exposure Route Values		ADL (mg/kg)
		Ingestion (mg/kg)	Inhalation (mg/kg)	Class I (mg/kg)	Class II (mg/kg)	
58-89-9	gamma-HCH (Lindane)[n]	0.5[e]	--[c]	0.009	0.047	*
77-47-4	Hexachlorocyclopentadiene	550[b]	10[b]	400	2,200[d]	*
67-72-1	Hexachloroethane	78[b]	--[c]	0.5[b]	2.6	*
193-39-5	Indeno(1,2,3-c,d)pyrene	0.9[e]	--[c]	14	69	*
78-59-1	Isophorone	15,600[b]	4,600[d]	8[b]	8	*
72-43-5	Methoxychlor	390[b]	--[c]	160	780	*
74-83-9	Methyl bromide (Bromomethane)	110[b]	10[b]	0.2[b]	1.2	*
75-09-2	Methylene chloride (Dichloromethane)	85[e]	13[e]	0.02[e]	0.2	*
95-48-7	2-Methylphenol (o - Cresol)	3,900[b]	--[c]	15[b]	15	*
91-20-3	Naphthalene	3,100[b]	--[c]	84[b]	420	*
98-95-3	Nitrobenzene	39[b]	92[b]	0.1[b,f]	0.1	0.26

457

CAS No.	Chemical Name	Exposure Route-Specific Values for Soils		Soil Component of the Groundwater Ingestion Exposure Route Values		
		Ingestion (mg/kg)	Inhalation (mg/kg)	Class I (mg/kg)	Class II (mg/kg)	ADL (mg/kg)
86-30-6	N-Nitrosodiphenylamine	130[e]	—[c]	1[c]	5.6	*
621-64-7	N-Nitrosodi-n-propylamine	0.09[e,f]	—[c]	0.00005[e,f]	0.00005	0.66
108-95-2	Phenol	47,000[b]	—[c]	100[b]	100	*
1918-02-1	Picloram[o]	5,500[b]	—[c]	2	20	NA
1336-36-3	Polychlorinated biphenyls (PCBs)[n]	1; 10[h]	—[c,h]	—[h]	—[h]	*
129-00-0	Pyrene	2,300[b]	—[c]	4,200[b]	21,000	*
122-34-9	Simazine[o]	390[b]	—[c]	0.04	0.37	NA
100-42-5	Styrene	16,000[b]	1,500[d]	4	18	*
127-18-4	Tetrachloroethylene (Perchloroethylene)	12[e]	11[e]	0.06	0.3	*
108-88-3	Toluene	16,000[b]	650[d]	12	29	*

CAS No.	Chemical Name	Exposure Route-Specific Values for Soils		Soil Component of the Groundwater Ingestion Exposure Route Values		ADL (mg/kg)
		Ingestion (mg/kg)	Inhalation (mg/kg)	Class I (mg/kg)	Class II (mg/kg)	
8001-35-2	Toxaphene[n]	0.6[e]	89[e]	31	150	*
120-82-1	1,2,4-Trichlorobenzene	780[b]	3,200[b]	5	53	*
71-55-6	1,1,1-Trichloroethane	—[c]	1,200[d]	2	9.6	*
79-00-5	1,1,2-Trichloroethane	310[b]	1,800[d]	0.02	0.3	*
79-01-6	Trichloroethylene	58[e]	5[e]	0.06	0.3	*
108-05-4	Vinyl acetate	78,000[b]	1,000[b]	170[b]	170	*
75-01-4	Vinyl chloride	0.3[e]	0.03[e]	0.01[f]	0.07	*
108-38-3	m-Xylene	160,000[b]	420[d]	210	210	*
95-47-6	o-Xylene	160,000[b]	410[d]	190	190	*
106-42-3	p-Xylene	160,000[b]	460[d]	200	200	*

		Exposure Route-Specific Values for Soils		Soil Component of the Groundwater Ingestion Exposure Route Values		
CAS No.	Chemical Name	Ingestion (mg/kg)	Inhalation (mg/kg)	Class I (mg/kg)	Class II (mg/kg)	ADL (mg/kg)
1330-20-7	Xylenes (total)	160,000[b]	410[d]	150	150	*
	Ionizable Organics					
65-85-0	Benzoic Acid	310,000[b]	---[c]	400[b,i]	400[i]	*
95-57-8	2-Chlorophenol	390[b]	53,000[d]	4[b,i]	4[i]	*
120-83-2	2,4-Dichlorophenol	230[b]	---[c]	1[b,i]	1[i]	*
51-28-5	2,4-Dinitrophenol	160[b]	---[c]	0.2[b,f]	0.2	3.3
88-85-7	Dinoseb[o]	3[c,j]	---[c]	0.34[b,i]	3.4[i]	*
87-86-5	Pentachlorophenol	630[b]	---[c]	0.03[f,i]	0.14[i]	2.4
93-72-1	2,4,5-TP (Silvex)	7,800[b]	---[c]	11[i]	55[i]	*
95-95-4	2,4,5-Trichlorophenol	7,800[b]	---[c]	270[b,i]	1,400[i]	*
88-06-2	2,4,6 Trichlorophenol	58[e]	200[e]	0.2[e,f,i]	0.77[i]	0.43

CAS No.	Chemical Name	Exposure Route-specific Values for Soils		Soil Component of the Groundwater Ingestion Exposure Route Values		ADL (mg/kg)
		Ingestion (mg/kg)	Inhalation (mg/kg)	Class I (mg/L)	Class II (mg/L)	
	Inorganics					
7440-36-0	Antimony	31[b]	---[c]	0.006[m]	0.024[m]	*
7440-38-2	Arsenic[l,n]	0.4[e,t]	750[e]	0.05[m]	0.2[m]	*
7440-39-3	Barium	5,500[b]	690,000[b]	2.0[m]	2.0[m]	*
7440-41-7	Beryllium	0.1[e,t]	1,300[e]	0.004[m]	0.5[m]	*
7440-42-8	Boron	7,000[b]	---[g]	2.0[m]	2.0[m]	*
7440-43-9	Cadmium[l,n]	78[b,t]	1,800[e]	0.005[m]	0.05[m]	*
16887-00-6	Chloride	---[c]	---[c]	200[m]	200[m]	*
7440-47-3	Chromium, total	390[b]	270[e]	0.1[m]	1.0[m]	*
16065-83-1	Chromium, ion, trivalent	78,000[b]	---[c]	---[g]	---[g]	*
18540-29-9	Chromium, ion, hexavalent	390[b]	270[e]	---	---	*
7440-48-4	Cobalt	4,700[b]	---[c]	1.0[m]	1.0[m]	*

CAS No.	Chemical Name	Exposure Route-specific Values for Soils		Soil Component of the Groundwater Ingestion Exposure Route Values		ADL (mg/kg)
		Ingestion (mg/kg)	Inhalation (mg/kg)	Class I (mg/L)	Class II (mg/L)	
7440-50-8	Copper[a]	2,900[b]	—[c]	0.65[m]	0.65[m]	*
57-12-5	Cyanide (amenable)	1,600[b]	—[c]	0.2[q]	0.6[q]	*
7782-41-4	Fluoride	4,700[b]	—[c]	4.0[m]	4.0[m]	*
15438-31-0	Iron	—[c]	—[c]	5.0[m]	5.0[m]	*
7439-92-1	Lead	400[k]	—[c]	0.0075[m]	0.1[m]	*
7439-96-5	Manganese	3,700[b]	69,000[b]	0.15[m]	10.0[m]	*
7439-97-6	Mercury[l,n]	23[b,s]	10[b,i]	0.002[m]	0.01[m]	*
7440-02-0	Nickel[l]	1,600[b]	13,000[c]	0.1[m]	2.0[m]	*
14797-55-8	Nitrate as N[p]	130,000[b]	—[c]	10.0[q]	100[q]	*
7782-49-2	Selenium[l,n]	390[b]	—[c]	0.05[m]	0.05[m]	*

CAS No.	Chemical Name	Exposure Route-specific Values for Soils		Soil Component of the Groundwater Ingestion Exposure Route Values		ADL (mg/kg)
		Ingestion (mg/kg)	Inhalation (mg/kg)	Class I (mg/L)	Class II (mg/L)	
7440-22-4	Silver	390[b]	---[c]	0.05[m]	---	*
14808-79-8	Sulfate	---[c]	---[c]	400[m]	400[m]	*
7440-28-0	Thallium	6.3[b,u]	---[c]	0.002[m]	0.02[m]	*
7440-62-2	Vanadium	550[b]	---[c]	0.049[m]	---	*
7440-66-6	Zinc[l]	23,000[b]	---[c]	5.0[m]	10[m]	*

"*" indicates that the ADL is less than or equal to the specified remediation objective.
NA means not available; no PQL or EQL available in USEPA analytical methods.

Chemical Name and Soil Remediation Objective Notations

a. Soil remediation objectives based on human health criteria only.

b. Calculated values correspond to a target hazard quotient of 1.

d. No toxicity criteria available for the route of exposure.

Soil saturation concentration (C_{sat}) = the concentration at which the absorptive limits of the soil particles, the solubility limits of the available soil moisture, and saturation of soil pore air have been reached. Above the soil saturation concentration, the assumptions regarding vapor transport to air and/or dissolved phase transport to groundwater (for chemicals which are liquid at ambient soil temperatures) have been violated, and alternative modeling approaches are required.

e. Calculated values correspond to a cancer risk level of 1 in 1,000,000.

f. Level is at or below Contract Laboratory Program required quantitation limit for Regular Analytical Services (RAS).

Chemical-specific properties are such that this route is not of concern at any soil contaminant concentration.

h. A preliminary goal of 1 ppm has been set for PCBs based on *Guidance on Remedial Actions for Superfund Sites with PCB Contamination*, EPA/540G-90/007, and on USEPA efforts to manage PCB contamination. See 40 CFR 761.120 - USEPA "PCB Spill Cleanup Policy." This regulation goes on to say that the remediation goal for an unrestricted area is 10 ppm and 25 ppm for a restricted area, provided both have at least 10 inches of clean cover.

i. Soil remediation objective for pH of 6.8. If soil pH is other than 6.8, refer to Appendix B, Tables C and D of this Part.

j. Ingestion soil remediation objective adjusted by a factor of 0.5 to account for dermal route.

k. A preliminary remediation goal of 400 mg/kg has been set for lead based on *Revised Interim Soil Lead Guidance for CERCLA Sites and RCRA Corrective Action Facilities*, OSWER Directive #9355.4-12.

l. Potential for soil-plant-human exposure.

m. The person conducting the remediation has the option to use: 1) TCLP or SPLP test results to compare with the remediation objectives listed in this Table; or 2) the total amount of contaminant in the soil sample results to compare with pH specific remediation objectives listed in Appendix B, Table C or D of this Part. (See Section 742.510.) If the person conducting the remediation wishes to calculate soil remediation objectives based on background concentrations, this should be done in accordance with Subpart D of this Part.

n. The Agency reserves the right to evaluate the potential for remaining contaminant concentrations to pose significant threats to crops, livestock, or wildlife.

o. For agrichemical facilities, remediation objectives for surficial soils which are based on field application rates may be more appropriate for currently registered pesticides. Consult the Agency for further information.

p. For agrichemical facilities, soil remediation objectives based on site-specific background concentrations of Nitrate as N may be more appropriate. Such determinations shall be conducted in accordance with the procedures set forth in Subparts D and I of this Part.

q. The TCLP extraction must be done using water at a pH of 7.0.

r. Value based on dietary Reference Dose.

s. Value based on Reference Dose for Mercuric chloride (CAS No. 7487-94-7).

t. Note that Table value is likely to be less than background concentration for this chemical; screening or remediation concentrations using the procedures of Subpart D of this Part may be more appropriate.

u. Value based on Reference Dose for thallium sulfate (CAS No. 7446-18-6).

Section 742. APPENDIX B: Tier 1 Tables and Illustrations

Section 742.Table B: Tier 1 Soil Remediation Objectives[a] for Industrial/Commercial Properties

| CAS No. | Chemical Name | Exposure Route-Specific Values for Soils | | | | | | Soil Component of the Groundwater Ingestion Exposure Route Values | | | |
| | | Industrial-Commercial | | Construction Worker | | | | | | |
		Ingestion (mg/kg)	Inhalation (mg/kg)	Ingestion (mg/kg)	Inhalation (mg/kg)			Class I (mg/kg)	ClassII (mg/kg)	ADL (mg/kg)
83-32-9	Acenaphthene	120,000[b]	-----[c]	120,000[b]	-----[c]			570[b]	2,900	*
67-64-1	Acetone	200,000[b]	100,000[d]	200,000[b]	100,000[d]			16[b]	16	*
15972-60-8	Alachlor[a]	72[c]	-----[c]	1,600[c]	-----[c]			0.04	0.2	NA
116-06-3	Aldicarb[a]	2,000[b]	-----[c]	200[b]	-----[c]			0.013	0.07	NA
309-00-2	Aldrin	0.3[c]	6.6[c]	6.1[b]	9.3[a]			0.5[a]	2.5	*
120-12-7	Anthracene	610,000[b]	-----[c]	610,000[b]	-----[c]			12,000[b]	59,000	*
1912-24-9	Atrazine[b]	72,000[b]	-----[c]	7,100[c]	-----[c]			0.066	0.33	NA
71-43-2	Benzene	200[c]	1.5[c]	4,300[c]	2.1[c]			0.03	0.17	*

465

| | | Exposure Route-Specific Values for Soils | | | | Soil Component of the Groundwater Ingestion Exposure Route Values | | |
| | | Industrial-Commercial | | Construction Worker | | | | |
CAS No.	Chemical Name	Ingestion (mg/kg)	Inhalation (mg/kg)	Ingestion (mg/kg)	Inhalation (mg/kg)	Class I (mg/kg)	Class II (mg/kg)	ADL (mg/kg)
56-55-3	Benzo(a)anthracene	8[e]	----[c]	170[e]	----[c]	2	8	*
205-99-2	Benzo(b)fluoranthene	8[e]	----[c]	170[e]	----[c]	5	25	*
207-08-9	Benzo(k)fluoranthene	78[e]	----[c]	1,700[e]	----[c]	49	250	*
50-32-8	Benzo(a)pyrene	0.8[e]	----[c]	17[e]	----[c]	8	82	*
111-44-4	Bis(2-chloroethyl)ether	5[e]	0.47[c]	75[e]	0.66[e]	0.0004[e,f]	0.0004	0.66
117-81-7	Bis(2-ethylhexyl)phthalate	410[e]	31,000[d]	4,100[e]	31,000[d]	3,600	31,000[d]	*
75-27-4	Bromodichloromethane (Dichlorobromomethane)	92[e]	3,000[d]	2,000[e]	3,000[d]	0.6	0.6	*
75-25-2	Bromoform	720[e]	100[e]	16,000[e]	140[e]	0.8	0.8	*
71-36-3	Butanol	200,000[b]	10,000[d]	200,000[b]	10,000[d]	17[b]	17	NA
85-68-7	Butyl benzyl phthalate	410,000[b]	930[d]	410,000[b]	930[d]	930[d]	930[d]	*
86-74-8	Carbazole	290[e]	----[c]	6,200[e]	----[c]	0.6[e]	2.8	NA

| CAS No. | Chemical Name | Exposure Route-Specific Values for Soils | | | | Soil Component of the Groundwater Ingestion Exposure Route Values | | |
| | | Industrial-Commercial | | Construction Worker | | | | |
		Ingestion (mg/kg)	Inhalation (mg/kg)	Ingestion (mg/kg)	Inhalation (mg/kg)	Class I (mg/kg)	Class II (mg/kg)	ADL (mg/kg)
1563-66-2	Carbofuran[a]	10,000[b]	----[c]	1,000[b]	----[c]	0.22	1.1	NA
75-15-0	Carbon disulfide	200,000[b]	720[d]	20,000[b]	9.0[b]	32[b]	160	*
56-23-5	Carbon tetrachloride	44[e]	0.64[e]	410[b]	0.90[e]	0.07	0.33	*
57-74-9	Chlordane	4[e]	38[e]	12[b]	53[e]	10	48	*
106-47-8	4 - Chloroaniline (p-Chloroaniline)	8,200[b]	----[c]	820[b]	----[c]	0.7[b]	0.7	1.3
108-90-7	Chlorobenzene (Monochlorobenzene)	41,000[b]	210[b]	4,100[b]	1.3[b]	1	6.5	*
124-48-1	Chlorodibromomethane (Dibromochloromethane)	41,000[b]	1,300[d]	41,000[b]	1,300[d]	0.4	0.4	*
67-66-3	Chloroform	940[e]	0.54[e]	2,000[b]	0.76[e]	0.6	2.9	*
218-01-9	Chrysene	780[e]	----[c]	17,000[e]	----[e]	160	800	*
94-75-7	2,4-D	20,000[b]	----[e]	2,000[b]	----[e]	1.5	7.7	*

| CAS No. | Chemical Name | Exposure Route-Specific Values for Soils | | | | Soil Component of the Groundwater Ingestion Exposure Route Values | | |
| | | Industrial-Commercial | | Construction Worker | | | | |
		Ingestion (mg/kg)	Inhalation (mg/kg)	Ingestion (mg/kg)	Inhalation (mg/kg)	Class I (mg/kg)	Class II (mg/kg)	ADL (mg/kg)
75-99-0	Dalapon	61,000[b]	----[c]	6,100[b]	----[c]	0.85	8.5	1.2
72-54-8	DDD	24[e]	----[c]	520[c]	----[c]	16[e]	80	*
72-55-9	DDE	17[e]	----[c]	370[c]	----[c]	54[e]	270	*
50-29-3	DDT	17[e]	1,500[e]	100[b]	2,100[e]	32[e]	160	*
53-70-3	Dibenzo(a,h)anthracene	0.8[e]	----[c]	17[e]	----[c]	2	7.6	*
96-12-8	1,2-Dibromo-3-chloropropane	4[e]	17[b]	89[e]	0.11[b]	0.002	0.002	*
106-93-4	1,2-Dibromoethane (Ethylene dibromide)	0.07[e]	0.32[e]	1.5[e]	0.45[e]	0.0004	0.004	0.005
84-74-2	Di-n-butyl phthalate	200,000[b]	2,300[d]	200,000[b]	2,300[d]	2,300[d]	2,300[d]	*
95-50-1	1,2-Dichlorobenzene (o - Dichlorobenzene)	180,000[b]	560[d]	18,000[b]	310[b]	17	43	*
106-46-7	1,4-Dichlorobenzene (p - Dichlorobenzene)	----[c]	17,000[b]	----[c]	340[b]	2	11	*

| | | Exposure Route-Specific Values for Soils | | | | Soil Component of the Groundwater Ingestion Exposure Route Values | | |
| | | Industrial-Commercial | | Construction Worker | | | | |
CAS No.	Chemical Name	Ingestion (mg/kg)	Inhalation (mg/kg)	Ingestion (mg/kg)	Inhalation (mg/kg)	Class I (mg/kg)	Class II (mg/kg)	ADL (mg/kg)
91-94-1	3,3'-Dichlorobenzidine	13[e]	----[c]	280[e]	----[c]	0.007[e,f]	0.033	1.3
75-34-3	1,1-Dichloroethane	200,000[b]	1,700[d]	200,000[b]	130[b]	23[b]	110	*
107-06-2	1,2-Dichloroethane (Ethylene dichloride)	63[e]	0.70[e]	1,400[e]	0.99[e]	0.02	0.1	*
75-35-4	1,1-Dichloroethylene	18,000[b]	1,500[d]	1,800[b]	1,500[d]	0.06	0.3	*
156-59-2	cis-1,2-Dichloroethylene	20,000[b]	1,200[d]	20,000[b]	1,200[d]	0.4	1.1	*
156-60-5	trans-1,2-Dichloroethylene	41,000[b]	3,100[d]	41,000[b]	3,100[d]	0.7	3.4	*
78-87-5	1,2-Dichloropropane	84[e]	23[b]	1,800[e]	0.50[b]	0.03	0.15	*
542-75-6	1,3-Dichloropropene (1,3-Dichloropropylene, cis + trans)	33[e]	0.23[e]	610[b]	0.33[e]	0.004[e]	0.02	0.005
60-57-1	Dieldrin[a]	0.4[e]	2.2[e]	7.8[e]	3.1[e]	0.004[e]	0.02	0.0013
84-66-2	Diethyl phthalate	1,000,000[b]	2,000[d]	1,000,000[b]	2,000[d]	470[b]	470	*

| CAS No. | Chemical Name | Exposure Route-Specific Values for Soils | | | | Soil Component of the Groundwater Ingestion Exposure Route Values | | |
| | | Industrial-Commercial | | Construction Worker | | | | |
		Ingestion (mg/kg)	Inhalation (mg/kg)	Ingestion (mg/kg)	Inhalation (mg/kg)	Class I (mg/kg)	Class II (mg/kg)	ADL (mg/kg)
105-67-9	2,4-Dimethylphenol	41,000[b]	---[c]	41,000[b]	---[c]	9[b]	9	*
121-14-2	2,4-Dinitrotoluene	8.4[e]	---[c]	180[e]	---[c]	0.0008[e,f]	0.0008	0.013
606-20-2	2,6-Dinitrotoluene	8.4[e]	---[c]	180[e]	---[c]	0.0007[e,f]	0.0007	0.0067
117-84-0	Di-n-octyl phthalate	41,000[c]	10,000[d]	4,100[b]	10,000[d]	10,000[d]	10,000[d]	*
115-29-7	Endosulfan	12,000[b]	---[c]	1,200[b]	---[c]	18[b]	90	*
145-73-3	Endothall[a]	41,000[c]	---[c]	4,100[b]	---[c]	0.4	0.4	NA
72-20-8	Endrin	610[b]	---[c]	61[b]	---[c]	1	5	*
100-41-4	Ethylbenzene	200,000[b]	400[d]	20,000[b]	58[b]	13	19	*
206-44-0	Fluoranthene	82,000[b]	---[c]	82,000[b]	---[c]	4,300[b]	21,000	*
86-73-7	Fluorene	82,000[b]	---[c]	82,000[b]	---[c]	560[b]	2,800	*
76-44-8	Heptachlor	1[e]	11[e]	28[e]	16[e]	23	110	*

| | | Exposure Route-Specific Values for Soils | | | | Soil Component of the Groundwater Ingestion Exposure Route Values | | |
| | | Industrial-Commercial | | Construction Worker | | | | |
CAS No.	Chemical Name	Ingestion (mg/kg)	Inhalation (mg/kg)	Ingestion (mg/kg)	Inhalation (mg/kg)	Class I (mg/kg)	Class II (mg/kg)	ADL (mg/kg)
1024-57-3	Heptachlor epoxide	0.6[e]	9.2[e]	2.7[b]	13[e]	0.7	3.3	*
118-74-1	Hexachlorobenzene	4[e]	1.8[e]	78[e]	2.6[e]	2	11	*
319-84-6	alpha-HCH (alpha-BHC)	0.9[e]	1.5[e]	20[e]	2.1[e]	0.0005[e,f]	0.003	0.002
58-89-9	gamma-HCH (Lindane)[a]	4[e]	----[c]	96[e]	----[c]	0.009	0.047	*
77-47-4	Hexachlorocyclopentadiene	14,000[b]	16[b]	14,000[b]	1.1[b]	400	2,200[d]	*
67-72-1	Hexachloroethane	2,000[b]	----[c]	2,000[b]	----[c]	0.5[b]	2.6	*
193-39-5	Indeno(1,2,3-c,d)pyrene	8[e]	----[c]	170[e]	----[c]	14	69	*
78-59-1	Isophorone	410,000[b]	4,600[d]	410,000[b]	4,600[d]	8[b]	8	*
72-43-5	Methoxychlor	10,000[b]	----[c]	1,000[b]	----[c]	160	780	*
74-83-9	Methyl bromide (Bromomethane)	2,900[b]	15[b]	1,000[b]	3.9[b]	0.2[b]	1.2	*

471

| | | Exposure Route-Specific Values for Soils | | | | Soil Component of the Groundwater Ingestion Exposure Route Values | | |
| | | Industrial-Commercial | | Construction Worker | | | | |
CAS No.	Chemical Name	Ingestion (mg/kg)	Inhalation (mg/kg)	Ingestion (mg/kg)	Inhalation (mg/kg)	Class I (mg/kg)	Class II (mg/kg)	ADL (mg/kg)
75-09-2	Methylene chloride (Dichloromethane)	760[e]	24[e]	12,000[b]	34[e]	0.02[c]	0.2	*
95-48-7	2-Methylphenol (o - Cresol)	100,000[b]	----[c]	100,000[b]	----[c]	15[b]	15	*
86-30-6	N-Nitrosodiphenylamine	1,200[e]	----[c]	25,000[e]	----[c]	1[e]	5.δ	0.66
621-64-7	N-Nitrosodi-n-propylamine	0.8[e]	----[c]	18[e]	----[c]	0.00005[e,f]	0.00005	0.66
91-20-3	Naphthalene	82,000[b]	----[c]	8,200[b]	----[c]	84[b]	420	*
98-95-3	Nitrobenzene	1,000[b]	140[b]	1,000[b]	9.4[b]	0.1[b,f]	0.1	0.26
108-95-2	Phenol	1,000,000[b]	----[c]	120,000[b]	----[c]	100[b]	100	*
1918-02-1	Picloram[o]	140,000[b]	----[c]	14,000[b]	----[c]	2	20	NA
1336-36-3	Polychlorinated biphenyls (PCBs)[n]	1; 10; 25[h]	----[c,h]	1[h]	----[c,h]	----[h]	----[h]	*
129-00-0	Pyrene	61,000[b]	----[c]	61,000[b]	----[c]	4,200[b]	21,000	*

| CAS No. | Chemical Name | Exposure Route-Specific Values for Soils | | | | Soil Component of the Groundwater Ingestion Exposure Route Values | | |
| | | Industrial-Commercial | | Construction Worker | | | | |
		Ingestion (mg/kg)	Inhalation (mg/kg)	Ingestion (mg/kg)	Inhalation (mg/kg)	Class I (mg/kg)	Class II (mg/kg)	ADL (mg/kg)
122-34-9	Simazine[a]	10,000[b]	----[c]	1,000[b]	----[c]	0.04	0.37	NA
100-42-5	Styrene	410,000[b]	1,500[d]	41,000[b]	430[b]	4	18	*
127-18-4	Tetrachloroethylene (Perchloroethylene)	110[e]	20[e]	2,400[e]	28[e]	0.06	0.3	*
108-88-3	Toluene	410,000[b]	650[d]	410,000[b]	42[b]	12	29	*
8001-35-2	Toxaphene[a]	5.2[c]	170[e]	110[e]	240[e]	31	150	*
120-82-1	1,2,4-Trichlorobenzene	20,000[b]	3,200[d]	2,000[b]	920[d]	5	53	*
71-55-6	1,1,1-Trichloroethane	----[c]	1,200[d]	----[c]	1,200[d]	2	9.6	*
79-00-5	1,1,2-Trichloroethane	8,200[b]	1,800[d]	8,200[b]	1,800[d]	0.02	0.3	*
79-01-6	Trichloroethylene	520[e]	8.9[e]	1,200[b]	12[e]	0.06	0.3	*
108-05-4	Vinyl acetate	1,000,000[b]	1,600[b]	200,000[b]	10[b]	170[b]	170	*

CAS No.	Chemical Name	Exposure Route-Specific Values for Soils				Soil Component of the Groundwater Ingestion Exposure Route Values		
		Industrial-Commercial		Construction Worker				
		Ingestion (mg/kg)	Inhalation (mg/kg)	Ingestion (mg/kg)	Inhalation (mg/kg)	Class I (mg/kg)	Class II (mg/kg)	ADL (mg/kg)
75-01-4	Vinyl chloride	3[a]	0.06[e]	65[a]	0.08[e]	0.01[f]	0.07	*
108-38-3	m-Xylene	1,000,000	420[d]	410,000[b]	420[d]	210	210	*
95-47-6	o-Xylene	1,000,000	410[d]	410,000[b]	410[d]	190	190	*
106-42-3	p-Xylene	1,000,000	460[d]	410,000[b]	460[d]	200	200	*
1330-20-7	Xylenes (total)	1,000,000[b]	410[d]	410,000[b]	410[d]	150	150	*
	Ionizable Organics							
65-85-0	Benzoic Acid	1,000,000[b]	-----[c]	820,000[b]	-----[c]	400[h,j]	400[j]	*
95-57-8	2-Chlorophenol	10,000[b]	53,000[d]	10,000[b]	53,000[d]	4[h,j]	20[j]	*
120-83-2	2,4-Dichlorophenol	6,100[b]	-----[c]	610[b]	-----[c]	1[h,j]	1[j]	*
51-28-5	2,4-Dinitrophenol	4,100[b]	-----[c]	410[b]	-----[c]	0.2[h,j,l]	0.2[j]	3.3
88-85-7	Dinoseb[o]	2,000[b]	-----[c]	200[b]	-----[c]	0.34[h,j]	3.4[l]	*

| | | Exposure Route-Specific Values for Soils | | | | Soil Component of the Groundwater Ingestion Exposure Route Values | | |
| | | Industrial-Commercial | | Construction Worker | | | | |
CAS No.	Chemical Name	Ingestion (mg/kg)	Inhalation (mg/kg)	Ingestion (mg/kg)	Inhalation (mg/kg)	Class I (mg/kg)	Class II (mg/kg)	ADL (mg/kg)
87-86-5	Pentachlorophenol	24[f,j]	----[c]	520[f,j]	----[c]	0.03[f,i]	0.14[i]	2.4
93-72-1	2,4,5-TP (Silvex)	16,000[b]	----[c]	1,600[b]	----[c]	11[i]	55[i]	*
95-95-4	2,4,5-Trichlorophenol	200,000[b]	----[c]	200,000[b]	----[c]	270[b,j]	1,400[j]	*
88-06-2	2,4,6-Trichlorophenol	520[e]	390[e]	11,000[e]	540[e]	0.2[e,f,i]	0.77[i]	0.43

CAS No.	Chemical Name	Exposure Route-Specific Values for Soils				Soil Component of the Groundwater Ingestion Exposure Route Values		
		Industrial-Commercial		Construction Worker				
		Ingestion (mg/kg)	Inhalation (mg/kg)	Ingestion (mg/kg)	Inhalation (mg/kg)	Class I (mg/L)	Class II (mg/L)	
	Inorganics							
7440-36-0	Antimony	820[b]	----[c]	82[b]	----[c]	0.006[m]	0.024[m]	*
7440-38-2	Arsenic[l,a]	3[e,i]	1,200[f]	61[b]	25,000[f]	0.05[m]	0.2[m]	*
7440-39-3	Barium	140,000[b]	910,000[b]	14,000[b]	870,000[b]	2.0[m]	2.0[m]	*
7440-41-7	Beryllium	1[e,i]	2,100[e]	29[b]	44,000[e]	0.004[m]	0.5[m]	*
7440-42-8	Boron	180,000[b]	1,000,000	18,000[b]	1,000,000	2.0[m]	2.0[m]	*
7440-43-9	Cadmium[l,h]	2,000[b,r]	2,800[e]	200[b,r]	59,000[e]	0.005[m]	0.05[m]	*
16887-00-6	Chloride	----[c]	----[c]	----[c]	----[c]	200[m]	200[m]	*
7440-47-3	Chromium, total	10,000[b]	420[e]	4,100[b]	8,800[e]	0.1[m]	1.0[m]	*
16065-83-1	Chromium, ion, trivalent	1,000,000[b]	----[c]	330,000[b]	----[c]	----[g]	----	*
18540-29-9	Chromium, ion, hexavalent	10,000[b]	420[e]	4,100[b]	8,800[e]	----	----	*

| | | Exposure Route-Specific Values for Soils | | | | Soil Component of the Groundwater Ingestion Exposure Route Values | | |
| | | Industrial-Commercial | | Construction Worker | | | | |
CAS No.	Chemical Name	Ingestion (mg/kg)	Inhalation (mg/kg)	Ingestion (mg/kg)	Inhalation (mg/kg)	Class I (mg/L)	Class II (mg/L)	
7440-48-4	Cobalt	120,000[b]	----[c]	12,000[b]	----[c]	1.0[m]	1.0[m]	*
7440-50-8	Copper[a]	82,000[b]	----[c]	8,200[b]	----[c]	0.65[m]	0.65[m]	*
57-12-5	Cyanide (amenable)	41,000[b]	----[c]	4,100[b]	----[c]	0.2[x]	0.6[q]	*
7782-41-4	Fluoride	120,000[b]	----[c]	12,000[b]	----[c]	4.0[m]	4.0[m]	*
15438-31-0	Iron	----[c]	----[c]	----[c]	----[c]	5.0[m]	5.0[m]	*
7439-92-1	Lead	400[k]	----[c]	400[k]	----[c]	0.0075[m]	0.1[m]	*
7439-96-5	Manganese	96,000[b]	91,000[b]	9,600[b]	8,700[b]	0.15[m]	10.0[m]	*
7439-97-6	Mercury[i,n]	610[b]	540,000[c]	61[b,s]	52,000[b]	0.002[m]	0.01[m]	*
7440-02-0	Nickel[i]	41,000[b]	21,000[c]	4,100[b]	440,000[c]	0.1[m]	2.0[m]	*
14797-55-8	Nitrate as N[p]	1,000,000[b]	----[c]	330,000[b]	----[c]	10.0[d]	100[d]	*
7782-49-2	Selenium[i,n]	10,000[b]	----[c]	1,000[b]	----[c]	0.05[m]	0.05[m]	*

| CAS No. | Chemical Name | Exposure Route-Specific Values for Soils | | | | Soil Component of the Groundwater Ingestion Exposure Route Values | | |
| | | Industrial-Commercial | | Construction Worker | | | | |
		Ingestion (mg/kg)	Inhalation (mg/kg)	Ingestion (mg/kg)	Inhalation (mg/kg)	Class I (mg/L)	Class II (mg/L)	
7440-22-4	Silver	10,000^b	----^{c}	1,000^b	----^{c}	0.05^m	-----	*
14808-79-8	Sulfate	----^{c}	----^{c}	----^{c}	----^{c}	400^m	400^m	*
7440-28-0	Thallium	160^{b,u}	----^{c}	160^{b,u}	----^{c}	0.002^m	0.02^m	*
7440-62-2	Vanadium	14,000^b	----^{c}	1,400^b	----^{c}	0.049^m	-----	*
7440-66-6	Zinc^l	610,000^b	----^{c}	61,000^b	----^{c}	5.0^m	10^m	*

"*" indicates that the ADL is less than or equal to the specified remediation objective.
NA means Not Available; no PQL or EQL available in USEPA analytical methods.

Chemical Name and Soil Remediation Objective Notations (2nd, 5th thru 8th Columns)

a   Soil remediation objectives based on human health criteria only.

b   Calculated values correspond to a target hazard quotient of 1.

c   No toxicity criteria available for this route of exposure.

d   Soil saturation concentration (C_{sat}) = the concentration at which the absorptive limits of the soil particles, the solubility limits of the available soil moisture, and saturation of soil pore air have been reached. Above the soil saturation concentration, the assumptions regarding vapor transport to air and/or dissolved phase transport to groundwater (for chemicals which are liquid at ambient soil temperatures) have been violated, and alternative modeling approaches are required.

e   Calculated values correspond to a cancer risk level of 1 in 1,000,000.

f   Level is at or below Contract Laboratory Program required quantitation limit for Regular Analytical Services (RAS).

g   Chemical-specific properties are such that this route is not of concern at any soil contaminant concentration.

h   A preliminary goal of 1 ppm has been set for PCBs based on *Guidance on Remedial Actions for Superfund Sites with PCB Contamination*, EPA/540G-90/007, and on USEPA efforts to manage PCB contamination. See 40 CFR 761.120 for USEPA "PCB Spill Cleanup Policy." This regulation goes on to say that the remediation goal for an unrestricted area is 10 ppm and 25 ppm for a restricted area, provided both have at least 10 inches of clean cover.

i   Soil remediation objective for pH of 6.8. If soil pH is other than 6.8, refer to Appendix B, Tables C and D in this Part.

j   Ingestion soil remediation objective adjusted by a factor of 0.5 to account for dermal route.

k   A preliminary remediation goal of 400 mg/kg has been set for lead based on *Revised Interim Soil Lead Guidance for CERCLA Sites and RCRA Corrective Action Facilities*, OSWER Directive #9355.4-12.

l   Potential for soil-plant-human exposure.

m   The person conducting the remediation has the option to use: (1) TCLP or SPLP test results to compare with the remediation objectives listed in this Table; or (2) the total amount of contaminant in the soil sample results to compare with pH specific remediation objectives listed in Appendix B, Table C or D of this Part. (See Section 742.510.) If the person conducting the remediation wishes to calculate soil remediation objectives based on background concentrations, this should be done in accordance with Subpart D of this Part.

n   The Agency reserves the right to evaluate the potential for remaining contaminant concentrations to pose significant threats to crops, livestock, or wildlife.

o   For agrichemical facilities, remediation objectives for surficial soils which are based on field application rates may be more appropriate for currently registered pesticides. Consult the Agency for further information.

p   For agrichemical facilities, soil remediation objectives based on site-specific background concentrations of Nitrate as N may be more appropriate. Such determinations shall be conducted in accordance with the located in Subparts D and I of this Part.

q   The TCLP extraction must be done using water at a pH of 7.0.

r   Value based on dietary Reference Dose.

s   Value based on Reference Dose for Mercuric chloride (CAS No. 7487-94-7).

t   Note that Table value is likely to be less than background concentration for this chemical; screening or remediation concentrations using the procedures of Subpart D of this Part.

u   Value based on Reference Dose for thallium sulfate (CAS No. 7446-18-6).

479

Section 742.Table C: pH Specific Soil Remediation Objectives for Inorganics and Ionizing Organics for the Soil
                         Component of the Groundwater Ingestion Route (Class I Groundwater)

Chemical (totals) (mg/kg)	pH 4.5 to 4.74	pH 4.75 to 5.24	pH 5.25 to 5.74	pH 5.75 to 6.24	pH 6.25 to 6.64	pH 6.65 to 6.89	pH 6.9 to 7.24	pH 7.25 to 7.74	pH 7.75 to 8.0
Inorganics									
Antimony	5	5	5	5	5	5	5	5	5
Arsenic	25	26	27	28	29	29	29	30	31
Barium	260	490	850	1,200	1,500	1,600	1,700	1,800	2,100
Beryllium	1.1	2.1	3.4	6.6	22	63	140	1,000	8,000
Cadmium	1.0	1.7	2.7	3.7	5.2	7.5	11	59	430
Chromium (+6)	70	62	54	46	40	38	36	32	28
Copper	330	580	2,100	11,000	59,000	130,000	200,000	330,000	330,000
Cyanide	40	40	40	40	40	40	40	40	40
Mercury	0.01	0.01	0.03	0.15	0.89	2.1	3.3	6.4	8.0
Nickel	20	36	56	76	100	130	180	700	3,800
Selenium	24	17	12	8.8	6.3	5.2	4.5	3.3	2.4
Silver	0.24	0.33	0.62	1.5	4.4	8.5	13	39	110

Chemical (totals) (mg/kg)	pH 4.5 to 4.74	pH 4.75 to 5.24	pH 5.25 to 5.74	pH 5.75 to 6.24	pH 6.25 to 6.64	pH 6.65 to 6.89	pH 6.9 to 7.24	pH 7.25 to 7.74	pH 7.75 to 8.0
Thallium	1.6	1.8	2.0	2.4	2.6	2.8	3.0	3.4	3.8
Vanadium	980	980	980	980	980	980	980	980	980
Zinc	1,000	1,800	2,600	3,600	5,100	6,200	7,500	16,000	53,000
Organics									
Benzoic Acid	440	420	410	400	400	400	400	400	400
2-Chlorophenol	4.0	4.0	4.0	4.0	3.9	3.9	3.9	3.6	3.1
2,4-Dichlorophenol	1.0	1.0	1.0	1.0	1.0	1.0	1.0	0.86	0.69
Dinoseb	8.4	4.5	1.9	0.82	0.43	0.34	0.31	0.27	0.25
Pentachlorophenol	0.54	0.32	0.15	0.07	0.04	0.03	0.02	0.02	0.02
2,4,5-TP (Silvex)	26	16	12	11	11	11	11	11	11
2,4,5-Trichlorophenol	400	390	390	370	320	270	230	130	64
2,4,6-Trichlorophenol	0.37	0.36	0.34	0.26	0.20	0.15	0.13	0.09	0.07

Section 742.Table D: pH Specific Soil Remediation Objectives for Inorganics and Ionizing Organics for the Soil Component of the Groundwater Ingestion Route (Class II Groundwater)

Chemical (totals) (mg/kg)	pH 4.5 to 4.74	pH 4.75 to 5.24	pH 5.25 to 5.74	pH 5.75 to 6.24	pH 6.25 to 6.64	pH 6.65 to 6.89	pH 6.9 to 7.24	pH 7.25 to 7.74	pH 7.75 to 8.0
Inorganics									
Antimony	20	20	20	20	20	20	20	20	20
Arsenic	100	100	100	110	110	120	120	120	120
Barium	260	490	850	1,200	1,500	1,600	1,700	1,800	2,100
Beryllium	140	260	420	820	2,800	7,900	17,000	130,000	1,000,000
Cadmium	10	17	27	37	52	75	110	590	4,300
Chromium (+6)	No Data	No Data	No Data	No Data	No Data	No Data	No Data	No Data	No Data
Copper	330	580	2,100	11,000	59,000	130,000	200,000	330,000	330,000
Cyanide	120	120	120	120	120	120	120	120	120
Mercury	0.05	0.06	0.14	0.75	4.4	10	16	32	40
Nickel	400	730	1,100	1,500	2,000	2,600	3,500	14,000	76,000
Selenium	24	17	12	8.8	6.3	5.2	4.5	3.3	2.4
Thallium	16	18	20	24	26	28	30	34	38
Zinc	2,000	3,600	5,200	7,200	10,000	12,000	15,000	32,000	110,000

Chemical (totals) (mg/kg)	pH 4.5 to 4.74	pH 4.75 to 5.24	pH 5.25 to 5.74	pH 5.75 to 6.24	pH 6.25 to 6.64	pH 6.65 to 6.89	pH 6.9 to 7.24	pH 7.25 to 7.74	pH 7.75 to 8.0
Organics									
Benzoic Acid	440	420	410	400	400	400	400	400	400
2-Chlorophenol	20	20	20	20	20	20	19	3.6	3.1
2,4-Dichlorophenol	1.0	1.0	1.0	1.0	1.0	1.0	1.0	0.86	0.69
Dinoseb	84	45	19	8.2	4.3	3.4	3.1	2.7	2.5
Pentachlorophenol	2.7	1.6	0.75	0.33	0.18	0.15	0.12	0.11	0.10
2,4,5-TP (Silvex)	130	79	62	57	55	55	55	55	55
2,4,5-Trichlorophenol	2,000	2,000	1,900	1,800	1,600	1,400	1,200	640	64
2,4,6-Trichlorophenol	0.37	0.36	0.34	0.26	0.20	0.15	0.13	0.09	0.07

Section 742.APPENDIX B: Tier 1 Tables and Illustrations

Section 742.TABLE E: Tier 1 Groundwater Remediation Objectives for the
                         Groundwater Component of the Groundwater Ingestion Route

CAS No.	Chemical Name	Groundwater Remediation Objective	
		Class I (mg/L)	Class II (mg/L)
83-32-9	Acenaphthene	0.42	2.1
67-64-1	Acetone	0.7	0.7
15972-60-8	Alachlor	0.002[c]	0.01[c]
116-06-3	Aldicarb	0.003[c]	0.015[c]
309-00-2	Aldrin	0.00004[a]	0.0002
120-12-7	Anthracene	2.1	10.5
1912-24-9	Atrazine	0.003[c]	0.015[c]
71-43-2 .	Benzene	0.005[c]	0.025[c]
56-55-3	Benzo(*a*)anthracene	0.00013[a]	0.00065
205-99-2	Benzo(*b*)fluoranthene	0.00018[a]	0.0009
207-08-9	Benzo(*k*)fluroanthene	0.00017[a]	0.00085
50-32-8	Benzo(*a*)pyrene	0.0002[a,c]	0.002[c]
111-44-4	Bis(2-chloroethyl)ether	0.01[a]	0.01
117-81-7	Bis(2-ethylhexyl)phthalate	0.006[a,c]	0.06[c]
75-27-4	Bromodichloromethane (Dichlorobromomethane)	0.00002[a]	0.00002
75-25-2	Bromoform	0.0002[a]	0.0002
71-36-3	Butanol	0.7	0.7
85-68-7	Butyl benzyl phthalate	1.4	7.0
86-74-8	Carbazole	---	---
1563-66-2	Carbofuran	0.04[c]	0.2[c]
75-15-0	Carbon disulfide	0.7	3.5
56-23-5	Carbon tetrachloride	0.005[c]	0.025[c]
57-74-9	Chlordane	0.002[c]	0.01[c]

CAS No.	Chemical Name	Groundwater Remediation Objective	
		Class I (mg/L)	Class II (mg/L)
108-90-7	Chlorobenzene (Monochlorobenzene)	0.1[c]	0.5[c]
124-48-1	Chlorodibromomethane (Dibromochloromethane)	0.14	0.14
67-66-3	Chloroform	0.00002[a]	0.0001
218-01-9	Chrysene	0.0015[a]	0.0075
94-75-7	2,4-D	0.07[c]	0.35[c]
75-99-0	Dalapon	0.2[c]	2.0[c]
72-54-8	DDD	0.00011[a]	0.00055
72-55-9	DDE	0.00004[a]	0.0002
50-29-3	DDT	0.00012[a]	0.0006
53-70-3	Dibenzo(a,h)anthracene	0.0003[a]	0.0015
96-12-8	1,2-Dibromo-3-chloropropane	0.0002[c]	0.0002[c]
106-93-4	1,2-Dibromoethane (Ethylene dibromide)	0.00005[a,c]	0.0005[c]
84-74-2	Di-n-butyl phthalate	0.7	3.5
95-50-1	1,2-Dichlorobenzene (o - Dichlorobenzene)	0.6[c]	1.5[c]
106-46-7	1,4-Dichlorobenzene (p - Dichlorobenzene)	0.075[c]	0.375[c]
91-94-1	3,3'-Dichlorobenzidine	0.02[a]	0.1
75-34-3	1,1-Dichloroethane	0.7	3.5
107-06-2	1,2-Dichloroethane (Ethylene dichloride)	0.005[c]	0.025[c]
75-35-4	1,1-Dichloroethylene[b]	0.007[c]	0.035[c]
156-59-2	cis-1,2-Dichloroethylene	0.07[c]	0.2[c]
156-60-5	$trans$-1,2-Dichloroethylene	0.1[c]	0.5[c]
78-87-5	1,2-Dichloropropane	0.005[c]	0.025[c]
542-75-6	1,3-Dichloropropene (1,3-Dichloropropylene, $cis + trans$)	0.001[a]	0.005

CAS No.	Chemical Name	Groundwater Remediation Objective	
		Class I (mg/L)	Class II (mg/L)
60-57-1	Dieldrin	0.00002[a]	0.0001
84-66-2	Diethyl phthalate	5.6	5.6
121-14-2	2,4-Dinitrotoluene[a]	0.00002	0.00002
606-20-2	2,6-Dinitrotoluene[a]	0.0001	0.0001
88-85-7	Dinoseb	0.007[c]	0.07[c]
117-84-0	Di-n-octyl phthalate	0.14	0.7
115-29-7	Endosulfan	0.042	0.21
145-73-3	Endothall	0.1[c]	0.1[c]
72-20-8	Endrin	0.002[c]	0.01[c]
100-41-4	Ethylbenzene	0.7[c]	1.0[c]
206-44-0	Fluoranthene	0.28	1.4
86-73-7	Fluorene	0.28	1.4
76-44-8	Heptachlor	0.0004[c]	0.002[c]
1024-57-3	Heptachlor epoxide	0.0002[c]	0.001[c]
118-74-1	Hexachlorobenzene	0.00006[a]	0.0003
319-84-6	alpha-HCH (alpha-BHC)	0.00003[a]	0.00015
58-89-9	gamma-HCH (Lindane)	0.0002[c]	0.001[c]
77-47-4	Hexachlorocyclopentadiene	0.05[c]	0.5[c]
67-72-1	Hexachloroethane	0.007	0.035
193-39-5	Indeno(1,2,3-c,d)pyrene	0.00043[a]	0.00215
78-59-1	Isophorone	1.4	1.4
72-43-5	Methoxychlor	0.04[c]	0.2[c]
74-83-9	Methyl bromide (Bromomethane)	0.0098	0.049
75-09-2	Methylene chloride (Dichloromethane)	0.005[c]	0.05[c]
91-20-3	Naphthalene[2]	0.025	0.039
98-95-3	Nitrobenzene[2]	0.0035	0.0035

CAS No.	Chemical Name	Groundwater Remediation Objective	
		Class I (mg/L)	Class II (mg/L)
1918-02-1	Picloram	0.5[c]	5.0[c]
1336-36-3	Polychlorinated biphenyls (PCBs)[a]	0.0005[c]	0.0025[c]
129-00-0	Pyrene	0.21	1.05
122-34-9	Simazine	0.004[c]	0.04[c]
100-42-5	Styrene	0.1[c]	0.5[c]
93-72-1	2,4,5-TP (Silvex)	0.05[c]	0.25[c]
127-18-4	Tetrachloroethylene (Perchloroethylene)	0.005[c]	0.025[c]
108-88-3	Toluene	1.0[c]	2.5[c]
8001-35-2	Toxaphene	0.003[c]	0.015[c]
120-82-1	1,2,4-Trichlorobenzene	0.07[c]	0.7[c]
71-55-6	1,1,1-Trichloroethane[2]	0.2[c]	1.0[c]
79-00-5	1,1,2-Trichloroethane	0.005[c]	0.05[c]
79-01-6	Trichloroethylene	0.005[c]	0.025[c]
108-05-4	Vinyl acetate	7.0	7.0
75-01-4	Vinyl chloride	0.002[c]	0.01[c]
1330-20-7	Xylenes (total)	10.0[c]	10.0[c]
	Ionizable Organics		
65-85-0	Benzoic Acid	28	28
106-47-8	4-Chloroaniline (p-Chloroaniline)	0.028	0.028
95-57-8	2-Chlorophenol	0.035	0.175
120-83-2	2,4-Dichlorophenol	0.021	0.021
105-67-9	2,4-Dimethylphenol	0.14	0.14
51-28-5	2,4-Dinitrophenol	0.014	0.014
95-48-7	2-Methylphenol (o - Cresol)	0.35	0.35
86-30-6	N-Nitrosodiphenylamine	0.01[a]	0.05

CAS No.	Chemical Name	Groundwater Remediation Objective	
		Class I (mg/L)	Class II (mg/L)
621-64-7	N-Nitrosodi-n-propylamine	0.01[a]	0.01
87-86-5	Pentachlorophenol	0.001[a,c]	0.005[c]
108-95-2	Phenol	0.1[c]	0.1[c]
95-95-4	2,4,5-Trichlorophenol	0.7	3.5
88-06-2	2,4,6 Trichlorophenol	0.0064[a]	0.032
	Inorganics		
7440-36-0	Antimony	0.006[c]	0.024[c]
7440-38-2	Arsenic	0.05[c]	0.2[c]
7440-39-3	Barium	2.0[c]	2.0[c]
7440-41-7	Beryllium	0.004[c]	0.5[c]
7440-42-8	Boron	2.0[c]	2.0[c]
7440-43-9	Cadmium	0.005[c]	0.05[c]
16887-00-6	Chloride	200[c]	200[c]
7440-47-3	Chromium, total	0.1[c]	1.0[c]
18540-29-9	Chromium, ion, hexavalent	---	---
7440-48-4	Cobalt	1.0[c]	1.0[c]
7440-50-8	Copper	0.65[c]	0.65[c]
57-12-5	Cyanide	0.2[c]	0.6[c]
7782-41-4	Fluoride	4.0[c]	4.0[c]
15438-31-0	Iron	5.0[c]	5.0[c]
7439-92-1	Lead	0.0075[c]	0.1[c]
7439-96-5	Manganese	0.15[c]	10.0[c]
7439-97-6	Mercury	0.002[c]	0.01[c]
7440-02-0	Nickel	0.1[c]	2.0[c]
14797-55-8	Nitrate as N	10.0[c]	100[c]
7782-49-2	Selenium	0.05[c]	0.05[c]
7440-22-4	Silver	0.05[c]	---
14808-79-8	Sulfate	400[c]	400[c]

| CAS No. | Chemical Name | Groundwater Remediation Objective | |
		Class I (mg/L)	Class II (mg/L)
7440-28-0	Thallium	0.002ᶜ	0.02ᶜ
7440-62-2	Vanadium[2]	0.049	---
7440-66-6	Zinc	5.0ᶜ	10ᶜ

Chemical Name and Groundwater Remediation Objective Notations

ᵃ   The groundwater Health Advisory concentration is equal to ADL for carcinogens.

ᵇ   Oral Reference Dose and/or Reference Concentration under review by USEPA. Listed values subject to change.

ᶜ   Value listed is also the Groundwater Quality Standard for this chemical pursuant to 35 Ill. Adm. Code 620.410 for Class I Groundwater or 35 Ill. Adm. Code 620.420 for Class II Groundwater.

Section 742.APPENDIX B: Tier 1 Tables and Illustrations

Section 742.TABLE F: Values Used to Calculate the Tier 1 Soil Remediation Objectives for the Soil Component of the Groundwater Ingestion Route

CAS No.	Chemical Name	GW$_{obj}$ Concentration used to Calculate Tier 1 Soil Rememdiation Objectives[a]	
		Class I (mg/L)	Class II (mg/L)
83-32-9	Acenaphthene	2.0[b]	10
67-64-1	Acetone	4.0[b]	4.0
15972-60-8	Alachlor	0.002[c]	0.01[c]
116-06-3	Aldicarb	0.003[c]	0.015[c]
309-00-2	Aldrin	5.0E-6[b]	2.5E-5
120-12-7	Anthracene	10[b]	50
1912-24-9	Atrazine	0.003[c]	0.015[c]
71-43-2	Benzene	0.005[c]	0.025[c]
56-55-3	Benzo(*a*)anthracene	0.0001[b]	0.0005
205-99-2	Benzo(*b*)fluoranthene	0.0001[b]	0.0005
207-08-9	Benzo(*k*)fluroanthene	0.001[b]	0.005
50-32-8	Benzo(*a*)pyrene	0.0002[a,c]	0.002[c]
111-44-4	Bis(2-chloroethyl)ether	8.0E-5[b]	8.0E-5
117-81-7	Bis(2-ethylhexyl)phthalate	0.006[a,c]	0.06[c]
75-27-4	Bromodichloromethane (Dichlorobromomethane)	0.1[b]	0.1
75-25-2	Bromoform	0.1[b]	0.01
71-36-3	Butanol	4.0[b]	4.0
85-68-7	Butyl benzyl phthalate	7.0[b]	35
86-74-8	Carbazole	0.004[b]	0.02
1563-66-2	Carbofuran	0.04[c]	0.2[c]
75-15-0	Carbon disulfide	4.0[b]	20
56-23-5	Carbon tetrachloride	0.005[c]	0.025[c]
57-74-9	Chlordane	0.002[c]	0.01[c]

CAS No.	Chemical Name	GW$_{obj}$ Concentration used to Calculate Tier 1 Soil Rememdiation Objectives[a]	
		Class I (mg/L)	Class II (mg/L)
108-90-7	Chlorobenzene (Monochlorobenzene)	0.1[c]	0.5[c]
124-48-1	Chlorodibromomethane (Dibromochloromethane)	0.06[b]	0.06
67-66-3	Chloroform	0.1[b]	0.5
218-01-9	Chrysene	0.1[b]	0.05
94-75-7	2,4-D	0.07[c]	0.35[c]
75-99-0	Dalapon	0.2[c]	2.0[c]
72-54-8	DDD	0.0004[b]	0.002
72-55-9	DDE	0.0003[b]	0.0015
50-29-3	DDT	0.0003[b]	0.0015
53-70-3	Dibenzo(a,h)anthracene	1.0E-5[b]	5.0E-5
96-12-8	1,2-Dibromo-3-chloropropane	0.0002[c]	0.0002[c]
106-93-4	1,2-Dibromoethane (Ethylene dibromide)	0.00005[a,c]	0.0005[c]
84-74-2	Di-n-butyl phthalate	4.0[b]	20
95-50-1	1,2-Dichlorobenzene (o - Dichlorobenzene)	0.6[c]	1.5[c]
106-46-7	1,4-Dichlorobenzene (p - Dichlorobenzene)	0.075[c]	0.375[c]
91-94-1	3,3'-Dichlorobenzidine	0.0002[b]	0.001
75-34-3	1,1-Dichloroethane	4.0[b]	20
107-06-2	1,2-Dichloroethane (Ethylene dichloride)	0.005[c]	0.025[c]
75-35-4	1,1-Dichloroethylene	0.007[c]	0.035[c]
156-59-2	cis-1,2-Dichloroethylene	0.07[c]	0.2[c]
156-60-5	$trans$-1,2-Dichloroethylene	0.1[c]	0.5[c]
78-97-5	1,2-Dichloropropane	0.005[c]	0.025[c]
542-75-6	1,3-Dichloropropene (1,3-Dichloropropylene, cis + $trans$)	0.0005[b]	0.0025

CAS No.	Chemical Name	GW$_{obj}$ Concentration used to Calculate Tier 1 Soil Rememdiation Objectives[a]	
		Class I (mg/L)	Class II (mg/L)
60-57-1	Dieldrin	5.0E-6[b]	2.5E-5
84-66-2	Diethyl phthalate	30[b]	30
121-14-2	2,4-Dinitrotoluene	0.0001[b]	0.0001
606-20-2	2,6-Dinitrotoluene	0.0001	0.0001
88-85-7	Dinoseb	0.007[c]	0.07[c]
117-84-0	Di-*n*-octyl phthalate	0.7[b]	3.5
115-29-7	Endosulfan	0.2[b]	1.0
145-73-3	Endothall	0.1[c]	0.1[c]
72-20-8	Endrin	0.002[c]	0.01[c]
100-41-4	Ethylbenzene	0.7[c]	1.0[c]
206-44-0	Fluoranthene	1.0[b]	5.0
86-73-7	Fluorene	1.0[b]	5.0
76-44-8	Heptachlor	0.0004[c]	0.002[c]
1024-57-3	Heptachlor epoxide	0.0002[c]	0.001[c]
118-74-1	Hexachlorobenzene	0.001[b]	0.005
319-84-6	*alpha*-HCH (*alpha*-BHC)	1.0E-5[b]	5.0E-5
58-89-9	*gamma*-HCH (Lindane)	0.0002[c]	0.001[c]
77-47-4	Hexachlorocyclopentadiene	0.05[c]	0.5[c]
67-72-1	Hexachloroethane	0.007	0.035
193-39-5	Indeno(1,2,3-*c,d*)pyrene	0.0001[b]	0.0005
78-59-1	Isophorone	1.4	1.4
72-43-5	Methoxychlor	0.04[c]	0.2[c]
74-83-9	Methyl bromide (Bromomethane)	0.05[b]	0.25
75-09-2	Methylene chloride (Dichloromethane)	0.005[c]	0.05[c]
91-20-3	Naphthalene	1.0[b]	5.0
98-95-3	Nitrobenzene	0.02[b]	0.02

CAS No.	Chemical Name	GW$_{obj}$ Concentration used to Calculate Tier 1 Soil Rememdiation Objectives[a]	
		Class I (mg/L)	Class II (mg/L)
1918-02-1	Picloram	0.5[c]	5.0[c]
1336-36-3	Polychlorinated biphenyls (PCBs)	---	---
129-00-0	Pyrene	1.0[b]	5.0
122-34-9	Simazine	0.004[c]	0.04[c]
100-42-5	Styrene	0.1[c]	0.5[c]
93-72-1	2,4,5-TP (Silvex)	0.05[c]	0.25[c]
127-18-4	Tetrachloroethylene (Perchloroethylene)	0.005[c]	0.025[c]
108-88-3	Toluene	1.0[c]	2.5[c]
8001-35-2	Toxaphene	0.003[c]	0.015[c]
120-82-1	1,2,4-Trichlorobenzene	0.07[c]	0.7[c]
71-55-6	1,1,1-Trichloroethane[2]	0.2[c]	1.0[c]
79-00-5	1,1,2-Trichloroethane	0.005[c]	0.05[c]
79-01-6	Trichloroethylene	0.005[c]	0.025[c]
108-05-4	Vinyl acetate	40[b]	40
75-01-4	Vinyl chloride	0.002[c]	0.01[c]
1330-20-7	Xylenes (total)	10.0[c]	10.0[c]
	Ionizable Organics		
65-85-0	Benzoic Acid	100[b]	100
106-47-8	4-Chloroaniline (p-Chloroaniline)	0.1[b]	0.1
95-57-8	2-Chlorophenol	0.2[b]	1.0
120-83-2	2,4-Dichlorophenol	0.1[b]	0.1
105-67-9	2,4-Dimethylphenol	0.7[b]	0.7
51-28-5	2,4-Dinitrophenol	0.04[b]	0.04
95-48-7	2-Methylphenol (o - Cresol)	2.0[b]	2.0
86-30-6	N-Nitrosodiphenylamine	0.02[b]	0.1

CAS No.	Chemical Name	GW$_{obj}$ Concentration used to Calculate Tier 1 Soil Rememdiation Objectives[a]	
		Class I (mg/L)	Class II (mg/L)
621-64-7	*N*-Nitrosodi-*n*-propylamine	1.0E-5[b]	1.0E-5
87-86-5	Pentachlorophenol	0.001[a,c]	0.005[c]
108-95-2	Phenol	0.1[c]	0.1[c]
95-95-4	2,4,5-Trichlorophenol	4.0[b]	20
88-06-2	2,4,6-Trichlorophenol	0.008[b]	0.04
	Inorganics		
7440-36-0	Antimony	0.006[c]	0.024[c]
7440-38-2	Arsenic	0.05[c]	0.2[c]
7440-39-3	Barium	2.0[c]	2.0[c]
7440-41-7	Beryllium	0.004[c]	0.5[c]
7440-42-8	Boron	2.0[c]	2.0[c]
7440-43-9	Cadmium	0.005[c]	0.05[c]
16887-00-6	Chloride	200[c]	200[c]
7440-47-3	Chromium, total	0.1[c]	1.0[c]
18540-29-9	Chromium, ion, hexavalent	---	---
7440-48-4	Cobalt	1.0[c]	1.0[c]
7440-50-8	Copper	0.65[c]	0.65[c]
57-12-5	Cyanide	0.2[c]	0.6[c]
7782-41-4	Fluoride	4.0[c]	4.0[c]
15438-31-0	Iron	5.0[c]	5.0[c]
7439-92-1	Lead	0.0075[c]	0.1[c]
7439-96-5	Manganese	0.15[c]	10.0[c]
7439-97-6	Mercury	0.002[c]	0.01[c]
7440-02-0	Nickel	0.1[c]	2.0[c]
14797-55-8	Nitrate as N	10.0[c]	100[c]
7782-49-2	Selenium	0.05[c]	0.05[c]
7440-22-4	Silver	0.05[c]	---
14808-79-8	Sulfate	400[c]	400[c]

CAS No.	Chemical Name	GW$_{obj}$ Concentration used to Calculate Tier 1 Soil Rememdiation Objectives[a]	
		Class I (mg/L)	Class II (mg/L)
7440-28-0	Thallium	0.002[c]	0.02[c]
7440-62-2	Vanadium	0.049	---
7440-66-6	Zinc	5.0[c]	10[c]

Chemical Name and Groundwater Remediation Objective Notations

[a] The Equation S17 is used to calculate the Soil Remediation Objective for the Soil Component of the Groundwater Ingestion Route; this equation requires calculation of the Target Soil Leachate Concentration (C_w) from Equation S18: $C_w = DF \times GW_{obj}$.

[b] Value listed is the Water Health Based Limit (HBL) for this chemical from Soil Screening Guidance: User's Guide, incorporated by reference at Section 742.210; for carcinogens, the HBL is equal to a cancer risk of 1.0E-6, and for noncarcinogens is equal to a Hazard Quotient of 1.0. NOTE: These GW$_{obj}$ concentrations are not equal to the Tier 1 Groundwater Remediation Objectives for the Direct Ingestion of Groundwater Component of the Groundwater Ingestion Route, listed in Section 742.Appendix B, Table E.

[c] Value listed is also the Groundwater Quality Standard for this chemical pursuant to 35 Ill. Adm. Code 620.410 for Class I Groundwater or 35 Ill. Adm. Code 620.420 for Class II Groundwater.

Section 742.APPENDIX C: Tier 2 Tables and Illustrations

Section 742.Illustration A: Tier 2 Evaluation for Soil

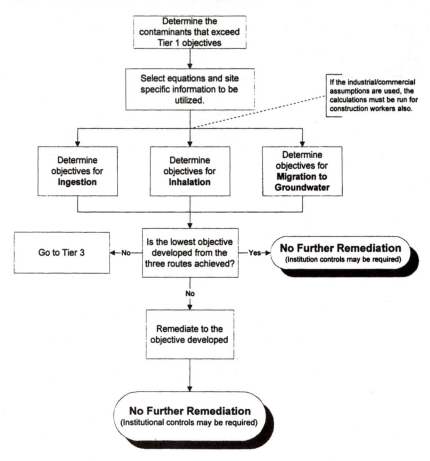

Section 742.APPENDIX C: Tier 2 Tables and Illustrations

Section 742.Illustration B: Tier 2 Evaluation for Groundwater

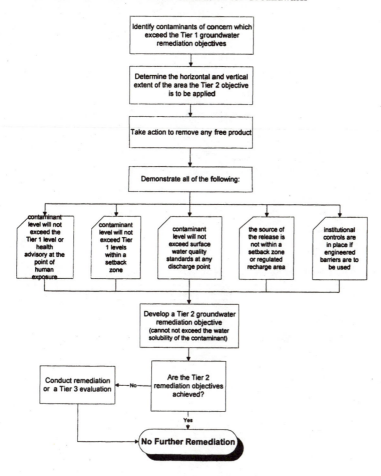

Section 742.APPENDIX C: Tier 2 Tables and Illustrations

Section 742.Illustration C: U.S. Department of Agriculture Soil Texture Classification

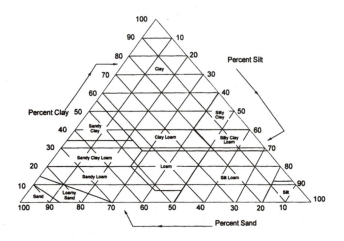

Criteria Used with the Field Method for Determining Soil Texture Classes

Criterion	Sand	Sandy loam	Loam	Silt loam	Clay loam	Clay
1. Individual grains visible to eye	Yes	Yes	Some	Few	No	No
2. Stability of dry clods	Do not form	Do not form	Easily broken	Moderately easily broken	Hard and stable	Very hard and stable
3. Stability of wet clods	Unstable	Slightyl stable	Moderately stable	Stable	Very stable	Very stable
4. Stability of "ribbon" when wet soil rubbed between thumb and fingers	Does not form	Does not form	Does not form	Broken appearance	Thin, will break	Very long, flexible

Particle Size, mm

	0.002		0.05	0.10	0.25	0.5	1.0		2.0	
Clay		Silt	Very Fine	Fine	Med.	Coarse	Very Coarse			Gravel
					Sand					

Section 742.APPENDIX C: Tier 2 Tables and Illustrations

Section 742.Table A: SSL Equations

Equations for Soil Ingestion Exposure Route		
Remediation Objectives for Noncarcinogenic Contaminants (mg/kg)	$$\dfrac{THQ \cdot BW \cdot AT \cdot 365\frac{d}{yr}}{\dfrac{1}{RfD_o} \cdot 10^{-6}\frac{kg}{mg} \cdot EF \cdot ED \cdot IR_{soil}}$$	S1
Remediation Objectives for Carcinogenic Contaminants - Residential (mg/kg)	$$\dfrac{TR \cdot AT_c \cdot 365\frac{d}{yr}}{SF_o \cdot 10^{-6}\frac{kg}{mg} \cdot EF \cdot IF_{soil-adj}}$$	S2
Remediation Objectives for Carcinogenic Contaminants - Industrial/ Commercial/ Construction Worker (mg/kg)	$$\dfrac{TR \cdot BW \cdot AT_c \cdot 365\frac{d}{yr}}{SF_o \cdot 10^{-6}\frac{kg}{mg} \cdot EF \cdot ED \cdot IR_{soil}}$$	S3
Equations for Inhalation Exposure Route (Volatiles)		
Remediation Objectives for Noncarcinogenic Contaminants - Residential, Industrial/Commercial (mg/kg)	$$\dfrac{THQ \cdot AT \cdot 365\frac{d}{yr}}{EF \cdot ED \cdot \left(\dfrac{1}{RfC} \cdot \dfrac{1}{VF}\right)}$$	S4

Remediation Objectives for Noncarcinogenic Contaminants - Construction Worker (mg/kg)	$$\dfrac{THQ \cdot AT \cdot 365 \frac{d}{yr}}{EF \cdot ED \cdot \left(\dfrac{1}{RfC} \cdot \dfrac{1}{VF'}\right)}$$	S5
Remediation Objectives for Carcinogenic Contaminants - Residential, Industrial/Commercial (mg/kg)	$$\dfrac{TR \cdot AT_c \cdot 365 \frac{d}{yr}}{URF \cdot 1{,}000 \frac{ug}{mg} \cdot EF \cdot ED \cdot \dfrac{1}{VF}}$$	S6
Remediation Objectives for Carcinogenic Contaminants - Construction Worker (mg/kg)	$$\dfrac{TR \cdot AT_c \cdot 365 \frac{d}{yr}}{URF \cdot 1{,}000 \frac{ug}{mg} \cdot EF \cdot ED \cdot \dfrac{1}{VF'}}$$	S7
Equation for Derivation of the Volatilization Factor - Residential, Industrial/Commercial, VF (m³/kg)	$$VF = \frac{Q}{C} \cdot \frac{\left(3.14 \cdot D_A \cdot T\right)^{1/2}}{\left(2 \cdot \rho_b \cdot D_A\right)} \cdot 10^{-4}\,\frac{m^2}{cm^2}$$	S8
Equation for Derivation of the Volatilization Factor - Construction Worker, VF' (m³/kg)	$$VF' = \frac{VF}{10}$$	S9
Equation for Derivation of Apparent Diffusivity, D_A (cm²/s)	$$D_A = \frac{\left(\theta_a^{3.33} \cdot D_i \cdot H'\right) + \left(\theta_w^{3.33} \cdot D_w\right)}{\eta^2} \cdot \frac{1}{\left(\rho_b \cdot K_d\right) + \theta_w + \left(\theta_a \cdot H'\right)}$$	S10

Equations for Inhalation Exposure Route (Fugitive Dusts)		
Remediation Objectives for Noncarcinogenic Contaminants - Residential, Industrial/Commercial (mg/kg)	$$\dfrac{THQ \cdot AT \cdot 365\frac{d}{yr}}{EF \cdot ED \cdot \left(\dfrac{1}{RfC} \cdot \dfrac{1}{PEF}\right)}$$	S11
Remediation Objectives for Noncarcinogenic Contaminants - Construction Worker (mg/kg)	$$\dfrac{THQ \cdot AT \cdot 365\frac{d}{yr}}{EF \cdot ED \cdot \left(\dfrac{1}{RfC} \cdot \dfrac{1}{PEF}\right)}$$	S12
Remediation Objectives for Carcinogenic Contaminants - Residential, Industrial/Commercial (mg/kg)	$$\dfrac{TR \cdot AT_c \cdot 365\frac{d}{yr}}{URF \cdot 1,000\frac{ug}{mg} \cdot EF \cdot ED \cdot \dfrac{1}{PEF}}$$	S13
Remediation Objectives for Carcinogenic Contaminants - Construction Worker (mg/kg)	$$\dfrac{TR \cdot AT_c \cdot 365\frac{d}{yr}}{URF \cdot 1,000\frac{ug}{mg} \cdot EF \cdot ED \cdot \dfrac{1}{PEF}}$$	S14
Equation for Derivation of Particulate Emission Factor, PEF (m³/kg)	$$PEF = \dfrac{Q}{C} \cdot \dfrac{3,600\frac{s}{hr}}{0.036 \cdot (1-V) \cdot \left(\dfrac{U_m}{U_t}\right)^3 \cdot F(x)}$$	S15

501

Equations for the Soil Component of the Groundwater Ingestion Exposure Route	Equation for Derivation of Particulate Emission Factor, PEF' - Construction Worker (m³/kg)	$PEF' = \dfrac{PEF}{10}$ NOTE: PEF must be the industrial/commercial value	S16
	Remediation Objective (mg/kg)	$C_w \cdot \left[K_d + \dfrac{(\theta_w + \theta_a \cdot H')}{\rho_b} \right]$ NOTE: This equation can only be used to model contaminant migration not in the water bearing unit.	S17
	Target Soil Leachate Concentration, C_w (mg/L)	$C_w = DF \cdot GW_{obj}$	S18
	Soil-Water Partition Coefficient, K_d (cm³/g)	$K_d = K_{oc} \cdot f_{oc}$	S19
	Water-Filled Soil Porosity, θ_w (L_{water}/L_{soil})	$\theta_w = \eta \cdot \left(\dfrac{I}{K_s}\right)^{1/(2b+3)}$	S20
	Air-Filled Soil Porosity, θ_a (L_{air}/L_{soil})	$\theta_a = \eta - \theta_w$	S21
	Dilution Factor, DF (unitless)	$DF = 1 + \dfrac{K \cdot i \cdot d}{I \cdot L}$	S22

Groundwater Remediation Objective for Carcinogenic Contaminants, GW_{obj} (mg/L)	$$\dfrac{TR \bullet BW \bullet AT_c \bullet 365\frac{d}{yr}}{SF_o \bullet IR_w \bullet EF \bullet ED}$$	S23
Total Soil Porosity, η (L_{pore}/L_{soil})	$$\eta = 1 - \dfrac{\rho_b}{\rho_s}$$	S24
Equation for Estimation of Mixing Zone Depth, d (m)	$$d = (0.0112 \bullet L^2)^{0.5} + d_a\left[1 - \exp\left(\dfrac{-L \bullet I}{K \bullet i \bullet d_a}\right)\right]$$	S25
Mass-Limit Volatilization Factor for the Inhalation Exposure Route - Residential, Industrial/Commercial, VF (m³/kg)	$$VF_{M-L} = \dfrac{Q}{C} \bullet \dfrac{\left[T_{M-L} \bullet \left(3.15 \bullet 10^7 \frac{s}{yr}\right)\right]}{\rho_b \bullet d_s \bullet 10^6 \frac{g}{mg}}$$	S26
Mass-Limit Equations for Inhalation Exposure Route and Soil Component of the Groundwater Ingestion Exposure Route	NOTE: This equation may be used when area and depth of contaminant source are known or can be estimated reliably.	
Mass-Limit Volatilization Factor for Inhalation Exposure Route - Construction Worker, VF' - (m³/kg)	$$VF'_{M-L} = \dfrac{VF_{M-L}}{10}$$	S27
Mass-Limit Remediation Objective for Soil Component of the Groundwater Ingestion Exposure Route (mg/kg)	$$\dfrac{\left(C_w \bullet I_{M-L} \bullet ED_{M-L}\right)}{\rho_b \bullet d_s}$$	S28
	NOTE: This equation may be used when area and depth of contaminant source are known or can be estimated reliably.	
Equation for Derivation of the Soil Saturation Limit, C_{sat}	$$C_{sat} = \dfrac{S}{\rho_b} \bullet \left[(K_d \bullet \rho_b) + \theta_w + (H' \bullet \theta_a)\right]$$	S29

Section 742.APPENDIX C: Tier 2 Tables and Illustrations

Section 742.Table B: SSL Parameters

Symbol	Parameter	Units	Source	Parameter Value(s)
AT	Averaging Time for Noncarcinogens in Ingestion Equation	yr		Residential = 6 Industrial/Commercial = 25 Construction Worker = 0.115
AT	Averaging Time for Noncarcinogens in Inhalation Equation	yr		Residential = 30 Industrial/Commercial = 25 Construction Worker = 0.115
AT_c	Averaging Time for Carcinogens	yr	SSL	70
BW	Body Weight	kg		Residential = 15, noncarcinogens 70, carcinogens Industrial/Commercial = 70 Construction Worker = 70
C_{sat}	Soil Saturation Concentration	mg/kg	Appendix A, Table A or Equation S29 in Appendix C, Table A	Chemical-Specific or Calculated Value
C_w	Target Soil Leachate Concentration	mg/L	Equation S18 in Appendix C, Table A	Groundwater Standard, Health Advisory concentration, or Calculated Value
d	Mixing Zone Depth	m	SSL or Equation S25 in Appendix C, Table A	2 m or Calculated Value
d_a	Aquifer Thickness	m	Field Measurement	Site-Specific

Symbol	Parameter	Units	Source	Parameter Value(s)
d_s	Depth of Source	m	Field Measurement or Estimation	Site-Specific
D_A	Apparent Diffusivity	cm²/s	Equation S10 in Appendix C, Table A	Calculated Value
D_i	Diffusivity in Air	cm²/s	Appendix C, Table E	Chemical-Specific
D_w	Diffusivity in Water	cm²/s	Appendix C, Table E	Chemical-Specific
DF	Dilution Factor	unitless	Equation S22 in Appendix C, Table A	20 or Calculated Value
ED	Exposure Duration for Ingestion of Carcinogens	yr		Industrial/Commercial = 25 Construction Worker = 1
ED	Exposure Duration for Inhalation of Carcinogens	yr		Residential = 30 Industrial/Commercial = 25 Construction Worker = 1
ED	Exposure Duration for Ingestion of Noncarcinogens	yr		Residential = 6 Industrial/Commercial = 25 Construction Worker = 1
ED	Exposure Duration for Inhalation of Noncarcinogens	yr		Residential = 30 Industrial/Commercial = 25 Construction Worker = 1
ED	Exposure Duration for the Direct Ingestion of Groundwater	yr		Residential = 30 Industrial/Commercial = 25 Construction Worker = 1

Symbol	Parameter	Units	Source	Parameter Value(s)
ED_{M-L}	Exposure Duration for Migration to Groundwater Mass-Limit Equation S28	yr	SSL	70
EF	Exposure Frequency	d/yr		**Residential = 350** **Industrial/Commercial = 250** **Construction Worker = 30**
F(x)	Function dependent on U_a/U_i	unitless	SSL	0.194
f_{oc}	Organic Carbon Content of Soil	g/g	SSL or Field Measurement (See Appendix C, Table F)	Surface Soil = 0.006 Subsurface soil = 0.002, or Site-Specific
GW_{obj}	Groundwater Remediation Remediation Objective	mg/L	Appendix B, Table E, 35 IAC 620.Subpart F, or Equation S23 in Appendix C, Table A	Chemical-Specific or Calculated
H'	Henry's Law Constant	unitless	Appendix C, Table E	Chemical-Specific
i	Hydraulic Gradient	m/m	Field Measurement (See Appendix C, Table F)	Site-Specific
I	Infiltration Rate	m/yr	SSL	0.3
I_{M-L}	Infiltration Rate for Migration to Groundwater Mass-Limit Equation S28	m/yr	SSL	0.18

Symbol	Parameter	Units	Source	Parameter Value(s)
$IF_{soil-adj}$ (residential)	Age Adjusted Soil Ingestion Factor for Carcinogens	(mg-yr)/(kg-d)	SSL	114
IR_{soil}	Soil Ingestion Rate	mg/d		Residential = 200 Industrial/Commercial = 50 Construction Worker = 480
IR_w	Daily Water Ingestion Rate	L/d		Residential = 2 Industrial/Commercial = 1
K	Aquifer Hydraulic Conductivity	m/yr	Field Measurement (See Appendix C, Table F)	Site-Specific
K_d	Soil-Water Partition Coefficient	cm³/g or L/kg	Equation S19 in Appendix C, Table A	Calculated Value
K_{oc}	Organic Carbon Partition Coefficient	cm³/g or L/kg	Appendix C, Table E or Appendix C, Table I	Chemical-Specific
K_s	Saturated Hydraulic Conductivity	m/yr	Appendix C, Table K Appendix C, Illustration C	Site-Specific
L	Source Length Parallel to Groundwater Flow	m	Field Measurement	Site-Specific
PEF	Particulate Emission Factor	m³/kg	SSL or Equation S15 in Appendix C, Table A	Residential = 1.32 • 10⁹ or Site-Specific Industrial/Commercial = 1.24 • 10⁹ or Site-Specific
PEF'	Particulate Emission Factor adjusted for Agitation (construction worker)	m³/kg	Equation S16 in Appendix C, Table A using PEF (industrial/commercial)	1.24 • 10⁸ or Site-Specific

Symbol	Parameter	Units	Source	Parameter Value(s)
Q/C (used in VF equations)	Inverse of the mean concentration at the center of a square source	$(g/m^2\text{-}s)/(kg/m^3)$	Appendix C, Table H	Residential = 68.81 Industrial/Commercial = 85.81 Construction Worker = 85.81
Q/C (used in PEF equations)	Inverse of the mean concentration at the center of a square source	$(g/m^2\text{-}s)/(kg/m^3)$	SSL or Appendix C, Table H	Residential = 90.80 Industrial/Commercial = 85.81 Construction Worker = 85.81
RfC	Inhalation Reference Concentration	mg/m^3	IEPA (IRIS/HEAST*)	Toxicological-Specific (Note: for Construction Workers use subchronic reference concentrations)
RfD_o	Oral Reference Dose	$mg/(kg\text{-}d)$	IEPA (IRIS/HEAST*)	Toxicological-Specific (Note: for Construction Worker use subchronic reference doses)
S	Solubility in Water	mg/L	Appendix C, Table E	Chemical-Specific
SF_o	Oral Slope Factor	$(mg/kg\text{-}d)^{-1}$	IEPA (IRIS/HEAST*)	Toxicological-Specific
T	Exposure Interval	s		Residential = $9.5 \bullet 10^8$ Industrial/Commercial = $7.9 \bullet 10^8$ Construction Worker = $3.6 \bullet 10^6$
$T_{M\text{-}L}$	Exposure Interval for Mass-Limit Volatilization Factor Equation S26	yr	SSL	30
THQ	Target Hazard Quotient	unitless	SSL	1

Symbol	Parameter	Units	Source	Parameter Value(s)
TR	Target Cancer Risk	unitless		Residential = 10^{-6} at the point of human exposure Industrial/Commercial = 10^{-4} at the point of human exposure Construction Worker = 10^{-5} at the point of human exposure
U_m	Mean Annual Windspeed	m/s	SSL	4.69
URF	Inhalation Unit Risk Factor	$(\mu g/m^3)^{-1}$	IEPA (IRIS/HEAST*)	Toxicological-Specific
U_t	Equivalent Threshold Value of Windspeed at 7 m	m/s	SSL	11.32
V	Fraction of Vegetative Cover	unitless	SSL or Field Measurement	0.5 or Site-Specific
VF	Volatilization Factor	m^3/kg	Equation S8 in Appendix C, Table A	Calculated Value
VF'	Volatilization Factor adjusted for Agitation	m^3/kg	Equation S9 in Appendix C, Table A	Calculated Value
VF_{M-L}	Mass-Limit Volatilization Factor	m^3/kg	Equation S26 in Appendix C, Table A	Calculated Value
VF'_{M-L}	Mass-Limit Volatilization Factor adjusted for Agitation	m^3/kg	Equation S27 in Appendix C, Table A	Calculated Value

Symbol	Parameter	Units	Source	Parameter Value(s)
η	Total Soil Porosity	L_{pore}/L_{soil}	SSL or Equation S24 in Appendix C, Table A	0.43, or Gravel = 0.25 Sand = 0.32 Silt = 0.40 Clay = 0.36, or **Calculated Value**
θ_a	Air-Filled Soil Porosity	L_{air}/L_{soil}	SSL or Equation S21 in Appendix C, Table A	Surface Soil (top 1 meter) = 0.28 Subsurface Soil (below 1 meter) = 0.13, or Gravel = 0.05 Sand = 0.14 Silt - 0.24 Clay = 0.19, or **Calculated Value**
θ_w	Water-Filled Soil Porosity	L_{water}/L_{soil}	SSL or Equation S20 in Appendix C, Table A	Surface Soil (top 1 meter) = 0.15 Subsurface Soil (below 1 meter) = 0.30, or Gravel = 0.20 Sand = 0.18 Silt = 0.16 Clay = 0.17, or **Calculated Value**

Symbol	Parameter	Units	Source	Parameter Value(s)
ρ_b	Dry Soil Bulk Density	kg/L or g/cm³	SSL or Field Measurement (See Appendix C, Table F)	1.5, or Gravel = 2.0 Sand = 1.8 Silt = 1.6 Clay = 1.7, or Site-Specific
ρ_s	Soil Particle Density	g/cm³	SSL or Field Measurement (See Appendix C, Table F)	2.65, or Site-Specific
ρ_w	Water Density	g/cm³	SSL	1
$1/(2b+3)$	Exponential in Equation S20	unitless	Appendix C, Table K Appendix C, Illustration C	Site-Specific

a   HEAST = Health Effects Assessment Summary Tables. USEPA, Office of Solid Waste and Emergency Response. EPA/SQO/R-95/036. Updated Quarterly.

Section 742.Appendix C: Tier 2 Tables and Illustrations

Section 742.Table C: RBCA Equations

Equations for the combined exposures routes of soil ingestion	Remediation Objectives for Carcinogenic Contaminants (mg/kg)	**R1** $$\dfrac{TR \bullet BW \bullet AT_c \bullet 365 \frac{d}{yr}}{EF \bullet ED \bullet \left[\left[SF_o \bullet 10^{-6} \frac{kg}{mg} \bullet \left((IR_{soil} \bullet RAF_o) + (SA \bullet M \bullet RAF_d)\right)\right] + \left[SF_i \bullet IR_{air} \bullet \left(VF_{ss} + VF_p\right)\right]\right]}$$
inhalation of vapors and particulates, and dermal contact with soil	Remediation Objectives for Non-carcinogenic Contaminants (mg/kg)	**R2** $$\dfrac{THQ \bullet BW \bullet AT_n \bullet 365 \frac{d}{yr}}{EF \bullet ED \bullet \left[\dfrac{10^{-6} \frac{kg}{mg} \bullet \left[\left(IR_{soil} \bullet RAF_o\right) + \left(SA \bullet M \bullet RAF_d\right)\right]}{RfD_o} + \dfrac{IR_{air} \bullet \left(VF_{ss} + VF_p\right)}{RfD_i}\right]}$$
	Volatilization Factor for Surficial Soils, VF$_{ss}$ (kg/m^3)	**R3** $$VF_{ss} = \dfrac{2 \bullet W \bullet \rho_s \bullet 10^3 \frac{cm^3 \cdot kg}{m^3 \cdot g}}{U_{air} \bullet \delta_{air}} \bullet \sqrt{\dfrac{D_s^{eff} \bullet H'}{\pi \bullet \left[\theta_{ws} + (k_s \bullet \rho_s) + (H' \bullet \theta_{air})\right] \bullet \tau}}$$
	Whichever is less between R3 and R4	**R4** $$VF_{ss} = \dfrac{W \bullet \rho_s \bullet d \bullet 10^3 \frac{cm^3 \cdot kg}{m^3 \cdot g}}{U_{air} \bullet \delta_{air} \bullet \tau}$$

	Equation	
Volatilization Factor for Surficial Soils Regarding Particulates, VF_p (kg/m³)	$$VF_p = \frac{P_e \cdot W \cdot 10^3 \, \frac{cm^3 \cdot kg}{m^3 \cdot g}}{U_{air} \cdot \delta_{air}}$$	R5
Effective Diffusion Coefficient in Soil Based on Vapor-Phase Concentration D_s^{eff} (cm²/s)	$$D_s^{eff} = \frac{D^{air} \cdot \theta_{as}^{3.33}}{\theta_T^2} + \frac{D^{water} \cdot \theta_{ws}^{3.33}}{H' \cdot \theta_T^2}$$	R6
Equations for the ambient vapor inhalation (outdoor) route from subsurface soils — Remediation Objectives for Carcinogenic Contaminants (mg/kg)	$$\frac{RBSL_{air} \cdot 10^{-3}}{VF_{samb}}$$	R7
Remediation Objectives for Non-carcinogenic Contaminants (mg/kg)	$$\frac{RBSL_{air} \cdot 10^{-3}}{VF_{samb}}$$	R8

Carcinogenic Risk-Based Screening Level for Air, $RBSL_{air}$ (ug/m^3)	$$RBSL_{air} = \frac{TR \bullet BW \bullet AT_c \bullet 365\frac{d}{yr} \bullet 10^3\frac{ug}{mg}}{SF_i \bullet IR_{air} \bullet EF \bullet ED}$$	R9
Noncarcinogenic Risk-Based Screening Level for Air, $RBSL_{air}$ (ug/m^3)	$$RBSL_{air} = \frac{THQ \bullet RfD_i \bullet BW \bullet AT_n \bullet 365\frac{d}{yr} \bullet 10^3\frac{ug}{mg}}{IR_{air} \bullet EF \bullet ED}$$	R10
Volatilization Factor - Subsurface Soil to Ambient Air, VF_{samb} ($mg/m^3)/(mg/kg_{soil})$	$$VF_{samb} = \frac{H \bullet \rho_s \bullet 10^3\frac{cm^3 \cdot kg}{m^3 \cdot g}}{\left[\theta_{ws} + (k_s \bullet \rho_s) + (H \bullet \theta_{as})\right] \bullet \left[1 + \frac{(U_{air} \bullet \delta_{air} \bullet L_s)}{(D_s^{eff} \bullet W)}\right]}$$	R11

Equations for the Soil Component of the Groundwater		
Remediation Objective (mg/kg)	$$\dfrac{GW_{source}}{LF_{sw}}$$ NOTE: This equation can only be used to model contaminant migration not in the water bearing unit.	R12
Ingestion Exposure Route		
Groundwater at the source, GW$_{source}$ (mg/L)	$$GW_{source} = \dfrac{GW_{comp}}{C_{(x)}\big/C_{source}}$$	R13
Leaching Factor, LF$_{sw}$ (mg/L$_{water}$)/(mg/kg$_{soil}$)	$$LF_{sw} = \dfrac{\rho_s \cdot \dfrac{cm^3 \cdot kg}{L \cdot g}}{\left[\theta_{ws} + (k_s \cdot \rho_s) + (H' \cdot \theta_{as})\right] \cdot \left[1 + \dfrac{(U_{gw} \cdot \delta_{gw})}{(I \cdot W)}\right]}$$	R14
Steady-State Attenuation Along the Centerline of a Dissolved Plume, C$_{(x)}$/C$_{source}$	$$\dfrac{C_{(x)}}{C_{source}} = \exp\left[\left(\dfrac{X}{2\alpha_x}\right) \cdot \left(1 - \sqrt{1 + \dfrac{4\lambda \cdot \alpha_x}{U}}\right)\right] \cdot erf\left[\dfrac{S_w}{4 \cdot \sqrt{\alpha_y \cdot X}}\right] \cdot erf\left[\dfrac{S_d}{2 \cdot \sqrt{\alpha_z \cdot X}}\right]$$ NOTE: 1. This equation does not predict the contaminant flow within bedrock. 2. If the value of the First Order Degradation Constant (λ) is not readily available, then set $\lambda = 0$.	R15
Longitudinal Dispersivity, α_x (cm)	$$\alpha_x = 0.10 \cdot X$$	R16

Transverse Dispersivity, α_y (cm)	$\alpha_y = \dfrac{\alpha_x}{3}$	R17
Vertical Dispersivity, α_z (cm)	$\alpha_z = \dfrac{\alpha_x}{20}$	R18
Specific Discharge, U (cm/d)	$U = \dfrac{K \bullet i}{\theta_T}$	R19
Soil-Water Sorption Coefficient, k_s	$k_s = K_{oc} \bullet f_{oc}$	R20
Volumetric Air Content in Vadose Zone Soils, θ_{as} (cm^3_{air}/cm^3_{soil})	$\theta_{as} = \theta_T - \dfrac{(w \bullet \rho_s)}{\rho_w}$	R21
Volumetric Water Content in Vadose Zone Soils, θ_{ws} (cm^3_{water}/cm^3_{soil})	$\theta_{ws} = \dfrac{w \bullet \rho_s}{\rho_w}$	R22
Total Soil Porosity, θ_T (cm^3/cm^3_{soil})	$\theta_T = \theta_{as} + \theta_{ws}$	R23

Equations for the Groundwater Ingestion Exposure Route	Groundwater Darcy Velocity, U_{gw} (cm/yr)	$$U_{gw} = K \cdot i$$ R24
	Remediation Objective for Carcinogenic Contaminants (mg/L)	$$\frac{TR \cdot BW \cdot AT_c \cdot 365\frac{d}{yr}}{SF_o \cdot IR_w \cdot EF \cdot ED}$$ R25
	Dissolved Hydrocarbon Concentration along Centerline, $C_{(x)}$ (g/cm^3_{water})	$C_{(x)} =$ $$C_{source} \cdot exp\left[\left(\frac{X}{2\alpha_x}\right) \cdot \left(1 - \sqrt{1 + \frac{4\lambda \cdot \alpha_x}{U}}\right)\right] \cdot erf\left[\frac{S_w}{4 \cdot \sqrt{\alpha_y \cdot X}}\right] \cdot erf\left[\frac{S_d}{2 \cdot \sqrt{\alpha_z \cdot X}}\right]$$ R26

NOTE:

1. This equation does not predict the contaminant flow within bedrock.

2. If the value of the First Order Degradation Constant (λ) is not readily available, then set $\lambda = 0$.

Section 742.APPENDIX C: Tier 2 Tables and Illustrations

Section 742.Table D: RBCA Parameters

Symbol	Parameter	Units	Source	Parameter Value(s)
AT_c	Averaging Time for Carcinogens	yr	RBCA	70
AT_n	Averaging Time for Noncarcinogens	yr	RBCA	Residential = 30 Industrial/Commercial = 25 Construction Worker = 0.115
BW	Adult Body Weight	kg	RBCA	70
C_source	The greatest potential concentration of the contaminant of concern in the groundwater at the source of the contamination, based on the concentrations of contaminants in groundwater due to the release and the projected concentration of the contaminant migrating from the soil to the groundwater.	mg/L	Field Measurement	Site-Specific
C_(x)	Concentration of Contaminant in Groundwater at Distance X from the source	mg/L	Equation R26 in Appendix C, Table C	Calculated Value

Symbol	Parameter	Units	Source	Parameter Value(s)
$C_{(x)}/C_{source}$	Steady-State Attenuation Along the Centerline of a Dissolved Plume	unitless	Equation R15 in Appendix C, Table C	Calculated Value
d	Lower Depth of Surficial Soil Zone	cm	Field Measurement	100 or Site-Specific (not to exceed 100)
D^{air}	Diffusion Coefficient in Air	cm^2/s	Appendix C, Table E	Chemical-Specific
D^{water}	Diffusion Coefficient in Water	cm^2/s	Appendix C, Table E	Chemical-Specific
D_s^{eff}	Effective Diffusion Coefficient in Soil Based on Vapor-Phase Concentration	cm^2/s	Equation R6 in Appendix C, Table C	Calculated Value
ED	Exposure Duration	yr	RBCA	Residential = 30 Industrial/Commercial = 25 Construction Worker = 1
EF	Exposure Frequency	d/yr	RBCA	Residential = 350 Industrial/Commercial = 250 Construction Worker = 30
erf	Error Function	unitless	Appendix C, Table G	Mathematical Function

Symbol	Parameter	Units	Source	Parameter Value(s)
f_{oc}	Organic Carbon Content of Soil	g/g	RBCA or Field Measurement (See Appendix C, Table F)	Surface Soil = 0.006 Subsurface Soil = 0.002 or Site-Specific
GW_{comp}	Groundwater Objective at the Compliance Point	mg/L	Appendix B, Table E, 35 IAC 620.Subpart F, or Equation R25 in Appendix C, Table C	Site-Specific
GW_{source}	Groundwater Concentration at the Source	mg/L	Equation R13 in Appendix C, Table C	Calculated Value
H'	Henry's Law Constant	cm^3_{water}/cm^3_{air}	Appendix C, Table E	Chemical-Specific
i	Hydraulic Gradient	cm/cm (unitless)	Field Measurement (See Appendix C, Table F)	Site-Specific
I	Infiltration Rate	cm/yr	RBCA	30
IR_{air}	Daily Outdoor Inhalation Rate	m^3/d	RBCA	20
IR_{soil}	Soil Ingestion Rate	mg/d	RBCA	Residential = 100 Industrial/Commercial = 50 Construction Worker = 480
IR_w	Daily Water Ingestion Rate	L/d	RBCA	Residential = 2 Industrial/Commercial = 1

Symbol	Parameter	Units	Source	Parameter Value(s)
K	Aquifer Hydraulic Conductivity	cm/d for Equations R15, R19 and R26 cm/yr for Equation R24	Field Measurement (See Appendix C, Table F)	Site-Specific
K_{oc}	Organic Carbon Partition Coefficient	cm^3/g or L/kg	Appendix C, Table E or Appendix C, Table I	Chemical-Specific
k_s (non-ionizing organics)	Soil Water Sorption Coefficient	cm^3_{water}/g_{soil}	Equation R20 in Appendix C, Table C	Calculated Value
k_s (ionizing organics)	Soil Water Sorption Coefficient	cm^3_{water}/g_{soil}	Equation R20 in Appendix C, Table C	Chemical-Specific
k_s (inorganics)	Soil Water Sorption Coefficient	cm^3_{water}/g_{soil}	Appendix C, Table J	Chemical-Specific
L_s	Depth to Subsurface Soil Sources	cm	RBCA	100
LF_{sw}	Leaching Factor	$(mg/L_{water})/(mg/kg_{soil})$	Equation R14 in Appendix C, Table C	Calculated Value
M	Soil to Skin Adherence Factor	mg/cm^2	RBCA	0.5

Symbol	Parameter	Units	Source	Parameter Value(s)
Pe	Particulate Emission Rate	g/cm²-s	RBCA	$6.9 \cdot 10^{-14}$
RAF$_d$	Dermal Relative Absorption Factor	unitless	RBCA	0.5
RAF$_d$ (PNAs)	Dermal Relative Absorption Factor	unitless	RBCA	0.05
RAF$_d$ (inorganics)	Dermal Relative Absorption Factor	unitless	RBCA	0
RAF$_o$	Oral Relative Absorption Factor	unitless	RBCA	1.0
RBSL$_{air}$	Carcinogenic Risk-Based Screening Level for Air	µg/m³	Equation R9 in Appendix C, Table C	Chemical-, Media-, and Exposure Route-Specific
RBSL$_{air}$	Noncarcinogenic Risk-Based Screening Level for Air	µg/m³	Equation R10 in Appendix C, Table C	Chemical-, Media-, and Exposure Route-Specific
RfD$_i$	Inhalation Reference Dose	mg/kg-d	IEPA (IRIS/HEAST)	Toxicological-Specific
RfD$_o$	Oral Reference Dose	mg/(kg-d)	IEPA (IRIS/HEAST)	Toxicological-Specific (Note: for Construction Worker use subchronic reference doses)
SA	Skin Surface Area	cm²/d	RBCA	3,160

Symbol	Parameter	Units	Source	Parameter Value(s)
S_d	Source Width Perpendicular to Groundwater Flow Direction in Vertical Plane	cm	Field Measurement	For Migration to Groundwater Route: Use 200 or Site-Specific For Groundwater remediation objective: Use Site-Specific
S_w	Source Width Perpendicular to Groundwater Flow Direction in Horizontal Plane	cm	Field Measurement	Site-Specific
SF_i	Inhalation Cancer Slope Factor	$(mg/kg-d)^{-1}$	IEPA (IRIS/HEAST")	Toxicological-Specific
SF_o	Oral Slope Factor	$(mg/kg-d)^{-1}$	IEPA (IRIS/HEAST")	Toxicological-Specific
THQ	Target Hazard Quotient	unitless	RBCA	1
TR	Target Cancer Risk	unitless	RBCA	Residential = 10^{-6} at the point of human exposure Industrial/Commercial = 10^{-6} at the point of human exposure Construction Worker = 10^{-6} at the point of human exposure
U	Specific Discharge	cm/d	Equation R19 in Appendix C, Table C	Calculated Value

Symbol	Parameter	Units	Source	Parameter Value(s)
U_{air}	Average Wind Speed Above Ground Surface in Ambient Mixing Zone	cm/s	RBCA	225
U_{gw}	Groundwater Darcy Velocity	cm/yr	Equation R24 in Appendix C, Table C	Calculated Value
VF_p	Volatilization Factor for Surficial Soils Regarding Particulates	kg/m³	Equation R5 in Appendix C, Table C	Calculated Value
VF_{samb}	Volatilization Factor (Subsurface Soils to Ambient Air)	(mg/m³ air)/(mg/ kg$_{soil}$) or kg/m³	Equation R11 in Appendix C, Table C	Calculated Value
VF_{ss}	Volatilization Factor for Surficial Soils	kg/m³	Use Equations R3 and R4 in Appendix C, Table C	Calculated Value from Equation R3 or R4 (whichever is less)
W	Width of Source Area Parallel to Direction to Wind or Groundwater Movement	cm	Field Measurement	Site-Specific

Symbol	Parameter	Units	Source	Parameter Value(s)
w	Average Soil Moisture Content	g_{water}/g_{soil}	RBCA or Field Measurement (See Appendix C, Table F)	0.1, or Surface Soil (top 1 meter) = 0.1 Subsurface Soil (below 1 meter) = 0.2, or Site-Specific
X	Distance along the Centerline of the Groundwater Plume Emanating from a Source. The x direction is the direction of groundwater flow	cm	Field Measurement	Site-Specific
α_x	Longitudinal Dispersivity	cm	Equation R16 in Appendix C, Table C	Calculated Value
α_y	Transverse Dispersivity	cm	Equation R17 in Appendix C, Table C	Calculated Value
α_z	Vertical Dispersivity	cm	Equation R18 in Appendix C, Table C	Calculated Value
δ_{air}	Ambient Air Mixing Zone Height	cm	RBCA	200

Symbol	Parameter	Units	Source	Parameter Value(s)
δ_{gw}	Groundwater Mixing Zone Thickness	cm	RBCA	200
θ_{as}	Volumetric Air Content in Vadose Zone Soils	cm^3_{air}/cm^3_{soil}	RBCA or Equation R21 in Appendix C, Table C	Surface Soil (top 1 meter) = 0.28 Subsurface Soil (below 1 meter) = 0.13, or Gravel = 0.05 Sand = 0.14 Silt = 0.16 Clay = 0.17, or Calculated Value
θ_{ws}	Volumetric Water Content in Vadose Zone Soils	cm^3_{water}/cm^3_{soil}	RBCA or Equation R22 in Appendix C, Table C	Surface Soil (top 1 meter) = 0.15 Subsurface Soil (below 1 meter) = 0.30, or Gravel = 0.20 Sand = 0.18 Silt = 0.16 Clay = 0.17, or Calculated Value

Symbol	Parameter	Units	Source	Parameter Value(s)
θ_T	Total Soil Porosity	cm^3/cm^3_{soil}	RBCA or Equation R23 in Appendix C, Table C	0.43, or Gravel = 0.25 Sand = 0.32 Silt = 0.40 Clay = 0.36, or Calculated Value
λ	First Order Degradation Constant	d^{-1}	Appendix C, Table E	Chemical-Specific
π	pi			3.1416
ρ_s	Soil Bulk Density	g/cm^3	RBCA or Field Measurement (See Appendix C, Table F)	1.5, or Gravel = 2.0 Sand = 1.8 Silt = 1.6 Clay = 1.7, or Site-Specific
ρ_w	Water Density	g/cm^3	RBCA	1
τ	Averaging Time for Vapor Flux	s	RBCA	$9.46 \cdot 10^8$

[a] HEAST = Health Effects Assessment Summary Tables. USEPA, Office of Solid Waste and Emergency Response. EPA/540/R-95/036. Updated Quarterly.

Section 742.APPENDIX C: Tier 2 Tables and Illustrations

Section 742.Table E: Default Physical and Chemical Parameters

CAS No.	Chemical	Solubility in Water (S) (mg/L)	Diffusivity in Air (D$_i$) (cm^2/s)	Diffusivity in Water (D$_w$) (cm^2/s)	Dimensionless Henry's Law Constant (H') (25°C)	Organic Carbon Partition Coefficient (K$_{oc}$) (L/kg)	First Order Degradation Constant (λ) (d^{-1})
Neutral Organics							
83-32-9	Acenaphthene	4.24	0.0421	7.69E-6	0.00636	7,080	0.0034
67-64-1	Acetone	1,000,000	0.124	1.14E-5	0.00159	0.575	0.0495
15972-60-8	Alachlor	242	0.0198	5.69E-6	0.00000132	394	No Data
116-06-3	Aldicarb	6,000	0.0305	7.19E-6	0.0000000574	12	0.00109
309-00-2	Aldrin	0.18	0.0132	4.86E-6	0.00697	2,450,000	0.00059
120-12-7	Anthracene	0.0434	0.0324	7.74E-6	0.00267	29,500	0.00075
1912-24-9	Atrazine	70	0.0258	6.69E-6	0.0000000005	451	No Data
71-43-2	Benzene	1,750	0.088	9.80E-6	0.228	58.9	0.0009

CAS No.	Chemical	Solubility in Water (S) (mg/L)	Diffusivity in Air (D_i) (cm²/s)	Diffusivity in Water (D_w) (cm²/s)	Dimensionless Henry's Law Constant (H') (25°C)	Organic Carbon Partition Coefficient (K_{oc}) (L/kg)	First Order Degradation Constant (λ) (d⁻¹)
56-55-3	Benzo(a)anthracene	0.0094	0.0510	9.00E-6	0.000137	398,000	0.00051
205-99-2	Benzo(b)fluoranthene	0.0015	0.0226	5.56E-6	0.00455	1,230,000	0.00057
207-08-9	Benzo(k)fluoranthene	0.0008	0.0226	5.56E-6	0.000034	1,230,000	0.00016
65-85-0	Benzoic Acid	3,500	0.0536	7.97E-6	0.0000631	0.600	No Data
50-32-8	Benzo(a)pyrene	0.00162	0.043	9.00E-6	0.0000463	1,020,000	0.00065
111-44-4	Bis(2-chloroethyl)ether	17,200	0.0692	7.53E-6	0.000738	15.5	0.0019
117-81-7	Bis(2-ethylhexyl)phthalate	0.34	0.0351	3.66E-6	0.00000418	15,100,000	0.0018
75-27-4	Bromodichloromethane	6,740	0.0298	1.06E-5	0.0656	55.0	No Data
75-25-2	Bromoform	3,100	0.0149	1.03E-5	0.0219	87.1	0.0019
71-36-3	Butanol	74,000	0.0800	9.30E-6	0.000361	6.92	0.01283
85-68-7	Butyl Benzyl Phthalate	2.69	0.0174	4.83E-6	0.0000517	57,500	0.00385
86-74-8	Carbazole	7.48	0.0390	7.03E-6	0.000000626	3,390	No Data

CAS No.	Chemical	Solubility in Water (S) (mg/L)	Diffusivity in Air (D_i) (cm^2/s)	Diffusivity in Water (D_w) (cm^2/s)	Dimensionless Henry's Law Constant (H') (25°C)	Organic Carbon Partition Coefficient (K_{oc}) (L/Kg)	First Order Degradation Constant (λ) (d^{-1})
1563-66-2	Carbofuran	320	0.0249	6.63E-6	.00377	37	No Data
75-15-0	Carbon Disulfide	1,190	0.104	1.00E-5	1.24	45.7	No Data
56-23-5	Carbon Tetrachloride	793	0.0780	8.80E-6	1.25	174	0.0019
57-74-9	Chlordane	0.056	0.0118	4.37E-6	0.00199	120,000	0.00025
106-47-8	p-Chloroaniline	5,300	0.0483	1.01E-5	0.0000136	66.1	No Data
108-09-7	Chlorobenzene	472	0.0730	8.70E-6	0.152	219	0.0023
124-48-1	Chlorodibromomethane	2,600	0.0196	1.05E-5	0.0321	63.1	0.00385
67-66-3	Chloroform	7,920	0.104	1.00E-5	0.15	39.8	0.00039
95-57-8	2-Chlorophenol	22,000	0.0501	9.46E-6	0.016	388	No Data
218-01-9	Chrysene	0.0016	0.0248	6.21E-6	0.00388	398,000	0.00035
94-75-7	2,4-D	680	0.0231	7.31E-6	0.00000041	451	0.00385
72-54-8	4,4'-DDD	0.09	0.0169	4.76E-6	0.000164	1,000,000	0.000062

CAS No.	Chemical	Solubility in Water (S) (mg/L)	Diffusivity in Air (D_i) (cm²/s)	Diffusivity in Water (D_w) (cm²/s)	Dimensionless Henry's Law Constant (H') (25°C)	Organic Carbon Partition Coefficient (K_{oc}) (L/kg)	First Order Degradation Constant (λ) (d⁻¹)
72-55-9	4,4'-DDE	0.12	0.0144	5.87E-6	0.000861	4,470,000	0.000062
50-29-3	4,4'-DDT	0.025	0.0137	4.95E-6	0.000332	2,630,000	0.000062
75-99-0	Dalapon	900,000	0.0414	9.46E-6	0.00000264	5.8	0.005775
53-70-3	Dibenzo(a,h)anthracene	0.00249	0.0202	5.18E-6	0.000000603	3,800,000	0.00037
96-12-8	1,2-Dibromo-3-chloropropane	1,200	0.0212	7.02E-6	0.00615	182	0.001925
106-93-4	1,2-Dibromoethane	4,200	0.0287	8.06E-6	0.0303	93	0.005775
84-74-2	Di-n-butyl Phthalate	11.2	0.0438	7.86E-6	0.0000000385	33,900	0.03013
95-50-1	1,2-Dichlorobenzene	156	0.0690	7.90E-6	0.0779	617	0.0019
106-46-7	1,4-Dichlorobenzene	73.8	0.0690	7.90E-6	0.0996	617	0.0019
91-94-1	3,3-Dichlorobenzidine	3.11	0.0194	6.74E-6	0.000000164	724	0.0019

CAS No.	Chemical	Solubility in Water (S) (mg/L)	Diffusivity in Air (D$_i$) (cm²/s)	Diffusivity in Water (D$_w$) (cm²/s)	Dimensionless Henry's Law Constant (H') (25°C)	Organic Carbon Partition Coefficient (K$_{oc}$) (L/kg)	First Order Degradation Constant (λ) (d^{-1})
75-34-3	1,1-Dichloroethane	5,060	0.0742	1.05E-5	0.23	31.6	0.0019
107-06-2	1,2-Dichloroethane	8,520	0.104	9.90E-6	0.0401	17.4	0.0019
75-35-4	1,1-Dichloroethylene	2,250	0.0900	1.04E-5	1.07	58.9	0.0053
156-59-2	cis-1,2-Dichloroethylene	3,500	0.0736	1.13E-5	0.167	35.5	0.00024
156-60-5	trans-1,2-Dichloroethylene	6,300	0.0707	1.19E-5	0.385	52.5	0.00024
120-83-2	2,4-Dichlorophenol	4,500	0.0346	8.77E-6	0.00013	147	0.00027
78-87-5	1,2-Dichloropropane	2,800	0.0782	8.73E-6	0.115	43.7	0.00027
542-75-6	1,3-Dichloropropylene (cis + trans)	2,800	0.0626	1.00E-5	0.726	45.7	0.061
60-57-1	Dieldrin	0.195	0.0125	4.74E-6	0.000619	21,400	0.00032
84-66-2	Diethyl Phthalate	1,080	0.0256	6.35E-6	0.0000185	288	0.00619
105-67-9	2,4-Dimethylphenol	7,870	0.0584	8.69E-6	0.000082	209	0.0495
51-28-5	2,4-Dinitrophenol	2,790	0.0273	9.06E-6	0.0000182	0.01	0.00132

CAS No.	Chemical	Solubility in Water (S) (mg/L)	Diffusivity in Air (D$_l$) (cm²/s)	Diffusivity in Water (D$_w$) (cm²/s)	Dimensionless Henry's Law Constant (H') (25°C)	Organic Carbon Partition Coefficient (K$_{oc}$) (L/kg)	First Order Degradation Constant (λ) (d^{-1})
121-14-2	2,4-Dinitrotoluene	270	0.203	7.06E-6	0.0000038	95.5	0.00192
606-20-2	2,6-Dinitrotoluene	182	0.0327	7.26E-6	0.0000306	69.2	0.00192
88-85-7	Dinoseb	52	0.0215	6.62E-6	0.0000189	1,120	0.002817
117-84-0	Di-n-octyl Phthalate	0.02	0.0151	3.58E-6	0.00274	83,200,000	0.0019
115-29-7	Endosulfan	0.51	0.0115	4.55E-6	0.000459	2,140	0.07629
145-73-3	Endothall	21,000	0.0291	8.07E-6	0.0000000107	0.29	No Data
72-20-8	Endrin	0.25	0.0125	4.74E-6	0.000308	12,300	0.00032
100-41-4	Ethylbenzene	169	0.0750	7.80E-6	0.323	363	0.003
206-44-0	Fluoranthene	0.206	0.0302	6.35E-6	0.00066	107,000	0.00019
86-73-7	Fluorene	1.98	0.0363	7.88E-6	0.00261	13,800	0.000691
76-44-8	Heptachlor	0.18	0.0112	5.69E-6	60.7	1,410,000	0.13
1024-57-3	Heptachlor epoxide	0.2	0.0132	4.23E-6	0.00039	83,200	0.00063

CAS No.	Chemical	Solubility in Water (S) (mg/L)	Diffusivity in Air (D_i) (cm²/s)	Diffusivity in Water (D_w) (cm²/s)	Dimensionless Henry's Law Constant (H') (25°C)	Organic Carbon Partition Coefficient (K_{oc}) (L/kg)	First Order Degradation Constant (λ) (d⁻¹)
118-74-1	Hexachlorobenzene	6.2	0.0542	5.91E-6	0.0541	55,000	0.00017
319-84-6	alpha-HCH (alpha-BHC)	2.0	0.0142	7.34E-6	0.000435	1,230	0.0025
58-89-9	gamma-HCH (Lindane)	6.8	0.0142	7.34E-6	0.000574	1,070	0.0029
77-47-4	Hexachlorocyclo-pentadiene	1.8	0.0161	7.21E-6	1.11	200,000	0.012
67-72-1	Hexachloroethane	50	0.0025	6.80E-6	0.159	1,780	0.00192
193-39-5	Indeno(1,2,3-c,d)pyrene	0.000022	0.0190	5.66E-6	0.0000656	3,470,000	0.00047
78-59-1	Isophorone	12,000	0.0623	6.76E-6	0.000272	46.8	0.01238
7439-97-6	Mercury	---	0.0307	6.30E-6	0.467	---	No Data
72-43-5	Methoxychlor	0.045	0.0156	4.46E-6	0.000648	97,700	0.0019
74-83-9	Methyl Bromide	15,200	0.0728	1.21E-5	0.256	10.5	0.01824
75-09-2	Methylene Chloride	13,000	0.101	1.17E-5	0.0898	11.7	0.012
95-48-7	2-Methylphenol	26,000	0.0740	8.30E-6	0.0000492	91.2	0.0495

CAS No.	Chemical	Solubility in Water (S) (mg/L)	Diffusivity in Air (D_i) (cm²/s)	Diffusivity in Water (D_w) (cm²/s)	Dimensionless Henry's Law Constant (H') (25°C)	Organic Carbon Partition Coefficient (K_{oc}) (L/kg)	First Order Degradation Constant (λ) (d⁻¹)
91-20-3	Naphthalene	31.0	0.0590	7.50E-6	0.0198	2,000	0.0027
98-95-3	Nitrobenzene	2,090	0.0760	8.60E-6	0.000984	64.6	0.00176
86-30-6	N-Nitrosodiphenylamine	35.1	0.0312	6.35E-6	0.000205	1,290	0.01
621-64-7	N-Nitrosodi-n-propylamine	9,890	0.0545	8.17E-6	0.0000923	24.0	0.0019
87-86-5	Pentachlorophenol	1,950	0.0560	6.10E-6	0.000001	592	0.00045
108-95-2	Phenol	82,800	0.0820	9.10E-6	0.0000163	28.8	0.099
1918-02-1	Picloram	430	0.0255	5.28E-6	0.00000000166	1.98	No Data
1336-36-3	Polychlorinated biphenyls (PCBs)	0.7	--------ᵃ	--------ᵃ	--------ᵃ	309,000	No Data
129-00-0	Pyrene	0.135	0.0272	7.24E-6	0.000451	105,000	0.00018
122-34-9	Simazine	5	0.027	7.36E-6	0.0000000133	133	No Data
100-42-5	Styrene	310	0.0710	8.00E-6	0.113	776	0.0033
93-72-1	2,4,5-TP (Silvex)	31	0.0194	5.83E-6	0.0000000032	5,440	No Data

CAS No.	Chemical	Solubility in Water (S) (mg/L)	Diffusivity in Air (D_i) (cm²/s)	Diffusivity in Water (D_w) (cm²/s)	Dimensionless Henry's Law Constant (H') (25°C)	Organic Carbon Partition Coefficient (K_{oc}) (L/kg)	First Order Degradation Constant (λ) (d^{-1})
127-18-4	Tetrachloroethylene	200	0.0720	8.20E-6	0.754	155	0.00096
108-88-3	Toluene	526	0.0870	8.60E-6	0.272	182	0.011
8001-35-2	Toxaphene	0.74	0.0116	4.34E-6	0.000246	257,000	No Data
120-82-1	1,2,4-Trichlorobenzene	300	0.0300	8.23E-6	0.0582	1,780	0.0019
71-55-6	1,1,1-Trichloroethane	1,330	0.0780	8.80E-6	0.705	110	0.0013
79-00-5	1,1,2-Trichloroethane	4,420	0.0780	8.80E-6	0.0374	50.1	0.00095
79-01-6	Trichloroethylene	1,100	0.0790	9.10E-6	0.422	166	0.00042
95-95-4	2,4,5-Trichlorophenol	1,200	0.0291	7.03E-6	0.000178	1,600	0.00038
88-06-2	2,4,6-Trichlorophenol	800	0.0318	6.25E-6	0.000319	381	0.00038
108-05-4	Vinyl Acetate	20,000	0.0850	9.20E-6	0.021	5.25	No Data
57-01-4	Vinyl Chloride	2,760	0.106	1.23E-6	1.11	18.6	0.00024
108-38-3	m-Xylene	161	0.070	7.80E-6	0.301	407	0.0019

CAS No	Chemical	Solubility in Water (S) (mg/L)	Diffusivity in Air (D_i) (cm²/s)	Diffusivity in Water (D_w) (cm²/s)	Dimensionless Henry's Law Constant (H') (25°C)	Organic Carbon Partition Coefficient (K_{oc}) (L/kg)	First Order Degradation Constant (λ) (d⁻¹)
95-47-6	o-Xylene	178	0.087	1.00E-5	0.213	363	0.0019
106-42-3	p-Xylene	185	0.0769	8.44E-6	0.314	389	0.0019
1330-20-7	Xylenes (total)	186	0.0720	9.34E-6	0.25	260	0.0019

Chemical Abstracts Service (CAS) registry number. This number in the format xxx-xx-x, is unique for each chemical and allows efficient searching on computerized data bases.

*Soil remediation objectives are determined pursuant to 40 CFR 761.120, as incorporated by reference at Section 732.104 (the USEPA "PCB Spill Cleanup Policy"), for most sites; persons remediating sites should consult with BOL if calculation of Tier 2 soil remediation objectives is desired.

Section 742.APPENDIX C: Tier 2 Tables and Illustrations

Section 742.Table F: Methods for Determining Physical Soil Parameters

Methods for Determining Physical Soil Parameters		
Parameter	Sampling Location[a]	Method
ρ_b (soil bulk density)	Surface	ASTM - D 1556-90 Sand Cone Method[b]
		ASTM - D 2167-94 Rubber Balloon Method[b]
		ASTM - D 2922-91 Nuclear Method[b]
	Subsurface	ASTM - D 2937-94 Drive Cylinder Method[b]
ρ_s (soil particle density)	Surface or Subsurface	ASTM - D 854-92 Specific Gravity of Soil[b]
w (moisture content)	Surface or Subsurface	ASTM - D 4959-89 (Reapproved 1994) Standard[b]
		ASTM - D 4643-93 Microwave Oven[b]
		ASTM - D2216-92 Laboratory Determination[b]
		ASTM - D3017-88 (Reapproved 1993) Nuclear Method[b]
		Equivalent USEPA Method (e.g., sample preparation procedures described in methods 3541 or 3550)
f_{oc} (organic carbon content)	Surface or Subsurface	Nelson and Sommers (1982)
		ASTM - D 2974-87 (Reapproved 1995) Moisture, Ash, and Organic Matter[b]
		USEPA Method 9060A Total Organic Content

Methods for Determining Physical Soil Parameters		
Parameter	Sampling Location[a]	Method
η or θ_T (total soil porosity)	Surface or Subsurface (calculated)	Equation S24 in Appendix C, Table A for SSL Model, or Equation R23 in Appendix C, Table C for RBCA Model
θ_a or θ_{as} (air-filled soil porosity)	Surface or Subsurface (calculated)	Equation S21 in Appendix C, Table A for SSL Model, or Equation R21 in Appendix C, Table C for RBCA Model
θ_w or θ_{ws} (water-filled soil porosity)	Surface or Subsurface (calculated)	Equation S20 in Appendix C, Table A for SSL Model, or Equation R22 in Appendix C, Table C for RBCA Model
K (hydraulic conductivity)	Surface or Subsurface	ASTM - D 5084-90 Flexible Wall Permeameter
		Pump Test
		Slug Test
i (hydraulic gradient)	Surface or Subsurface	Field Measurement

[a] This is the location where the sample is collected
[b] As incorporated by reference in Section 742.120.

Section 742.APPENDIX C: **Tier 2 Tables and Illustrations**

Section 742.Table G: **Error Function (erf)**

$$erf(\beta) = \frac{2}{\sqrt{\pi}} \int_0^\beta e^{-\varepsilon^2} d\varepsilon$$

β	erf (β)
0	0
0.05	0.056372
0.1	0.112463
0.15	0.167996
0.2	0.222703
0.25	0.276326
0.3	0.328627
0.35	0.379382
0.4	0.428392
0.45	0.475482
0.5	0.520500
0.55	0.563323
0.6	0.603856
0.65	0.642029
0.7	0.677801
0.75	0.711156
0.8	0.742101
0.85	0.770668
0.9	0.796908
0.95	0.820891
1.0	0.842701
1.1	0.880205
1.2	0.910314

1.3	0.934008
1.4	0.952285
1.5	0.966105
1.6	0.976348
1.7	0.983790
1.8	0.989091
1.9	0.992790
2.0	0.995322
2.1	0.997021
2.2	0.998137
2.3	0.998857
2.4	0.999311
2.5	0.999593
2.6	0.999764
2.7	0.999866
2.8	0.999925
2.9	0.999959
3.0	0.999978

742.APPENDIX C: Tier 2 Tables and Illustrations

Section 742.Table H: Q/C Values by Source Area

Source (Acres)	Area Q/C Value $(g/m^2\text{-s per } kg/m^3)$
0.5	97.78
1	85.81
2	76.08
5	65.75
10	59.16
30	50.60

Section 742.TABLE I: K_{oc} Values for Ionizing Organics as a Function of pH (cm^3/g or L/kg)

pH	Benzoic Acid	2-Chloro-phenol	2,4-Dichloro-phenol	Pentachloro-phenol	2,4,5-Trichloro-phenol	2,4,6-Trichloro-phenol	Dinoseb	2,3,5-TP (Silvex)
4.5	1.07E+01	3.98E+02	1.59E+02	1.34E+04	2.37E+03	1.06E+03	3.00E+03	1.28E+04
4.6	9.16E+00	3.98E+02	1.59E+02	1.24E+04	2.37E+03	1.05E+03	2.71E+03	1.13E+04
4.7	7.79E+00	3.98E+02	1.59E+02	1.13E+04	2.37E+03	1.05E+03	2.41E+03	1.01E+04
4.8	6.58E+00	3.98E+02	1.59E+02	1.02E+04	2.37E+03	1.05E+03	2.12E+03	9.16E+03
4.9	5.54E+00	3.98E+02	1.59E+02	9.05E+03	2.37E+03	1.04E+03	1.85E+04	8.40E+03
5.0	4.64E+00	3.98E+02	1.59E+02	7.96E+03	2.36E+03	1.03E+03	1.59E+04	7.76E+03
5.1	3.88E+00	3.98E+02	1.59E+02	6.93E+03	2.36E+03	1.02E+03	1.36E+04	7.30E+03
5.2	3.25E+00	3.98E+02	1.59E+02	5.97E+03	2.35E+03	1.01E+03	1.15E+04	6.91E+03
5.3	2.72E+00	3.98E+02	1.59E+02	5.10E+03	2.34E+03	9.99E+02	9.66E+03	6.60E+03
5.4	2.29E+00	3.98E+02	1.58E+02	4.32E+03	2.33E+03	9.82E+02	8.10E+03	6.36E+03
5.5	1.94E+00	3.97E+02	1.58E+02	3.65E+03	2.32E+03	9.62E+02	6.77E+03	6.16E+03
5.6	1.65E+00	3.97E+02	1.58E+02	3.07E+03	2.31E+03	9.38E+02	5.65E+03	6.00E+03
5.7	1.42E+00	3.97E+02	1.58E+02	2.58E+03	2.29E+03	9.10E+02	4.73E+03	5.88E+03
5.8	1.24E+00	3.97E+02	1.58E+02	2.18E+03	2.27E+03	8.77E+02	3.97E+03	5.78E+03
5.9	1.09E+00	3.97E+02	1.57E+02	1.84E+03	2.24E+03	8.39E+02	3.35E+03	5.70E+03

pH	Benzoic Acid	2-Chloro-phenol	2,4-Dichloro-phenol	Pentachloro-phenol	2,4,5-Trichloro-phenol	2,4,6-Trichloro-phenol	Dinoseb	2,3,5-TP (Silvex)
6.0	9.69E-01	3.96E+02	1.57E+02	1.56E+03	2.21E+03	7.96E+02	2.84E+03	5.64E+03
6.1	8.75E-01	3.96E+02	1.57E+02	1.33E+03	2.17E+03	7.48E+02	2.43E+03	5.59E+03
6.2	7.99E-01	3.96E+02	1.56E+02	1.15E+03	2.12E+03	6.97E+02	2.10E+03	5.55E+03
6.3	7.36E-01	3.95E+02	1.55E+02	9.98E+02	2.06E+03	6.44E+02	1.83E+03	5.52E+03
6.4	6.89E-01	3.94E+02	1.54E+02	8.77E+02	1.99E+03	5.89E+02	1.62E+03	5.50E+03
6.5	6.51E-01	3.93E+02	1.53E+02	7.81E+02	1.91E+03	5.33E+02	1.45E+03	5.48E+03
6.6	6.20E-01	3.92E+02	1.52E+02	7.03E+02	1.82E+03	4.80E+02	1.32E+03	5.46E+03
6.7	5.95E-01	3.90E+02	1.50E+02	6.40E+02	1.71E+03	4.29E+02	1.21E+03	5.45E+03
6.8	5.76E-01	3.88E+02	1.47E+02	5.92E+02	1.60E+03	3.81E+02	1.12E+03	5.44E+03
6.9	5.60E-01	3.86E+02	1.45E+02	5.52E+02	1.47E+03	3.38E+02	1.05E+03	5.43E+03
7.0	5.47E-01	3.83E+02	1.41E+02	5.21E+02	1.34E+03	3.00E+02	9.96E+02	5.43E+03
7.1	5.38E-01	3.79E+02	1.38E+02	4.96E+02	1.21E+03	2.67E+02	9.52E+02	5.42E+03
7.2	5.32E-01	3.75E+02	1.33E+02	4.76E+02	1.07E+03	2.39E+02	9.18E+02	5.42E+03
7.3	5.25E-01	3.69E+02	1.28E+02	4.61E+02	9.43E+02	2.15E+02	8.90E+02	5.42E+03
7.4	5.19E-01	3.62E+02	1.21E+02	4.47E+02	8.19E+02	1.95E+02	8.68E+02	5.41E+03
7.5	5.16E-01	3.54E+02	1.14E+02	4.37E+02	7.03E+02	1.78E+02	8.50E+02	5.41E+03
7.6	5.13E-01	3.44E+02	1.07E+02	4.29E+02	5.99E+02	1.64E+02	8.36E+02	5.41E+03

pH	Benzoic Acid	2-Chloro-phenol	2,4-Dichloro-phenol	Pentachloro-phenol	2,4,5-Trichloro-phenol	2,4,6-Trichloro-phenol	Dinoseb	2,3,5-TP (Silvex)
7.7	5.09E-01	3.33E+02	9.84E+01	4.23E+02	5.07E+02	1.53E+02	8.25E+02	5.41E+03
7.8	5.06E-01	3.19E+02	8.97E+01	4.18E+02	4.26E+02	1.44E+02	8.17E+02	5.41E+03
7.9	5.06E-01	3.04E+02	8.07E+01	4.14E+02	3.57E+02	1.37E+02	8.10E+02	5.41E+03
8.0	5.06E-01	2.86E+02	7.17E+01	4.10E+02	2.98E+02	1.31E+02	8.04E+02	5.41E+03

Section 742.TABLE J: Values to be Substituted for k_s when Evaluating Inorganics as a Function of pH (cm^3_{water}/g_{soil})

pH	As	Ba	Be	Cd	Cr (+3)	Cr (+6)	Hg	Ni	Ag	Se	Tl	Zn
4.9	2.5E+	1.1E+	2.3E+	1.5E+	1.2E+	3.1E+	4.0E-02	1.6E+	1.0E-01	1.8E+	4.4E+	1.6E+
5.0	2.5E+	1.2E+	2.6E+	1.7E+	1.9E+	3.1E+	6.0E-02	1.8E+	1.3E-01	1.7E+	4.5E+	1.8E+
5.1	2.5E+	1.4E+	2.8E+	1.9E+	3.0E+	3.0E+	9.0E-02	2.0E+	1.6E-01	1.6E+	4.6E+	1.9E+
5.2	2.6E+	1.5E+	3.1E+	2.1E+	4.9E+	2.9E+	1.4E-01	2.2E+	2.1E-01	1.5E+	4.7E+	2.1E+
5.3	2.6E+	1.7E+	3.5E+	2.3E+	8.1E+	2.8E+	2.0E-01	2.4E+	2.6E-01	1.4E+	4.8E+	2.3E+
5.4	2.6E+	1.9E+	3.8E+	2.5E+	1.3E+	2.7E+	3.0E-01	2.6E+	3.3E-01	1.3E+	5.0E+	2.5E+
5.5	2.6E+	2.1E+	4.2E+	2.7E+	2.1E+	2.7E+	4.6E-01	2.8E+	4.2E-01	1.2E+	5.1E+	2.6E+
5.6	2.6E+	2.2E+	4.7E+	2.9E+	3.5E+	2.6E+	6.9E-01	3.0E+	5.3E-01	1.1E+	5.2E+	2.8E+
5.7	2.7E+	2.4E+	5.3E+	3.1E+	5.5E+	2.5E+	1.0E-00	3.2E+	6.7E-01	1.1E+	5.4E+	3.0E+
5.8	2.7E+	2.6E+	6.0E+	3.3E+	8.7E+	2.5E+	1.6E-00	3.4E+	8.4E-01	9.8E+	5.5E+	3.2E+
5.9	2.7E+	2.8E+	6.9E+	3.5E+	1.3E+	2.4E+	2.3E-00	3.6E+	1.1E+	9.2E+	5.6E+	3.4E+
6.0	2.7E+	3.0E+	8.2E+	3.7E+	2.0E+	2.3E+	3.5E-00	3.8E+	1.3E+	8.6E+	5.8E+	3.6E+
6.1	2.7E+	3.1E+	9.9E+	4.0E+	3.0E+	2.3E+	5.1E-00	4.0E+	1.7E+	8.0E+	5.9E+	3.9E+
6.2	2.8E+	3.3E+	1.2E+	4.2E+	4.2E+	2.2E+	7.5E-00	4.2E+	2.1E+	7.5E+	6.1E+	4.2E+
6.3	2.8E+	3.5E+	1.6E+	4.4E+	5.8E+	2.2E+	1.1E+	4.5E+	2.7E+	7.0E+	6.2E+	4.4E+
6.4	2.8E+	3.6E+	2.1E+	4.8E+	7.7E+	2.1E+	1.6E+	4.7E+	3.4E+	6.5E+	6.4E+	4.7E+
6.5	2.8E+	3.7E+	2.8E+	5.2E+	9.9E+	2.0E+	2.2E+	5.0E+	4.2E+	6.1E+	6.6E+	5.1E+
6.6	2.8E+	3.9E+	3.9E+	5.7E+	1.2E+	2.0E+	3.0E+	5.4E+	5.3E+	5.7E+	6.7E+	5.4E+

pH	As	Ba	Be	Cd	Cr (+3)	Cr (+6)	Hg	Ni	Ag	Se	Tl	Zn
6.7	2.9E+	4.0E+	5.5E+	6.4E+	1.5E+	1.9E+	4.0E+	5.8E+	6.6E+	5.3E+	6.9E+	5.8E+
6.8	2.9E+	4.1E+	7.9E+	7.5E+	1.8E+	1.9E+	5.2E+	6.5E+	8.3E+	5.0E+	7.1E+	6.2E+
6.9	2.9E+	4.2E+	1.1E+	9.1E+	2.1E+	1.8E+	6.6E+	7.4E+	1.0E+	4.7E+	7.3E+	6.8E+
7.0	2.9E+	4.2E+	1.7E+	1.1E+	2.5E+	1.8E+	8.2E+	8.8E+	1.3E+	4.3E+	7.4E+	7.5E+
7.1	2.9E+	4.3E+	2.5E+	1.5E+	2.8E+	1.7E+	9.9E+	1.1E+	1.6E+	4.1E+	7.6E+	8.3E+
7.2	3.0E+	4.4E+	3.8E+	2.0E+	3.1E+	1.7E+	1.2E+	1.4E+	2.0E+	3.8E+	7.8E+	9.5E+
7.3	3.0E+	4.4E+	5.7E+	2.8E+	3.4E+	1.6E+	1.3E+	1.8E+	2.5E+	3.5E+	8.0E+	1.1E+
7.4	3.0E+	4.5E+	8.6E+	4.0E+	3.7E+	1.6E+	1.5E+	2.5E+	3.1E+	3.3E+	8.2E+	1.3E+
7.5	3.0E+	4.6E+	1.3E+	5.9E+	3.9E+	1.6E+	1.6E+	3.5E+	3.9E+	3.1E+	8.5E+	1.6E+
7.6	3.1E+	4.6E+	2.0E+	8.7E+	4.1E+	1.5E+	1.7E+	4.9E+	4.8E+	2.9E+	8.7E+	1.9E+
7.7	3.1E+	4.7E+	3.0E+	1.3E+	4.2E+	1.5E+	1.8E+	7.0E+	5.9E+	2.7E+	8.9E+	2.4E+
7.8	3.1E+	4.9E+	4.6E+	1.9E+	4.3E+	1.4E+	1.9E+	9.9E+	7.3E+	2.5E+	9.1E+	3.1E+
7.9	3.1E+	5.0E+	6.9E+	2.9E+	4.3E+	1.4E+	1.9E+	1.4E+	8.9E+	2.4E+	9.4E+	4.0E+
8.0	3.1E+	5.2E+	1.0E+	4.3E+	4.3E+	1.4E+	2.0E+	1.9E+	1.1E+	2.2E+	9.6E+	5.3E+

Section 742.APPENDIX C: Tier 2 Tables and Illustrations

Section 742.TABLE K: **Parameter Estimates for Calculating Water-Filled Soil Porosity (θ_w)**

Soil Texture[a]	Saturated Hydraulic Conductivity, K_s (m/yr)	$1/(2b+3)$[b]
Sand	1,830	0.090
Loamy Sand	540	0.085
Sandy Loam	230	0.080
Silt Loam	120	0.074
Loam	60	0.073
Sandy Clay Loam	40	0.058
Silt Clay Loam	13	0.054
Clay Loam	20	0.050
Sandy Clay	10	0.042
Silt Clay	8	0.042
Clay	5	0.039

[a]   The appropriate texture classification is determined by a particle size analysis by ASTM D2488-93 as incorporated by reference in Section 742.210 and the U.S. Department of Agriculture Soil Textural Triangle shown in Appendix C, Illustration C.

[b]   Where b is the soil-specific exponential parameter (unitless)

IT IS SO ORDERED.

Section 41 of the Environmental Protection Act (415 ILCS 5/41(1994)) provides for the appeal of final Board opinions and orders to the Illinois Appellate Court within 35 days of the date of service of this order. The Rules of the Supreme Court of Illinois establish filing requirement. (See also 35 Ill.Adm. Code 101.246 "Motions for Reconsideration.")

I, Dorothy M. Gunn, Clerk of the Illinois Pollution Control Board, hereby certify that the above opinion and order was adopted on the _5th_ day of _June_, 1997, by a vote of _7-0_.

Dorothy M. Gunn, Clerk
Illinois Pollution Control Board

Appendix 2

Memorandum of Understanding Between IEPA and USEPA Region 5

Memorandum of Understanding
between
the Illinois Environmental Protection Agency
and
the United States Environmental Protection Agency Region 5
on
the Illinois Site Remediation Program,
the Illinois Tiered Approach to Corrective Action Objectives,
and
the Environmental Remediation Programs
administered by
the Region 5 Waste, Pesticides, and Toxics Division
under
the Resource Conservation and Recovery Act (RCRA)
and
the Toxic Substances Control Act (TSCA)

I.   Introduction

The Illinois Environmental Protection Agency ("Illinois EPA") and the United States Environmental Protection Agency, Region 5 ("Region 5") entered a Memorandum of Agreement ("MOA") under the Resource Conversation and Recovery Act ("RCRA") Subtitle C, effective January 3 1, 1986. Illinois EPA and Region 5 have periodically modified that MOA to reflect authorization changes. Among other things, the RCRA MOA established operating procedures for general RCRA program coordination and communication under Subtitle C between Illinois EPA and Region 5. Illinois EPA and Region 5 do not have a general operating MOA under Subtitle I, but have maintained a continuous workirig relationship under successive co-operative agreements since 1987.

On April 6, 1995 the Illinois EPA and Region 5 entered Superfund Memorandum of Agreement, Addendum No. I. That agreement specifies how the Illinois EPA Pre-Notice Site Cleanup Program, precursor of the Site Remediation Program referenced in this MOU, intersects with administration of the Superfund program by Region 5 and Illinois EPA.

Effective December 21 ,1 995, the Environmental Protection Act of the State of Illinois was amended to add Title XVII: Site Remediation Program(415 Illinois Compiled-Statutes 5/58-58.12). Title XVII was amended effective June 30, 1996. The Illinois EPA and Region 5 have agreed to establish this Memorandum of Understanding ("MOU") for the following purposes:

(1)   to encourage voluntary environmental cleanup, which is protective of human health and the environment, at contaminated locations in Illinois;

(2)   to establish how the State of Illinois Site Remediation Program intersects with RCRA and the Toxic Substances Control Act ("TSCA"), as administered by the Waste, Pesticides, and Toxics Division of Region 5; and

(3)   to recognize the Illinois EPA's use of the Tiered Approach to Corrective Action Objectives (35 Ill. Adm_ Code 742) for sites subject to RCRA or the **TSCA**[1].

This MOU **is not** intended to alter any other existing agreements between Region 5 and Illinois EPA, including the Memorandum of Agreement authorizing administration of the State's RCRA Subtitle C program.

II.   Background

The Illinois EPA and Region 5 recognize that revitalization of contaminated property provides a significant benefit to both the environment and the economy. This is especially true for 'brownfields". The term "brownfields" refers to properties which are abandoned, idled, or under-used industrial and commercial facilitieswhere expansion or redevelopment is complicated by real or perceived environmental contamination. Some of the contaminated properties *in* Illinois, including some brownfields, are subject to environmental cleanup requirements which are established by Federal laws (e.g.,closure, post-closure+ and corrective action under RCRA; PCB Cleanup Policy under TSCA; the National Oil and Hazardous Substances Pollution Contingency Plan under the Comprehensive Environmental Response, Compensation and Liability Act ("CERCLA")).

Both Illinois EPA and Region 5 are mandated to protect human health and the environment and both play a critical role in Illinois in the cleanup and redevelopment of brownfields. Each Agency acknowledges the potentiai benefits that can be achieved by clarifying the liabilities associated with brownfields as a result of environmental cleanup requirements in both State and Federal laws. Both agencies recognize each other as key partners in addressing the perceived uncertainties in the financing, transfer and development of brownfields. Both agencies seek to facilitate the productive use of their authorities and resources in ways that are mutually complementary and are not redundant-. Both Region 5 and Illinois EPA acknowledge their mutual respect, positive working relationship and commitment to the successful implementation of the MOU. In particular, both agencies seek to protect human health and the environment by:

(1)   Promoting appropriate voluntary investigations and *cleanups* of browfields in Illinois.

(2)   Developing partnerships between Region 5, Illinois EPA, other Federal, State, local governmental agencies and other stakeholders, including representatives from the **private** sector and citizen/community groups, for the cleanup and redevelopment of brownfields.

(3)   Providing information and technical assistance to the key stakeholders to allow for informed decision making by property owners, prospective purchasers, lenders, public and private developers, citizens, municipalities, counties and elected officials.

'Facilities which **perfom** PCB cleanups under this MOU must, at this time be limited to TACO Tier.1 cleanup due to regulatory limitations under the preemption provisions of Section 18 of TSCA and the applicable PCB disposal rules and polices (e.g. U.S. EPA's Spill Cleanup **Policy, 40** CFR 761 Subpart G). Upon adoption of the **pending** amendments to TSCA PCB rules, Region V EPA anticipates modifying this MOU to include PCB cleanups under Tiers 2 and 3 of **TACO.**

(4)   Ensuring remediation of sites that protects human health and the environment and promoting revitalization of contaminated property for an appropriate use.

(5)   Promoting processes by which corrective action activities and consistent cleanup objectives are carried out.

III.   **Illinois** EPA Administration of **Title XVII**

Illinois EPA's administrative responsibilities under Title XVII are divided into several subject matters, two of which directly pertain to the purposes of this MOU. First, Illinois EPA is directed to administer a program that provides standards and procedures for remediation **activities for sites** voluntarily entering the Site Remediation Program.-(See Sections 58.6,58.7,58.8,and 58.10). These standards and procedures are set forth in 35 Ill. Adm. Code 740. Second, Illinois EPA is directed to establish, through the Illinois Pollution Control Board, risk-based remediation objectives. (See Section 58.5). These standards are incorporated in 35 Ill. Adm. Code 742.

A. Site Remediation Program (35 Ill. Adm. Code 740)

Under Title XVII, any "remediation applicant"* who proceeds under the Title may choose to have the Illinois EPA review and approve any of the remediation objectives for any or all of the "regulated substances of concern"² by submitting plans and reports to Illinois EPA. Illinois EPA then carries out its review in conformance with Title XVII and its rules. Illinois EPA may approve, disapprove, or approve with conditions, a plan or report. Under Title XVII, Illinois EPA administers the Site Remediation Program using 35 111. Adm Code 7.4O Part 740, in turn, requires remediation objectives to be established in accordance with 3.5 Ill. Adm: Code 742.. Part 740 allows sites to enter the Site Remediation Program to the extent allowed by federal law, federal authorization, or by other federal approval, such as through this MOU.

In the case of Illinois EPA approving, or approving with conditions, a plan or report, Illinois EPA prepares a document known as a "No Further Remediation Letter." Within 45 days of a remediation applicant's receipt of such a letter, the remediation applicant must submit the letter to the Office of the Recorder or the Registrar of Titles of the County in which the site is located. When the letter is accepted and recorded in accordance with Illinois law so that it forms a permanent part of the chain of title for the site, the letter becomes effective. The remediation applicant then submits a copy of the letter, as recorded; to the Illinois EPA.

The Illinois EPA's issuance of the No Further Remediation Letter signifies a release from further responsibilities under the State of Illinois Environmental Protection Act in performing the

---

*"Remediation Applicant" means any person seeking to perform or performing investigative or **remedial** activities under Title XVII, including the owner or operator of the site or persons authorized by law or consent to act *an* behalf of or in lieu of the owner or operator of the site.

²**"Regulated** substance of concern" means any contaminant that is expected to be present at the site **bared upon** past and current land uses and associated releases that are known to the "Remediation Applicant" based **upon reasonable** inquiry.

approved remedial action and shall be considered prima facie evidence that the site does not constitute a threat to human health and the environment and does not require further remediation under that act, so long as the site is maintained and utilized in accordance with the terms and conditions of the No Further Remediation Letter.

B. Tiered Approach to Corrective Action Objectives ("TACO") (35 Ill. Adm. Code 742)

TACO establishes a comprehensive tiered approach to the development of remediation objectives at sites evaluating cleanup needs in Illinois. This approach sets forth five independent methodologies for use, singly or in combination, in developing methodologies. The centerpiece of TACO is a set of Tier I baseline objectives for residential and commercial uses that were drawn directly from the technical concepts and principles established by: USEPA's final "Soil Screening Guidance: User's Guide", EPA/540/R-96/018,PB96-963505 (April 1996)). TACO is used by the Illinois EPA in developing remediation objectives for remediation activities under the following programs:

(1)   Leaking Underground Storage Tanks (35 Ill. Adm. Code 73 I and 732);

(2)   Site Remediation Program (35 Ill. Adm Code 740); and

(3)   RCRA Part B Permits and Closure Plans (35 Ill. m. Code 724 and 725).

IV.   Eligibility for Site Remediation Program Under 35 Ill. Adm. Code 740

This agreement approves the use of 35 Ill. Adm. Code 740 with regards to contaminated properties in Illinois subject to RCRA or TSCA except for the following

(1)   facilities which are required to have RCRA permits[4] issued by either (i) Illinois EPA, (ii) U.S. EPA, or (iii) both agencies;

(2)   sites at which investigation or remedial action has been required by a Federal court order or an order issued by the U.S. EPA. Such orders include orders or consent agreement and consent orders issued under:

- Section 3008(a), 3008(h), 3013,7003, or 9003(h) of RCRA;
- Section 16 of TSCA; and
- Sections 106,107,120, and 122 of CERCLA;

(3)   units, and associated releases from such units, at which treatment, storage, or disposal of hazardous waste has occurred after November 19,1980,and whose owners and operators

---

[4]RCRA Subtitle C permits for the treatment. storage or disposal of hazardous waste shall require corrective action for all releases of hazardous waste or constituents from any solid waste management unit at the permitted facility, regardless of the time at which waste was placed in the unit. Illinois EPA is authorized by U.S. EPA to issue, administer, and enforce such permits. U.S. EPA may also enforce such permits.

are required to (and have not yet) plan, conduct and certify closure and, if necessary, post-closure monitoring and maintenance pursuant to Subtitle C of RCRA;

(4) properties which are the subject of an order or a consent agreement and consent order proposed to be issued by Region 5 under section 3008(a), 3008(h), 3013,7003 or 9003(h) of RCRA; or section 16 of TSCA;

(5) properties approved by, or **seeking the** approval of U.S. EPA under TSCA (40 CFR Part 761, Subpart D) for the disposal or commercial storage or polychlorinated biphenyls (PCBs);

(6) sites listed in the CERCLA National Priorities List (40 CFR Part 300, Appendix B); and

(7) sites subject to 35 Ill. Adm. Code 807,810-817, or 830-832 that have not satisfied all development, operation, and closure requirements (including postclosure) applicable under 35 Ill. Adm. Code 807,810-817, or 830-832.

V. Principles

A. Although nothing in this MOU constitutes a release from liability under applicable Federal law, generally Region 5 does not anticipate taking **any** federal environmental cleanup action under RCRA or TSCA at a site, or portion thereof where the Illinois EPA has approved a remediation as having met the requirements of 35 Ill. Adm. Code 742 through:

(1) a 'No Further Remediation" letter issued pursuant to 35 Ill. Adm. Code 73 1,732 or 740;

(2) a Part B permit issued pursuant to 35 Ill. Adm. Code 724; or

(3) a closure certification approval issued pursuant to 35 Ill. Adm. Code 724 or 725.

This principle shall not apply if Region 5 determines that there may be an imminent and substantial endangerment to public health, welfare or the environment at a site, or portion thereof, where Illinois EPA has approved a remediation as having met the requirements of 35 Ill. **Adm.** Code 742. This principle shall not apply if the letter, permit **or** approval ceases to be in effect. If, following the issuance of the No Further Remediation Letter, permit or approval by Illinois EPA, conditions at the site previously unknown to Illinois EPA and/or Region 5 indicate that the response action undertaken is not protective of human health and the environment., Illinois EPA and Region 5 reserve the right to take necessary response action to protect human health and the environment.

B. Pursuant to this MOU, Region 5 approves the use of 35 Ill. Adm Code 740 for sites subject to RCRA or TSCA only at eligible sites. In this light, Region 5 acknowledges the use of 35 Ill. Adm_ Code 740, in conjunction with the applicable requirements of 35 111. Adm. Code 731 or 732, for remediatian of sites subject to RCRA Subtitle I, as long as the remediation meets the requirements of 35 111. Adm. Code 742.

VI. Reporting

Upon request, Illinois EPA will provide to Region 5 the following:

(1) The name and location of sites- with regard to which remediation applicants are seeking Illinois EPA review and approval pursuant to 35 Ill. Adm. Code 740; and

(2) The Illinois EPA review status of applications, and the status of remediation applicants' compliance with plans or reports approved, disapproved, or approved with conditions, by Illinois EPA pursuant to 35 Ill. Adm. Code 740.

To the extent practicable, for those sites identified by the Illinois EPA pursuant to VI,(l), Region 5 will provide notice to Illinois EPA in an enforcement confidential manner when U.S. EPA is proposing to issue anenvironmental cleanup order under Section 3008(a), 3008(h), 3013,7003, or 9003(h) of RCRA; or Section 16 of TSCA.

VII. Reservation of Rights

Notwithstanding any provision in this MOU, Region 5 and Illinois EPA reserve any and all rights or authority that they respectively have and nothing in any provision of this MOU limits or affects the authority or ability of either Agency to take any action authorized by law.

This MOU will be reviewed on an annual basis by Region 5 and Illinois EPA. In addition, at the request of either Agency, this MOU may be reevaluated and modified as appropriate.

VIII. Signatures

This MOU has been developed by mutual cooperation and consent, and hereby becomes an integral part of Illinois **EPA**'s and **Region 5**'s working relationship. The effective date of this MOU is July 1, 1997.

For the Illinois Environmental Protection Agency

[signature]

Director
Illinois Environmental Protection Agency

Date _6/23/97_

For the U.S. Environmental Protection Agency

[signature]

Acting Regional Administrator
U.S. Environmental Protection Agency, Region 5

Date _June 13, 1997_

Appendix 3
Acronyms

AICPA	American Institute of Certified Public Accountants
ASTM	American Society for Testing and Materials
CDBG	Community development block grants
CERCLA	Comprehensive Environmental Response, Compensation and Liability Act
CERCLIS	Comprehensive Environmental Response, Compensation and Liability Information System
CGL	Comprehensive general liability
COC	Constituent of concern
CRE	Counselors of Real Estate
DOC	Dissolved organic carbon
EAV	Equalized assessed value
FASB	Financial Accounting Standards Board
FIRREA	Financial Institutions Reform, Recovery and Enforcement Act
FID	Flame ionization detector
GAO	General Accounting Office
GC	Gas chromatography
GC-MS	Gas chromatography–mass spectrometry
GL	General liability
G.O.	General obligation (bonds)
HPLC	High-performance liquid chromatography
HUD	Department of Housing and Urban Development
ICP	Inductively coupled plasma
IDB	Industrial development bonds
IEPA	Illinois Environmental Protection Agency

MAI	Member of Appraisal Institute
MOA	Memorandum of agreement
MOU	Memorandum of understanding
NFR	No further remediation
NFRP	No further remedial action planned
NOI	Net operating income
OVA	Organic vapor analyzer
PAH	Polycyclic aromatic hydrocarbons
PCE	Perchlorethylene
PCP	Pentachlorophenol
PID	Photo ionization detector
PMD	Planned manufacturing district
PPA	Prospective purchaser agreement
PRP	Potentially responsible party
RBCA	Risk-based corrective action (also pronounced "Rebecca")
RCRA	Resource Conservation and Recovery Act
REC	Recognized environmental condition
REEL	Real estate environmental liability
SEC	Securities and Exchange Commission
SEIDS	Site Environmental Information Data System (Illinois)
SIR	Self-insured retention level
SVOC	Semivolatile organic compound
TACO	Tiered Approach to Corrective Action Objectives
TAL	Target Analyte List
TCD	Thermal conductivity detector
TCE	Trichlorethylene
TIF	Tax increment financing
TPP	Total priority pollutant
USPAP	Uniform Standards of Professional Approval Practice
UST	Underground storage tank
VCP	Voluntary cleanup program
VOC	Volatile organic compound

Index

About the Authors

Robert N. Rafson, P.E., is a licensed professional engineer, a graduate of the University of Wisconsin, and a partner in Greenfield Partners, Ltd., a company that purchases and redevelops environmentally contaminated properties. He is also president of Rafson Engineering, a firm that specializes in environmental engineering, and air pollution control design, testing, and training.

Harold J. Rafson is a chemical engineer and founder of a company that both invented and manufactured air pollution control equipment. In a chemical engineering career that spanned 50 years, he also worked for many years in food processes, engineering, and quality control, and environmental engineering. He holds bachelor's and master's degrees in chemical engineering from Brooklyn Polytechnic Institute.